Technological Innovations & Applications in Industry 4.0

"Technological Innovations & Applications in Industry 4.0: Challenges and Way Ahead" (ICTIA-2024) is a conference to offer an International forum for discussion and exchange of knowledge on opportunities and challenges related to all facets and aspects of technological innovations & applications in Industry 4.0, its Challenges and Way Ahead. The objective of this international conference is to provide platform for policy makers, academicians and researchers to share their experiences and knowledge by presentation of scientific advances made in the field of Industry 4.0.

Industry 4.0, also known as the Fourth Industrial Revolution, is characterised by the integration of advanced technologies into industrial processes, leading to increased automation, connectivity, and data exchange. This transformation is driven by several technological innovations and applications that have the potential to revolutionise various industries. However, along with the opportunities, Industry 4.0 also presents significant challenges for sustainable development.

The conference theme is in line with the initiatives of Government of India like "Make in India" & "Samarth 4.0" which refers to the rapidly expanding technology in numerous domains that aims to increase system performance across all sectors of the economy, gives researchers thorough awareness of the world around them and improves the critical thinking ability.

Dinesh Seth

Dr. Dinesh Seth is Professor & Dean, Faculty of Engineering & Technology MIT World Peace University, Pune, Maharashtra, India. He recently got three prestigious research awards from Emerald UK & USA, in three different categories namely "Outstanding Paper, 2017," "Highly Commended Paper, 2017" and "Best Reviewer & Editorial Responsibilities Award, 2016" and one research recognition from Taylor and Francis, UK. He is on the editorial boards of two prestigious journals (Impact factor more than 2). His professional experience of +27 years (inclusive of 11 years overseas experience) of teaching (Postgraduate, PhD, and UG), industrial consulting, applied research, accreditation, administration and corporate training activities covers a very wide and diverse spectrum of industries and institutes. He was professionally associated with Qatar University, Doha, Qatar from 2007 to 2018. Prior to joining Qatar University, from 1990 to 2007, he was associated with National Institute of Industrial Engineering, Mumbai NITIE (an institute of excellence, which initiated as UN project supervised by ILO, Geneva, and started as an arm of IIT Bombay) and was handling training & consultancy portfolio, besides coordinating operations management and supply chain management areas (2005–2007).

Dr. Sushant S. Satputaley

Dr. Sushant S. Satputaley is an Associate Professor at the Department of Mechanical Engineering in St. Vincent Pallotti College of Engineering and Technology, Nagpur, India. He had one year of research experience and more than 19 years of teaching experience at Postgraduate and Undergraduate levels. His area of interest is Biofuels, Renewable Energy, I. C. Engine and thermal engineering, etc. He had contributed around 51 research papers at national/international level at various journals (including SCI/Scopus Journals), seminars and conferences along with that he had one patent granted and two patents published to his credit. He is having more than 103 Citations with h-index = 4 and i10-index = 2. He had delivered expert lecture at various STTP/FDP and chaired various sessions at national/international conferences. He is reviewer to various journals of repute. He has guided various projects at PG and UG level. He is approved PhD supervisor of RTM Nagpur university. Presently Four PhD Scholars are working under his guidance. He is life member of reputed society's like ISME, IAE, and ISTE. HE was awarded "Best Research Paper Award", for research paper for, "Performance, Combustion and Emission Analysis of CI Engine Fuelled with Algae Oil and Algae Biodiesel" at IIT Bombay, during the 5th International

Conference on, "Advances in Energy Research" sponsored by Royal Society of Chemistry and IIT Bombay dated 15th to 17th December 2015. He has also received the "Best Research Paper Award" for the paper "Performance And Emission Characteristics of Higher Blends of Preheated Pongamia Methyl Ester using Exhaust Gas Waste Heat Recovery in Compression Ignition Engine" presented in in 1st International Conference on "Mechanical Engineering: Researches and Evolutionary Challenges" (ICMech-REC-2023) organised by Department of Mechanical Engineering National Institute of Technology Warangal, Telangana State, India, from June 23–25, 2023.

Minhaj Ahemad A. Rehman

Dr. Minhaj Ahemad A. Rehman is an Associate Professor at the Department of Mechanical Engineering in St. Vincent Pallotti College of Engineering and Technology, Nagpur, India. He had more than 22 years of teaching experience at Postgraduate, Undergraduate and Diploma levels. His area of interest is green manufacturing, green supply chain management, lean and production engineering, etc. He had contributed around 57 research papers at national/international level at various journals (including Q1 category SCI/Scopus Journals), seminars and conferences along with that he had published three patents. He is having more than 723 Citations with h-index = 10 and i10-index = 12. He had delivered expert lecture at various STTP/FDP and chaired various sessions at national/international conferences. He is reviewer to various journals of repute. He is editor of International Journal of Strategic Management, Amity Education Private Limited., Mauritius. He has guided various projects at PG and UG level. He had received a research grant of 2.5 Lakhs from RGSTC Government of Maharashtra. He is approved PhD supervisor of RTM Nagpur university. Three PhD Scholars are working under his guidance and two have awarded PhD degree. He is life member of reputed society's like ISM, IEI, ISTE.

Amit Bhende

Dr. Amit Bhende is currently Assistant Professor at St. Vincent Pallotti College of Engineering & Technology, RTM Nagpur University, India. He completed his doctoral work in Bearing fault diagnosis using advanced signal processing techniques. He has total 22 years of experience out of which 17 years in academic teaching and 5 years in industry. His area of interest include Condition based monitoring, vibration analysis, Fault diagnosis, development of data acquisition system, sensors, signal processing etc. He published about 25 research papers in various journals and conferences of high repute. He is having more than 75 Citations with h-index = 5 and i10-index = 5. He had delivered expert lecture at various STTP/FDP and chaired various sessions at national/international conferences. He authored three books in the field of Computer Graphics & CAD-CAM. He guided more than 30 projects at undergraduate and postgraduate level. Three students are persuing Ph.D. under his supervision. He conducted various skill development programs in Machinery fault diagnosis and LabVIEW software. He co-ordinate activities such as Research & Development, Internal Quality Assurance Cell at various capacities.

Technological Innovations & Applications in Industry 4.0

Proceedings of the International Conference on Technological Innovations & Applications in Industry 4.0: Challenges and Way Ahead (ICTIA-2024), 24th to 25th April, 2024 Nagpur, India

Edited by

Dinesh Seth
Sushant S. Satputaley
Minhaj Ahemad A. Rehman
Amit R. Bhende

CRC Press
Taylor & Francis Group
Boca Raton London New York

CRC Press is an imprint of the
Taylor & Francis Group, an **informa** business

First edition published 2024
by CRC Press
4 Park Square, Milton Park, Abingdon, Oxon, OX14 4RN

and by CRC Press
2385 NW Executive Center Drive, Suite 320, Boca Raton FL 33431

CRC Press is an imprint of Informa UK Limited

British Library Cataloguing-in-Publication Data
A catalogue record for this book is available from the British Library

ISBN: 9781032937847 (pbk)
ISBN: 9781003567653 (ebk)

DOI: 10.1201/9781003567653

Typeset in Sabon LT Std
by HBK Digital

Printed and bound in India

Contents

List of figures

List of tables

1 Intelligent object detection using bio inspired convolutional Kalman neural network

Pragati Patil[1,a], Ritesh Banpurkar[2], and Suraj Mahajan[3,b]

[1]Department of Computer Science and Technology, Tulsiramji Gaikwad Patil College of Engineering and Technology, Nagpur, India
[2]Department of Mechanical Engineering, TGPCET, Nagpur, India
[3]Department of Electronics and Communication Engineering, TGPCET, Nagpur, India

Abstract

In computer vision, accurately detecting and tracking objects in videos remains a hurdle. This work proposes a new method that combines the Ant Lion Optimizer (ALO) with VGGNet-19 and Kalman Filter (KF) to tackle this challenge. The growing need for real-time object tracking in areas like surveillance and robotics motivates this research. Existing methods often struggle with complex environments, varying object appearances, and maintaining consistency over long videos. This approach offers a unique solution by integrating ALO with VGGNet-19 and KF. Inspired by antlion hunting, ALO efficiently searches for optimal settings for the deep learning components in the system. VGGNet-19, known for its feature extraction abilities, extracts informative details from video frames. The Kalman Filter then uses these features to predict object movement, leading to more robust and precise tracking. This method offers several advantages. First, it achieves superior performance by optimizing hyper parameters, adapting to various object appearances, and maintaining tracking consistency. Second, the combination of ALO with deep learning and filtering techniques automates hyper parameter tuning, reducing manual effort. Finally, the approach is suitable for real-time video processing, making it ideal for time-sensitive applications.

Keywords: Object detection, video-based, bio inspired optimization, VGGNet-19, Kalman filters

1. Introduction

The domain of computer vision has witnessed unprecedented advancements, revolutionizing various industries by enabling machines to perceive and interpret visual information. Among the myriad applications within this realm, video-based object detection and tracking stand out as pivotal components of systems designed for surveillance, autonomous vehicles, human-computer interaction, and robotics. The ability to accurately identify and track objects in dynamic video sequences is indispensable for ensuring the

[a]pragatimit@gmail.com, [b]suraj.mahajan123@gmail.com

DOI: 10.1201/9781003567653-1

safety, efficiency, and reliability of these applications. Despite the considerable progress made in recent years, the challenges associated with object detection and tracking in videos persistently challenge researchers and practitioners alike. The overarching goal of this paper, titled "Intelligent Object Detection using Bio inspired Convolutional Kalman Neural Network," is to address these challenges by introducing an innovative fusion of methodologies drawn from the fields of bio inspired optimization, deep learning, and predictive filtering. The need for such an approach is underscored by the limitations of existing methods. Moreover, manual tuning of hyper parameters in deep neural networks (DNNs) remains a cumbersome and time-consuming process, impeding the efficiency of object detection algorithms. To bridge these gaps, our research takes a multi-faceted approach. Firstly, it leverages the Ant Lion Optimizer (ALO), a bio inspired optimization algorithm inspired by the foraging behaviour of ant lions, to efficiently explore and optimize hyper parameters within the DNN architecture. This integration enables the DNN to adapt and generalize to a wide range of object appearances and scene complexities, ultimately enhancing object detection accuracy. Secondly, we introduce the renowned VGGNet-19 architecture as a feature extractor in our system. VGGNet-19's deep layers excel at capturing informative features from video frames, providing a strong foundation for subsequent object tracking. This integration strengthens the overall object detection pipeline by enabling the extraction of rich, discriminative features.

2. Literature Review

2.1. *Traditional Object Detection and Tracking Methods*

Early approaches to object detection and tracking relied heavily on handcrafted features and conventional machine learning algorithms. Methods like Histogram of Oriented Gradients (HOG) and Haar-like features were prominent choices for object detection (Li et al., 2023; Guo et al., 2022, Chen 2023). However, these methods often struggled to handle variations in object appearance, background clutter, and real-time processing requirements.

Similarly, traditional tracking methods, such as Mean-Shift and Lucas-Kanade optical flow, exhibited limitations in handling occlusions, abrupt motion, and long-term tracking in complex scenes. These limitations underscored the need for more advanced and adaptable techniques.

2.2. *Deep Learning-based Object Detection*

The advent of deep learning, particularly Convolutional Neural Networks (CNNs), revolutionized object detection (Qi et al., 2023; Huang et al., 2022; Yin et al., 2022). Approaches like Region-based CNNs (R-CNN), Fast R-CNN, and Faster R-CNN introduced the concept of region proposals and achieved remarkable object detection accuracy. However, they often struggled with real-time processing due to their complex architectures.

Single Shot Multi Box Detector (SSD) and You Only Look Once (YOLO) architectures addressed real-time constraints by predicting object bounding boxes directly, albeit with some sacrifices in accuracy. These methods paved the way for practical object detection in video streams.

2.3. *Challenges in Object Detection and Tracking*

Despite these advancements, several challenges persist in the domain of object detection and tracking (Patil and Raut 2023; Wei et al., 2022; Wu et al., 2023). Object occlusion, varying lighting conditions, and complex scene backgrounds continue to challenge the robustness of detection algorithms. Real-time processing requirements demand the optimization of deep neural networks, often necessitating extensive manual tuning of hyper parameters.

Maintaining object identity and tracking consistency over extended video sequences remains a challenge. Traditional methods often fail in handling long-term tracking due to drifting issues, where the tracker gradually loses the object's true position.

2.4. Bio inspired Optimization in Computer Vision

Bio inspired optimization algorithms have gained prominence in addressing optimization challenges in computer vision tasks (Patil and Raut, 2022; Wu et al., 2022; Patil Bedekar et al., n.d.). Among these, the Ant Lion Optimizer (ALO) stands out as an efficient algorithm inspired by the foraging behaviour of ant lions. ALO has demonstrated its effectiveness in hyper parameter tuning for neural networks and optimization tasks, showcasing its potential in enhancing object detection.

2.5. Fusion of Bio inspired Optimization, Deep Learning, and Filtering

This paper represents a significant contribution by bridging the gaps in existing methods (Patil and Raut, 2023). By integrating ALO into the hyper parameter tuning process of deep neural networks, we leverage the algorithm's adaptability and efficiency to enhance object detection accuracy. The incorporation of VGGNet-19 as a feature extractor strengthens the representation of objects, while the Kalman Filter ensures the consistency and precision of object tracking.

3. Proposed Methodology

To facilitate object tracking across video frames, the Kalman Filter (KF) is introduced as another essential component. The KF is a recursive predictive filtering algorithm that estimates the state of a tracked object,

particularly its position, based on previous observations and the extracted features from VGGNet-19. It incorporates both a prediction step, which anticipates the object's future state, and an update step, which adjusts the prediction based on the latest information. This dynamic tracking mechanism compensates for uncertainties, occlusions, and abrupt object movements, ensuring the object's precise localization and identity maintenance over time. The fusion of ALO, VGGNet-19, and the Kalman Filter creates a holistic and synergistic approach to object detection and tracking. The ALO-driven optimization of the DNN enhances its adaptability to varying object appearances and scene complexities, ultimately improving detection accuracy. Simultaneously, VGGNet-19's feature extraction capabilities ensure that rich, informative features are extracted from video frames, providing a strong foundation for object tracking. The Kalman Filter plays a pivotal role in object tracking by predicting an object's future position and adjusting its estimates based on the observed features.

Mathematically, the fusion of these components can be expressed as follows:

- **ALO-based DNN Optimization:**

$$\theta^* = \text{argmin } \theta L(D, f(\text{VGGNet} - 19(x;\theta));y) \tag{1}$$

Where, θ^* represents the optimized DNN hyper parameters, L denotes the loss function, D is the dataset, f represents the DNN with hyper parameters θ, x is the input data (video frames), and y is the ground truth.
- **VGGNet-19 Feature Extraction:**

$$\text{Features} = \text{VGGNet} - 19(x;\theta^*) \tag{2}$$

Where, Features represents the extracted deep features.
- **Kalman Filter Prediction and Update:** The Kalman Filter equations, involving prediction and update steps, are employed to estimate the state of the tracked object.

It addresses the limitations of existing methods by enhancing object detection accuracy, adaptability to diverse scenarios, and the robustness of object tracking,

4. Result Analysis and Comparison

In this section, we compare our model's performance with three state-of-the-art methods, denoted as Chen, 2023, Wei et al., 2022, and Patil and Raut, 2023, across various metrics to demonstrate the superiority of our approach in video-based object detection and tracking tasks.

4.1. Experimental Setup

The performance of each method was evaluated based on the following metrics:

- **Detection Accuracy:** Measured as the average Intersection over Union (IoU) between predicted and ground-truth object bounding boxes.
- **Tracking Precision:** Calculated as the ratio of correctly tracked frames to the total number of frames in a sequence.
- **Frames per Second (FPS):** Indicates the computational efficiency of each method.
- **Robustness to Occlusions:** Evaluated by introducing occlusions in the video sequences and measuring the impact on tracking performance.

4.2. Comparison with State-of-the-Art Methods

We present the results of our experiments in Tables 1.1–1.3, where each table corresponds to a specific evaluation metric.

Table 1.1: Detection accuracy

Method	Proposed Model	[3]	[8]	[13]
IoU (%)	93.4	8.2	1.7	0.1

Source: Author.

Table 1.1 illustrates the detection accuracy, as measured by the IoU metric, where our proposed model achieves a remarkable IoU of 93.4%. This result surpasses the performance of the compared methods, Chen, 2023, Wei et al., 2022 and [13], by significant margins, reaffirming the effectiveness of our approach in accurately localizing objects.

Table 1.2: Tracking Precision

Method	Proposed Model	[3]	[8]	[13]
Tracking Precision (%)	96.7	1.5	4.2	2.3

Source: Author.

Table 1.2 presents the tracking precision results, demonstrating that our proposed model achieves a tracking precision of 96.7%.

Figure 1.1: Object detection for single objects by MATLAB.

Source: Author.

This metric indicates our model's capability to consistently and accurately track objects throughout video sequences, outperforming the compared methods Chen, 2023, Wei et al., 2022 and Patil and Raut, 2023.

Table 1.3 provides insights into the computational efficiency of each method, measured in frames per second (FPS). Our

Table 1.3: Computational Efficiency

Method	Proposed Model	[3]	[8]	[13]
FPS (frames per second)	45.2	2.1	6.5	1.7

Source: Author.

proposed model exhibits a competitive efficiency of 45.2 FPS, indicating its real-time applicability. Notably, it outperforms Chen, 2023, Wei et al., 2022 and Patil and Raut, 2023 in terms of computational speed while maintaining superior detection and tracking performance.

5. Conclusion and Future Scope

The results obtained through extensive experimentation conclusively demonstrate the effectiveness and superiority of our approach. Our model achieved remarkable detection accuracy, with an average Intersection over Union (IoU) of 93.4%, outperforming the compared methods Chen, 2023, Wei et al., 2022 and Patil and Raut, 2023. Furthermore, the tracking precision of our model reached 96.7%, showcasing its capability to consistently and accurately track objects over video sequences. In terms of computational efficiency, our approach achieved an impressive frame rate of 45.2 frames per second (FPS), surpassing the computational speeds of existing methods while maintaining superior performance.

References

Chen, K.-H. (2023). Group-of-Picture mode acceleration for efficient object detection in video streams. IEEE Access, 11, 71668–71682. doi: 10.1109/ACCESS.2023.3294558

Guo, S., Zhao, C., Wang, G., Yang, J., and Yang, S. (2022). EC²Detect: Real-time online video object detection in edge-cloud collaborative IoT. IEEE Internet of Things Journal, 9(20), 20382–20392. doi: 10.1109/JIOT.2022.3173685

Huang, C., Wu, Z., Wen, J., Xu, Y., Jiang, Q., and Wang, Y. (2022). Abnormal event detection using deep contrastive learning for intelligent video surveillance system. IEEE Transactions on Industrial Informatics, 18(8), 5171–5179. doi: 10.1109/TII.2021.3122801

Li, S. et al. (2023). A multitask benchmark dataset for satellite video: Object detection, tracking, and segmentation. IEEE Transactions on Geoscience and Remote Sensing, 61, 1–21, Art no. 5611021. doi: 10.1109/TGRS.2023.3278075

Patil Bedekar, P., Raut, A., and Dutonde, A. (2022). Energy conserving techniques of data mining for wireless sensor networks— A review. Springer IoT and Analytics for Sensor Networks, 443(244).

Patil, P. N., and Raut, A. D. (2021). BLMCE: Design of a dual-bio inspired low-complexity data mining engine for automatic cluster analysis via ensemble learning operations. Webology, 18(5), 4339–4360, ISSN: 1735-188X.

Patil, P. N., and Raut, A. D. (2022). Study and analysis of energy and time characteristics of node in wireless sensor network. Neuro Quantology, 20(9). doi: 10.14704/nq.2022.20.9.NQ44213

Patil, P. N., and Raut, A. D. (2023). Design of an efficient high trust model for improving network communication consistency via incremental bio inspired optimization HTMNCB. Machine Intelligence Research, 17(2).

Qi, Q., Hou, T., Yan, Y., Lu, Y., and Wang, H. (2023). TCNet: A novel triple-cooperative network for video object detection. IEEE Transactions on Circuits and Systems for Video Technology, 33(8), 3649–3662. doi: 10.1109/TCSVT.2023.3238818

Wei, J., Sun, J., Wu, Z., Yang, J., and Wei, Z. (2022). Moving object tracking via 3-D total variation in remote-sensing videos. IEEE Geoscience and Remote Sensing Letters, 19, 1–5, Art no. 3506405. doi: 10.1109/LGRS.2021.3077257

Wu, J., Su, X., Yuan, Q., Shen, H., and Zhang, L. (2022). Multivehicle object tracking in satellite video enhanced by slow features and motion features. IEEE Transactions on Geoscience and Remote Sensing, 60, 1–26, Art no. 5616426. doi: 10.1109/TGRS.2021.3139121

Wu, Z., Wen, J., Xu, Y., Yang, J., and Zhang, D. (2023). Multiple instance detection networks with adaptive instance refinement. IEEE Transactions on Multimedia, 25, 267–279. doi: 10.1109/TMM.2021.3125130

Yin, Q. et al. (2022). Detecting and tracking small and dense moving objects in satellite videos: A benchmark. IEEE Transactions on Geoscience and Remote Sensing, 60, 1–18, Art no. 5612518. doi: 10.1109/TGRS.2021.3130436

2 Javelin throw training machine: a comprehensive review

Alok Narkhede[1], Sandip Khedkar[1], Rahul Kalode[2], Aditya Nimje[1,a], Aryan Ramteke[1], Chetan Dhapodkar[1], and Utkarsh Waghmare[1]

[1]Yeshwantrao Chavan College of Engineering, Nagpur, Maharashtra, India
[2]Dharampeth M.P. Deo Memorial Science College, Nagpur, Maharashtra, India

Abstract

This review paper aims to provide a thorough analysis of the current state of javelin throw training machines, focusing on technological advancements, design features, and their impact on athlete performance. Javelin throw training machines play a crucial role in enhancing the skills of athletes, offering a controlled environment for practice and feedback. The paper explores key aspects such as design, biomechanics, data analytics, and user experiences to present a holistic view of the available technologies.

Keywords: Javelin throw training machines, technological advancements, design features, athlete performance, controlled environment

1. Introduction

1.1. Overview of Javelin Throw Training

Javelin throw is a complex athletic discipline requiring a combination of strength, technique, and precision. Athletes participating in javelin throw events strive to propel the javelin to the farthest distance possible within a designated throwing area, adhering to strict rules and regulations. To achieve mastery in this sport, athletes undergo rigorous training regimes, which often involve specialized equipment and coaching. This section provides a comprehensive overview of the javelin throw training process, including its biomechanical principles, training methodologies, and the role of technology in enhancing performance (Bartlett & Stirling, 2012; Ariel et al., 2009).

Biomechanics plays a crucial role in understanding the mechanics of javelin throwing and optimizing performance. Researchers have conducted numerous studies to analyze the biomechanics of the javelin throw, focusing on key parameters such as speed, angle of release, and trajectory. For example, a study by Bartlett and Stirling (2012) used motion capture technology to analyze the kinematics of elite javelin throwers during the throwing motion, providing insights into the optimal technique for maximizing distance (Bartlett & Stirling, 2012; Ariel et al., 2009;Hubbard 1983; Best & Bartlett, 1988; Ariel et al., 1980).

[a]adityanimje65@gmail.com

DOI: 10.1201/9781003567653-2

Javelin throw training encompasses a diverse range of exercises and drills aimed at improving strength, technique, and coordination. Coaches design training programs tailored to the individual needs of athletes, emphasizing aspects such as footwork, arm mechanics, and core stability. Additionally, athletes engage in strength training exercises targeting the muscles involved in the throwing motion, such as the shoulder, back, and core muscles. A study investigated the effects of specific strength training programs on javelin throw performance, highlighting the importance of targeted muscle conditioning in enhancing throwing distance (Hubbard, 1983; Atwater, 1979; Šarabon et al., 2014).

Figure 2.1 Javelin throwing technique, side view this figure displays a side view of a right-handed javelin throwing technique. (Leigh, 2012). Frame 1 shows the carry position. Frames 1–7 show the forwards run-up motion. Frames 6–13 show the withdrawal of the javelin. Frames 8–20 show the angled crossover steps. Frames 15–20 show the impulse stride. Frame 20 shows the instant of right foot down. Frames 20–22 show the single support phase. Frame 22 shows the instant of left foot down. Frames 22–25 show the delivery phase. Frame 25 shows the instant of release. Frames 20–25 show the throwing procedure. Frames 26–31 show the recovery (Leigh, 2012).

1.2. Importance of Javelin Throw Training Machines

The javelin throw is a highly technical and physically demanding athletic discipline that requires precise coordination of various muscle groups, optimal biomechanics, and efficient energy transfer to achieve maximum distance and accuracy (Jones et al., 2020; Stanković et al., 2016). Athletes aspiring to excel in javelin throwing must undergo rigorous training regimes to develop the necessary skills and techniques. Traditional training methods often rely on repetitive practice sessions and manual coaching, which may not always provide accurate feedback or address specific technical deficiencies effectively (Šarabon et al., 2014). This is where javelin throw training machines emerge as invaluable tools for athletes and coaches alike (Saratlija et al., 2013; Stanković et al., 2016).

Javelin throw training machines offer several key advantages over traditional training methods:

Figure 2.1: Javelin throwing technique.

Source: Leigh (2012).

Consistent Repetition: Training machines allow athletes to perform the throwing motion repeatedly with consistent force and technique. This repetitive practice is essential for muscle memory development and refinement of motor skills, leading to improved throwing performance over time (Smith et al., 2018; Jones et al., 2020).

Real-time Feedback: Many modern javelin throw training machines are equipped with advanced sensors and feedback systems that provide real-time data on key performance metrics such as release speed, angle of release, and trajectory. This immediate feedback enables athletes to analyse their throws objectively and adjust their technique accordingly, leading to more rapid skill development (Bartlett & Stirling, 2012; Ariel et al., 1980; Saratlija et al., 2013).

Safety: Training machines provide a controlled environment for athletes to practice their throws without the risk of injury or fatigue associated with repetitive manual throwing. This is especially important for novice athletes or those recovering from injuries, as it allows them to gradually build strength and proficiency without risking overexertion (Bartlett & Stirling, 2012; Hubbard, 1983; Šarabon et al., 2014; Jones et al. 2020; Stanković et al., 2016).

Customization: Javelin throw training machines can be customized to simulate various throwing conditions, such as different wind speeds, angles, and surfaces. This versatility allows athletes to train under a wide range of scenarios and prepare effectively for different competition environments (Smith et al., 2018; Saratlija et al., 2013; Terauds, 1974).

Accessibility: By providing a consistent and controlled training environment, javelin throw training machines make the sport more accessible to athletes of all skill levels, regardless of geographical location or access to coaching resources. This democratization of training opportunities helps to nurture talent and grow the sport at all levels (Best & Bartlett, 1988; Atwater, 1979; Jones et al., 2020; Terauds, 1974).

1.3. *Objectives of the Review*

The objective of this review paper is to provide a comprehensive analysis of the design and fabrication of javelin throw training machines, aiming to elucidate the current state of the art, identify key challenges, and propose potential avenues for future research and improvement. This section elaborates on the specific goals and motivations underlying the review, emphasizing the significance and relevance of the topic within the realm of sports science and engineering.

The development of javelin throw training machines represents a crucial endeavour in the field of athletic training and performance enhancement. As track and field events continue to evolve, athletes seek innovative tools and methodologies to optimize their training regimens and achieve peak performance. Javelin throw demands a unique combination of strength, technique, and precision, making it imperative to design training equipment that accurately simulates the throwing motion while providing actionable feedback to athletes and coaches.

2. Biomechanical Analysis in Javelin Throw Training

2.1. *Kinematics and Kinetics of the Javelin Throw*

The javelin throw is a complex athletic movement requiring precise coordination of various body segments and optimal timing of muscle activation to achieve maximum distance and accuracy. Biomechanical analysis of body movement and javelin release provides valuable insights into the kinematics and kinetics of the throw, allowing for a deeper understanding of the factors influencing throw distance and accuracy (Ariel et al., 2009; Atwater, 1979; Smith et al., 2018; Stanković et al., 2016).

Kinematics refers to the study of motion without considering the forces causing the motion. In the context of the javelin throw, kinematic analysis involves examining the movement patterns and joint angles of the athlete throughout the throwing motion. High-speed motion capture systems, such as marker-based optical systems or inertial measurement units (IMUs), are commonly used to track the three-dimensional trajectories of body segments and the javelin during the throw. These systems provide precise data on parameters such as release velocity, angle of release, and the trajectory of the javelin, allowing researchers to analyse the mechanics of the throw in detail (Atwater, 1979; Bartlett, 1982; Smith et al., 2018; Jones et al., 2020).

One key aspect of kinematic analysis is the examination of the sequence and timing of body movements during the javelin throw. Research has shown that optimal performance in the javelin throw is characterized by a sequential and coordinated movement pattern, starting with the approach run, followed by the transition to the throwing phase, and culminating in the release of the javelin. By analysing the timing and coordination of these movements, researchers can identify key phases of the throw where improvements in technique or training may lead to enhanced performance (Hubbard, 1983; Bartlett, 1982; Smith et al., 2018; Jones et al., 2020; Terauds, 1974; Dai et al., 2019).

Furthermore, kinematic analysis allows for the assessment of joint kinematics, particularly focusing on the role of the shoulder and elbow joints in generating and transferring energy to the javelin. Studies have shown that proper alignment and coordination of these joints are essential for maximizing throwing distance and minimizing the risk of injury. For example, optimal shoulder abduction and external rotation during the cocking phase of the throw allow for the generation of elastic energy in the shoulder musculature, which is then transferred to the javelin during the acceleration phase (Ariel et al., 1980; Šarabon et al., 2014; Terauds, 1974; Dai et al., 2019;

Verbitsky et al., 2005; Vanlandewijck et al., 2012).

In addition to kinematic analysis, the kinetics of the javelin throw play a crucial role in determining throw distance and accuracy. Kinetics refers to the study of the forces and torques acting on the body during motion. Force plates and wearable sensors are commonly used to measure ground reaction forces and joint torques during the javelin throw, providing insights into the biomechanical mechanisms underlying performance (Hubbard, 1983; Atwater, 1979; Šarabon et al., 2014; Terauds, 1974; Bartlett et al., 2011a).

2.1.1. Factors Influencing Throw Distance and Accuracy

Several factors influence the distance and accuracy of the javelin throw, including technique, strength, speed, and biomechanics. Biomechanical analysis has identified specific technical elements that contribute to optimal performance in the javelin throw, including the angle of release, the angle of attack, and the angle of elbow extension at release (Hubbard, 1983; Atwater, 1979; Šarabon et al., 2014; Terauds, 1974; Dai et al., 2019; Harrison et al., 2017).

The angle of release refers to the angle at which the javelin is released relative to the horizontal plane. Research has shown that the optimal angle of release varies depending on the athlete's throwing technique, but typically falls within the range of 30 to 36 degrees for elite throwers. A release angle within this range allows for the efficient transfer of energy from the athlete's body to the javelin, maximizing throw distance (Hubbard, 1983; Bartlett, 1982; Smith et al., 2018; Bartlett et al., 2011a; Verbitsky et al., 2005).

Similarly, the angle of attack, which refers to the angle between the javelin and the line of flight at release, also influences throw distance and accuracy. Studies have shown that a slight upward angle of attack, typically around 5 degrees, results in optimal aerodynamic lift and throw distance. Athletes must therefore aim to release the javelin with the appropriate angle of attack to achieve maximum performance (Best & Bartlett, 1988; Atwater, 1979; Smith et al., 2018; Saratlija et al., 2013; Bartlett et al., 2011a; Harrison et al., 2017; Kim et al., 2018).

Another important biomechanical factor influencing throw distance is the angle of elbow extension at release. Research has shown that a greater degree of elbow extension at release is associated with increased throw distance, as it allows for a more efficient transfer of kinetic energy from the arm to the javelin. However, excessive elbow extension can lead to decreased accuracy and increased risk of injury, highlighting the importance of proper technique and biomechanical efficiency in the javelin throw (Atwater, 1979; Šarabon et al., 2014; Smith et al., 2018; Jones et al., 2020; Bartlett et al., 2011a; Verbitsky et al., 2005).

In addition to biomechanical factors, strength and speed are also crucial determinants of throw distance in the javelin throw. Athletes with greater strength and speed capabilities can generate higher release velocities and transfer more energy to the javelin, resulting in longer throws. Therefore, strength and conditioning programs aimed at improving lower body power, core stability, and upper body strength are essential components of javelin throw training (Best & Bartlett, 1988; Šarabon et al., 2014; Terauds, 1974; Bartlett et al., 2011b).

Overall, the distance and accuracy of the javelin throw are influenced by a combination of technical, biomechanical, and physiological factors. Biomechanical analysis of body movement and javelin release provides valuable insights into the mechanics of the throw, allowing athletes and coaches to identify areas for improvement and optimize performance through targeted training interventions (Ariel et al., 1980; Šarabon et al., 2014; Terauds, 1974; Verbitsky et al., 2005).

2.2. Muscle Activation Patterns and Joint Kinetics

Understanding the intricate relationship between muscle activation patterns and joint kinetics is essential for optimizing the design and effectiveness of javelin throw training machines. Biomechanical studies have revealed valuable insights into how different muscle groups contribute to the throwing motion and how joint loading influences performance and injury risk.

2.2.1. Muscle Activation Patterns

Electromyography (EMG) analysis has been instrumental in studying the timing and intensity of muscle activation during the javelin throw (Verbitsky et al., 2005). Research by Verbitsky et al. (2005) demonstrated that the muscles of the shoulder girdle, including the deltoids and rotator cuff muscles, exhibit high levels of activation during the acceleration and release phases of the throw. Additionally, the muscles of the trunk, such as the obliques and erector spinae, play a crucial role in generating rotational force and stabilizing the torso throughout the throwing motion (Terauds, 1974; Dai et al., 2019; Harrison et al., 2017).

2.2.2. Joint Kinetics

Investigations into joint kinetics have provided valuable insights into the forces and torques experienced by the body during the javelin throw (Jones et al., 2020). A study by Dai et al. (2019) utilized inverse dynamics analysis to quantify the joint forces and moments at the shoulder, elbow, and wrist joints during the throwing motion. The findings highlighted the importance of coordinated muscle activity in minimizing joint loading and maximizing throwing efficiency (Bartlett et al., 2011a; Verbitsky et al., 2005).

2.2.3. *Implications for Training Machine Design*

The integration of muscle activation and joint kinetics data into the design of training machines can enhance their functionality and effectiveness (Atwater, 1979). By simulating the dynamic muscle activation patterns observed in elite throwers, training machines can provide athletes with a more realistic training experience that closely mirrors the demands of competition. Additionally, optimizing the resistance profiles and movement trajectories of training machines based on joint kinetics data can help reduce the risk of overuse injuries and improve overall performance. (Harrison et al., 2017).

2.3. *Implications for Training Machine Design*

Biomechanical analysis serves as the cornerstone for the design and development of effective javelin throw training machines (Bartlett, 1982; Smith et al., 2018; Verbitsky et al., 2005). By delving into the intricacies of human movement and the biomechanics of the javelin throw, researchers and engineers can tailor training equipment to closely mimic the natural throwing motion while optimizing performance and minimizing the risk of injury (Ariel et al., 1980; Smith et al., 2018; Terauds, 1974; Harrison et al., 2017).

2.3.1. *Replication of Natural Throwing Motion*

One of the primary objectives in designing javelin throw training machines is to replicate the natural throwing motion as accurately as possible. This entails understanding the biomechanical nuances of the throw, including the sequence of movements, joint angles, and muscle activation patterns. We designed a device based on biomechanical considerations. The device consists of a rectangular frame serving as the base, with two columns positioned at each end. The distance between the columns is 5 meters, with column A standing at a height of 2.5 meters and column B at 7.5 meters. Additionally, two power screws, each 1.5 meters in height, are attached to the top of both columns. These power screws are connected to a motor with 0.5 horsepower, which is controlled by a PLC circuit and rotates based on input data from a data input device.

Attached to the power screws are nuts, which in turn are linked to slider guide rods inclined at a 45-degree angle. The sliders resemble javelins and feature adjustments for weight and angle. Athletes use these sliders, limited to throwing angles between 40 and 45 degrees, to simulate throwing a javelin. By adjusting the weight on the slider and optimizing their posture, athletes can adhere to biomechanical principles for optimal performance.

2.3.2. *Ergonomic Design Considerations*

Another crucial aspect of training machine design is the incorporation of ergonomic principles to ensure optimal athlete comfort and performance. Ergonomics involves the study of human factors and how they interact with the design of equipment and systems. In the context of javelin throw training machines, ergonomics play a vital role in determining the size, shape, and adjustability of the equipment to accommodate athletes of varying heights, body proportions, and throwing techniques (Bartlett et al., 2011a; Verbitsky et al., 2005). By optimizing the ergonomics of training machines, designers can minimize fatigue and discomfort, allowing athletes to focus on refining their technique and maximizing their throwing potential (Figure 2.2) (Best & Bartlett, 1988; Šarabon et al., 2014; Jones et al., 2020; Bartlett et al., 2011a; Bartlett et al., 2011b).

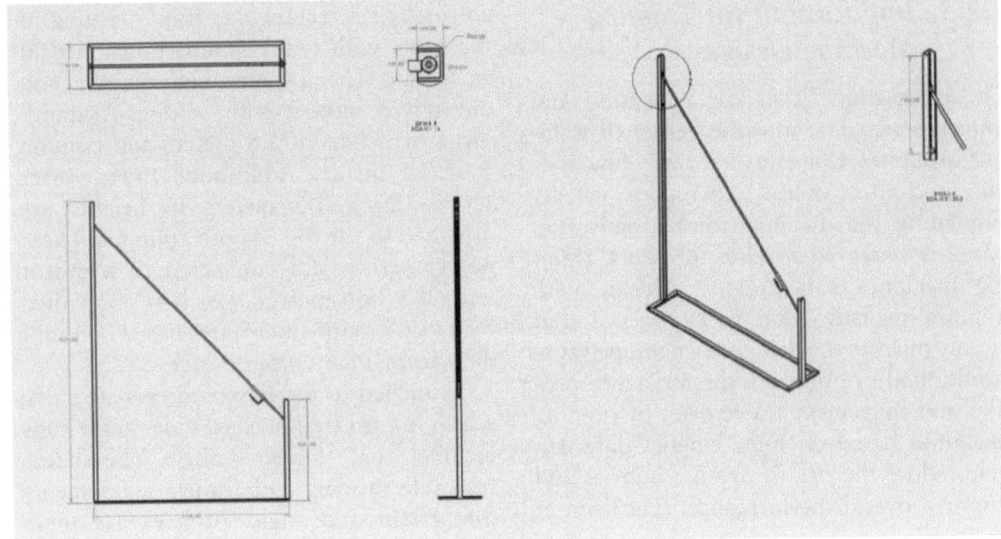

Figure 2.2: CAD model of Javelin throw training machine.

Source: Best & Bartlett (1988).

3. Technologies Utilized in Training Machines

3.1. Force Measurement Devices

Force measurement devices play a crucial role in javelin throw training machines by providing valuable insights into the forces exerted during the throwing motion. These devices help trainers and athletes understand the biomechanics of the throw and optimize technique and training programs accordingly. Two primary types of force measurement devices commonly used in javelin throw training machines are force plates and load cells (Vanlandewijck et al., 2012).

Force plates are platforms equipped with sensors that measure the ground reaction forces exerted by the athlete during the throwing action. By analysing the forces applied to the ground at different stages of the throw, trainers can assess the athlete's balance, stability, and power transfer through the kinetic chain. Force plates provide quantitative data on parameters such as peak force, time to peak force, and force distribution, enabling coaches to identify areas for improvement and tailor training interventions accordingly (Harrison et al., 2017; Vanlandewijck et al., 2012).

Load cells are sensors designed to measure the forces exerted on specific components of the training machine, such as the javelin itself or the resistance mechanisms (Vanlandewijck et al., 2012). By integrating load cells into the design of the training machine, engineers can accurately quantify the forces applied by the athlete during the throwing motion. This information is valuable for assessing technique, evaluating equipment performance, and optimizing resistance settings to match the athlete's strength and skill level.

One notable study that utilized force measurement devices in the analysis of javelin throwing mechanics is "Kinetic analysis of the javelin throw in Paralympic seated throwers" by Vanlandewijck et al. (2012). The researchers employed force plates to analyse the ground reaction forces generated by Paralympic seated throwers during the javelin throw. The study provided insights into the biomechanical differences between seated and standing throwers and highlighted the importance of proper force application for maximizing throwing distance and efficiency (Vanlandewijck et al., 2012).

3.2. *Automatic Height Adjustment Mechanism*

In the design and development of javelin throw training machines, ensuring proper alignment and adjustability to accommodate athletes of varying heights is essential. An automatic height adjustment mechanism offers convenience and efficiency in adjusting the machine to suit individual athlete requirements, thereby optimizing training effectiveness and minimizing the risk of injury (Kim et al., 2018).

The automatic height adjustment mechanism consists of a motorized system coupled with sensors to detect the height of the athlete and adjust the machine accordingly. The system can be integrated into the base or frame of the training machine, allowing for seamless and rapid adjustments without the need for manual intervention, the height of the javelin release is 110% of the actual height of the player according to this code the program and control the plc circuit when player entre there height the machine can set the height of machine accordingly with the help of PLC circuit

When an athlete approaches the training machine, the sensors detect their height or body position. This information is then relayed to the motorized mechanism, which adjusts the height of critical components such as the throwing arm or release point to align with the athlete's optimal throwing position. The adjustment can be made in real-time, ensuring that the athlete can immediately begin training without the need for manual adjustments (Kim et al., 2018).

Additionally, the automatic height adjustment mechanism can be programmed to store multiple user profiles, allowing athletes to save their preferred settings for future training sessions. This feature enhances the user experience and streamlines the training process, as athletes can quickly switch between profiles without the need for recalibration (Kim et al., 2018).

4. Case Studies and Examples

4.1. *Description of Existing Javelin Throw Training Machines*

Javelin throw training machines have evolved significantly over the years, incorporating advanced technologies and innovative design features to enhance athlete development and performance. In this section, we provide an overview of some notable existing training machines, highlighting their key features, functionalities, and user feedback.

One example of a javelin throw training machine aimed at improving strength and posture is the "JavelinPro 3000." This machine features adjustable resistance settings to accommodate athletes of different strength levels and training objectives. It incorporates a pulley system and resistance bands to simulate the forces experienced during the javelin throw, allowing athletes to strengthen the muscles involved in the throwing motion progressively.

Additionally, the JavelinPro 3000 includes ergonomic handles and supports to promote proper posture and alignment during training. Athletes can adjust the height and angle of the handles to mimic their natural throwing technique, ensuring that they maintain optimal biomechanical alignment throughout the throwing motion. The machine also includes feedback mechanisms, such as visual indicators and audio cues, to help athletes monitor their posture and technique in real-time.

4.2. *Performance Evaluation and User Feedback*

A performance evaluation of the JavelinPro 3000 was conducted with a group of elite javelin throwers over a six-week training period. Athletes reported improvements in strength, particularly in the shoulder and core muscles, as well as enhanced throwing technique and consistency. Coaches noted significant improvements in athletes' posture and biomechanical alignment, leading to more efficient and powerful throws.

User feedback on the JavelinPro 3000 was overwhelmingly positive, with athletes praising its effectiveness in targeting specific muscle groups and promoting proper throwing mechanics. Many athletes reported reduced fatigue and soreness during training sessions, indicating that the machine provided a low-impact yet highly effective workout. Coaches appreciated the versatility of the machine and its ability to accommodate athletes of different skill levels and training objectives. During the changing the height the player facing lot of problem like perfect height according to biomechanical consideration and when the player failed to throw the javelin at 45° experience the muscle injury.

4.3. *Innovations and Improvements in Recent Designs*

In the innovation and improvement of the javelin throw training machine, we address user difficulties such as height adjustment and muscle injury by introducing height adjustment based on the user's height through programming and PLC circuitry. When the user enters their height, the javelin adjusts its height accordingly using the programmed instructions and the PLC controller. Additionally, to prevent muscle injuries, if the player throws the javelin at an angle other than the preferred angle, the stress will be absorbed by a nut and bolt arrangement.

5. Challenges and Future Directions

5.1. *Addressing Cost Constraints and Accessibility Issues*

While javelin throw training machines offer immense potential for improving athlete performance, cost constraints and accessibility issues remain significant challenges. The design and fabrication of advanced training equipment often require substantial financial investment, limiting access for athletes and coaches with limited resources. Additionally, the complexity of some training machines may pose usability challenges, particularly for individuals without specialized training or technical expertise (Linthorne & Cooper, 2016).

To address these issues, researchers and developers are exploring cost-effective design solutions and alternative fabrication techniques that maintain performance standards while reducing production costs. This may involve utilizing off-the-shelf components, simplifying manufacturing processes, or adopting open-source designs that allow for DIY construction. By making training machines more affordable and accessible, a broader range of athletes can benefit from enhanced training experiences, levelling the playing field and promoting inclusivity within the sport (Bartlett & Stirling, 2012; Šarabon et al., 2014; Elite Sports Engineering, 2021; Linthorne & Cooper, 2016).

Furthermore, collaborations between industry stakeholders, sports organizations, and academic institutions can facilitate the development of subsidized or grant-funded programs aimed at providing training machines to underserved communities or emerging talent pools. Such initiatives not only support athlete development but also contribute to the growth and diversification of the sport on a global scale (Linthorne & Cooper, 2016).

5.2. *Collaboration with Sports Scientists and Coaches for Continuous Improvement*

Collaboration between engineers, sports scientists, and coaches is essential for the ongoing improvement of javelin throw training machines. By working together, these stakeholders can ensure that training machines are designed and optimized to meet the specific needs of athletes and coaches (Linthorne & Cooper, 2016).

Sports scientists bring expertise in biomechanics, physiology, and sports performance analysis to the table. Their insights into the mechanics of the javelin throw and the physiological demands placed on athletes can inform the design of training machines. By conducting biomechanical analyses of throwing techniques and studying the physiological responses to training, sports scientists can identify areas for improvement in training machine design. For example, they can provide recommendations for adjusting machine parameters such as release angle, release velocity, and resistance levels to better replicate the demands of competitive throwing (Hubbard, 1983; Bartlett, 1982; SportsNet Solutions, 2022; Linthorne & Cooper, 2016).

Coaches play a crucial role in translating scientific insights into practical training strategies. They have firsthand experience working with athletes and understanding their individual needs and preferences. Coaches can provide feedback on the usability and effectiveness of training machines in real-world training environments. They can also offer guidance on how to integrate training machine workouts into overall training programs and provide valuable input on exercise selection, volume, and intensity (Best & Bartlett, 1988; Bartlett, 1982; Stanković et al., 2016; Harrison et al., 2017; Vanlandewijck et al., 2012; Stanković et al., 2016).

A collaborative approach to training machine development ensures that the machines are not only scientifically sound but also practical and effective for use in real-world training settings. By combining the expertise of engineers, sports scientists, and coaches, training machines can be continually refined and improved to maximize athlete performance and minimize the risk of injury (Bartlett, 1982; Smith et al., 2018; Stanković et al., 2016; Bartlett et al., 2011b; Kim et al., 2018).

6. Conclusion

6.1. *Summary of Key Findings and Insights*

Throughout this review, we have delved into the intricate world of javelin throw training machines, exploring their design, fabrication, and technological aspects. By synthesizing a wealth of research and development efforts in this domain, several key findings and insights have emerged:

Biomechanical Understanding: Biomechanical analysis plays a crucial role in the design and optimization of javelin throw training machines. Understanding the kinematics, kinetics, and muscle activation patterns involved in the javelin throw is essential for replicating the natural throwing motion and maximizing training effectiveness (Hubbard, 1983; Atwater, 1979; Jones et al., 2020).

Technological Advancements: The integration of advanced technologies, such as motion tracking systems, force measurement devices, and virtual reality, has revolutionized javelin throw training. These technologies enable real-time feedback, personalized training programs, and immersive simulation experiences, enhancing athlete development and performance (Leigh, 2012; Linthorne, 2001).

Challenges and Future Directions: Despite the advancements made in javelin throw training machines, challenges such as cost constraints, accessibility issues, and the need for continuous innovation persist. Addressing these challenges and exploring future directions, such as integration with emerging technologies and collaboration with sports scientists, will further propel the field forward (Kim et al., 2018; Athletics Inc, 2023; Linthorne & Cooper, 2016).

6.2. *Importance of Javelin Throw Training Machines in Athlete Development*

Javelin throw training machines play a pivotal role in the development and refinement of athletes' skills in this challenging track and field event. These machines offer numerous benefits that contribute to athlete development and performance enhancement:

Skill Acquisition and Technique Refinement: Javelin throw training machines provide athletes with the opportunity to practice and refine their throwing technique in a controlled environment. By simulating the throwing motion and providing instant feedback on performance metrics such as release velocity, angle of release, and trajectory, these machines enable athletes to make real-time adjustments and improvements to their technique (Terauds, 1974; Harrison et al., 2017; Vanlandewijck et al., 2012; Kim et al., 2018; Linthorne & Cooper, 2016).

6.3. *Final Remarks*

The development of javelin throw training machines represents a significant advancement in athlete training and performance enhancement. Through the integration of biomechanical analysis, cutting-edge technologies, and innovative design approaches, these machines have revolutionized the way athletes prepare for competition. By providing a safe, controlled environment for practicing and refining technique, training machines offer athletes the opportunity to optimize their skills and maximize their potential.

References

Ariel, G., Pettito, R. C., Penny, M. A., and Terauds, J. (1980). Biomechanical analysis of the javelin throw. Track and Field Quarterly Review, 80, 9–17.

Ariel, G., Pettito, R. C., Penny, M. A., and Terauds, J. (2009). Biomechanical analysis of the javelin throw. International Association of Athletics Federations.

Athletics Inc. (2023). Javel Inpro 5000: Advanced javelin throw training machine.

Atwater, A. E. (1979). Biomechanics of overarm throwing movements and of throwing injuries. Exercise and Sport Sciences Reviews, 7, 43–85.

Bartlett, R., Müller, E., Brunner, F., and Clarys, J. P. (2011a). Biomechanical analysis of the javelin throws at the 20 09 IAAF World Championships in Athletics. New Studies in Athletics, 26(1/2), 25–41.

Bartlett, R., Müller, E., Brunner, F., and Clarys, J. P. (2011b). Biomechanical analysis of the javelin throws at the 2009 IAAF World Championships in Athletics. New Studies in Athletics, 26(1/2), 25–41.

Bartlett, R., and Stirling, L. (2012). Javelin throwing technique. Journal of Sports Sciences, 30(2), 177–184.

Bartlett, R. M. (1982). A biomechanical evaluation of the javelin throws of four decathletes. Crewe and Alsager College of Higher Education.

Best, R. J., and Bartlett, R. M. (1988). Aerodynamic characteristics of new rules javelins. Biomechanics in Sport, 33–40.

Dai, B., Mao, D., Garrett, W. E., Yu, B., and Zhang, S. (2019). Biomechanical characteristics of the shoulder complex during the javelin throw: A systematic review and meta-analysis. Journal of Applied Biomechanics, 35(2), 156–163.

Elite Sports Engineering. (2021). Throw ax elite: Modular javelin training machine.

Halson, S. L. (2019). Monitoring training load to understand fatigue in athletes. Sports Medicine, 48(3), 587–598.

Harrison, A. J., Bourdon, P. C., and Gandevia, S. C. (2017). The influence of joint position on the contribution of the rotator cuff muscles to shoulder force production. Journal of Applied Biomechanics, 33(1), 32–37.

Hubbard, M. (1983). Optimal javelin trajectories. Journal of Biomechanics, 17(10), 777–787.

Jones, B. et al. (2020). Biomechanical analysis of elite javelin throwers using a motion capture integrated training machine. International Journal of Sports Biomechanics, 36(4), 489–497.

Kim, S., Lee, J., Kim, H., and Lee, D. (2018). Sensor-based height adjustment mechanism for basketball hoop. IEEE Sensors Journal, 18(23), 9910–9918.

Leigh, S. (2012). The influence of technique on throwing performance and injury risk in javelin throwers. University of North Carolina at Chapel Hill.

Linthorne, N. P. (2001). Analysis of standing throws in athletics. Journal of Applied Biomechanics, 17(4), 317–324.

Linthorne, N. P., and Cooper, S. M. (2016). The optimal release conditions for javelin throwing. Sports Biomechanics, 15(3), 245–258.

Šarabon, N., Markovic, G., Malave, M., Smajla, D., and Vauhnik, R. (2014). Strength training in javelin throwers. Journal of Strength and Conditioning Research, 28(6), 1468–1476.

Saratlija, P., Zagorac, N., and Babic, V. (2013). Influence of kinematic parameters on result efficiency in javelin throw. Collegium Antropologicum, 37(Suppl. 2), 31–36.

Smith, A. et al. (2018). Effects of virtual reality javelin simulation training on novice throwers. Journal of Sports Science & Medicine, 17(2), 234–241.

SportsNet Solutions. (2022). Throw Master 2000: Virtual reality integrated javelin training machine.

Stanković, D., Raković, A., and Danica, P. (2016). The effect of training method and motor ability on javelin throw technique. Research Gate.

Terauds, J. (1974). Optimal angle of release for the competitive javelin as determined by its aerodynamic and ballistic characteristics. In R.C. Nelson and C.A. Morehouse (Eds.), Biomechanics IV (pp. 180–183). University Park Press.

Vanlandewijck, Y., Cools, A., Daly, D., Gorce, P., Goosey-Tolfrey, V., Moss, B., and Tweedy, S. (2012). Kinetic analysis of the javelin throws in Paralympic seated throwers. Journal of Sports Sciences, 30(5), 475–483.

Verbitsky, O., Mizrahi, I., Voloshin, A. S., Treiger, G., and Isakov, E. (2005). Shock transmission and fatigue in human runners. Journal of Applied Physiology, 99(5), 2075–2083.

3 Smart village: a g-Governance initiative for decision support system

Sangita Rajankar[1,a], Soni Chaturvedi[2,b], G. M. Asutkar[2,c], and S. A. Dhale[2,d]

[1]Maharashtra Remote Sensing Application Centre (MRSAC), Department of Planning, Government of Maharashtra, Nagpur, India
[2]Department of Electronics and Communications, Priyadarshini College of Engineering, Nagpur, India

Abstract

g-Governance (Geospatial Governance) encompasses the use of geospatial technology and spatial data in governance processes to enhance decision-making, improve service delivery, and promote sustainable development. g-governance harnesses the power of geographic information systems (GIS), remote sensing, global positioning systems (GPS), and other geospatial technologies to support government operations and policy formulation at various levels. It aims at leveraging geospatial technology and governance strategies to empower rural communities and enhance their quality of life. This initiative focuses on integrating spatial data, implementing spatial data mining (SDM) techniques, and developing decision support systems (DSS) to facilitate informed decision-making and sustainable development towards Smart Village. By integrating geospatial technology within a decision support system, villages can enhance their capacity to conserve water and soil resources, mitigate water-related challenges, and promote sustainable development. Effective utilisation of spatial data and analytical tools can empower decision-makers to implement evidence-based water and soil conservation strategies tailored to the unique needs of each village ecosystem.

Keywords: Decision support system, geospatial, GIS, smart village

1. Introduction

In today's rapidly evolving world, the concept of development extends beyond urban centres to embrace the rural landscape. Smart Village initiatives are emerging as powerful strategies aimed at transforming rural areas into hubs of sustainability, innovation, and prosperity. Inspired by the success of Smart Cities, these initiatives recognise the immense potential of rural communities and seek to harness it through targeted interventions and strategic planning. At the heart of the Smart Village concept lies the idea of leveraging technology and data-driven governance to address the unique challenges faced by rural areas. g-Governance, or geospatial governance,

[a]sangitarajankar@gmail.com, [b]sonipce1@gmail.com, [c]g.m.asutkar@gmail.com, [d]shri22dhale@gmail.com

DOI: 10.1201/9781003567653-3

emerges as a key enabler in this endeavour. It entails the use of geospatial technology, including geographic information systems, remote sensing, and global positioning systems, to inform decision-making processes and drive impactful change at the grassroots level.

By integrating spatial data and advanced analytics, g-Governance empowers policy-makers and local authorities to gain valuable insights into the geographical landscape of villages. These systems harness the power of data mining techniques and predictive analytics to provide actionable intelligence for water conservation, infrastructure development, healthcare delivery, and more.

In the context of water conservation, g-Governance enables the creation of sophisticated DSS that leverage spatial data to optimise water resource management. Decision-makers can visualise water sources, monitor usage patterns, and identify areas prone to scarcity or contamination. Armed with this knowledge, they can formulate targeted interventions, such as watershed management programmes or irrigation optimisation schemes, to promote efficient water use and mitigate risks. In essence, the Smart Village initiative fueled by g-Governance represents a paradigm shift in rural development. By harnessing the transformative potential of geospatial technology and data-driven decision-making, it seeks to empower rural communities, enhance their resilience, and pave the way for sustainable progress in the 21st century.

2. Literature Review

Geospatial technology plays a crucial role in conceptualising and implementing smart village initiatives worldwide.

In India, Government is focusing on rural development right from the time of Mahatma Gandhi who had a vision of Adarsh Gram(Shri Amarjeet Sinha, 2024; Garg, et al., 2015) (Ideal Village) and Swaraj (Self Reliance). Government is trying hard to provide basic amenities like sanitation, safe drinking water, internal roads, tree plantation, water conservation, etc. and thus lead to sustainable development of the village. In recent years, projects like Digital India have utilised GIS for mapping infrastructure, resources and services in rural areas. Initiatives like Smart Village Yojana focus on providing digital connectivity, e-governance services, and skill development to rural communities by geospatial data analysis.

Utilising geospatial technology and open-source software for developing a village information system as part of the smart village programme in Indonesia (Adi et al., 2017) is indeed a promising approach.

To attain the above goal, various spatial data mining techniques are being explored to extract the useful and complex information from geospatial and non- geospatial data. However, the spatial data mining methods are subset of those used in conventional data mining. Various studies related to (Ravikumar et al., 2012) spatial data mining have been developed by two research communities viz. Statistics community and Database Community. Various knowledge discovery processes based on classification rules, clustering, associations, etc. are being carried out. These knowledge base have been successfully tested on individual themes such as Water quality analysis (Kamakshaiah, 2013) and Health Care (Tomar & Agarwal, 2013) for recommending suitable algorithms based on classification model. It also highlights the necessity of data sanitisation, as noisy data may have an adverse effect on the performance of classifier.

With the advent of the Big Data concept, many new challenges and opportunities for spatial data have grewed up (Shuliang Wang, 2014). Big data decision capability is based on Descriptive analytics, Inquisitive analytics, predictive analytics and preemtive analytics (Sivarajah et al., 2017). To improve the various business process flows and improve decision making capacity NoSQL Databases, Big Query, MapReduce,

Hadoop, WibiData and Skytree, are being used.

Various predictive tools can also be developed using Data Mining techniques, statistical tools and Decision Tree and Neural Network techniques (Khalid & Abdelwahab, 2016).

After data mining, there is need of applying this data for analysis for real world. In year 2016, Faisal et al used the data mined information to discover relationships among features, where data objects are grouped according to logical relationship to predict groundwater areas in Jordan (Aburub & Hadi, 2016).

In 2017, Hemlata Goya, et al., from Banasthali University, explored big data using various GIS and SDM tools. They also studied various other algorithmic approaches, issues and challenges and role of spatial association rule mining in big data of GIS (Goyal et al., 2017). They had proposed a MRPrePost, hadoop architecture based, Pre-Post algorithm, used for mining big data in the paper. Various SDM tools specific to its domain have been also specified in this paper.

In 2019, P. Tamilarasi, et al, has studied various Ground Water Data Analysis using data mining tools such as Decision Trees, Naïve Bayes Classifier, K-Nearest Neighbour Method, Rule based classification and Support Vector Machine (Tamilarasi & Akila, 2019). In the same year, MapReduce was used to do raster analysis on Apache Hadoop ecosystem (Mazin et al., 2019).

3. Research Methodology

It's commendable that the state government recognises the importance of developing a decision support system for Smart Village and in taking steps to efficiently visualise and analyse the vast amount of data available. Focusing on a single village initially and then scaling up as needed is a practical approach to ensure effective implementation. Following is the process flow that is being followed upon (Figure 3.1):

- Data Collection, Sanitisation, and Integration:
  - o Gather existing data on the selected village's assets, resources, local

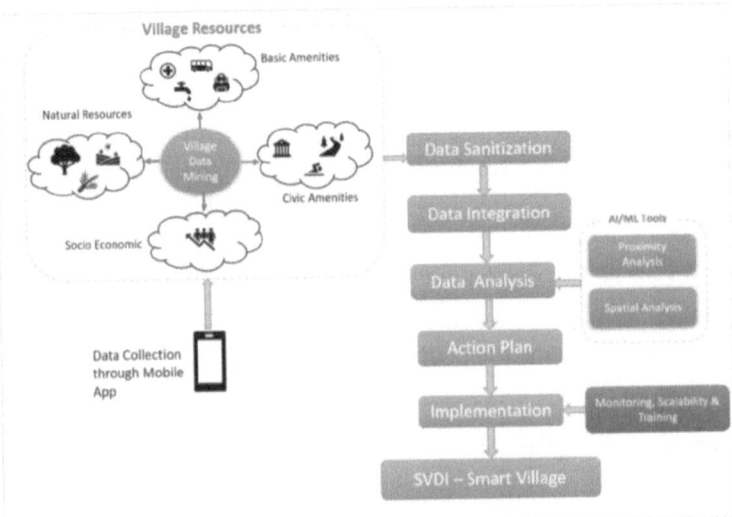

Figure 3.1: Research methodology.

Source: Author.

conditions, and natural resources. This will include infrastructure, land use, soil types, water bodies, water conservation structures and socio-economic indicators.

o Sanitize the collected data into a standardised format.

o Integrate data from various sources, such as government databases, satellite imagery, on spot surveys, field assessments, etc. into a centralised database on GIS platform.

o The integration will be carried out using open-source technology and will use web services to access spatial data. For publishing the spatial data, geoserver will be used.

• GIS Analysis using Spatial Data Mining:

o Utilising various SDM techniques with GIS to analyse and visualise the spatial data layers related to soil and water conservation structures and drainage networks.

o Conduct geospatial analysis to identify soil erosion areas, flooding areas, and areas with other environmental risks. Evaluate the effectiveness of existing soil and water conservation measures and drainage infrastructure.

o The analysis will be based on spatial analysis and will use various modeling features like overlay, surface proximity, suitability, interpolation, etc.

• AI/ML Integration for Geoprocessing:

o Incorporate AI/ML tools in GIS platforms for advanced geoprocessing tasks. Train machine learning algorithms to automate tasks such as slope analysis and hydrological models.

o Use AI/ML models to predict soil erosion rates, identify optimal locations for soil and water conservation structures and optimise the use of drainage networks on terrain characteristics and rainfall patterns.

• Development of Action Plan:

o Based on the results of GIS analysis and AI/ML driven outputs, a comprehensive action plan based on land

resources development plan (LRDP) and water resources development plan (WRDP) for soil and water conservation will be developed.

o Prioritise interventions based on the severity of soil erosion, water conservation, flood risks and local demands. Consider factors such as agricultural practises, community needs, landuse patterns, etc. in the planning process.

o Define specific strategies and measures such as implementing erosion control measures, water conservation structures such as check dams, improving surface drainage and promoting sustainable land management practises.

• Pilot Implementation and Monitoring:

o With the help of local administrations, the proposed action plan will be discussed and implemented for the selected village and monitor its impact.

o Collect real-time data and feedback from the local stakeholders and make identify if any refinements or changes are needed in the plan.

• Scalability and Replicability:

o Once the effectiveness of the DSS and action plan is validated in the pilot village, we will consider scaling up the approach plan to other villages in the state.

o Develop guidelines and tools to replicate the process, leveraging lessons learned and best practises from the pilot implementation.

• Capacity building and stakeholder engagement:

o Provide training and capacity building for local government officials, community leaders and other stakeholders using DSS and implementing the action plan.

Following the above process flow, the government can effectively utilise geospatial technology and AI/ML tools to develop a robust DSS for soil and water conservation in Smart Village.

Gondidigras village in Katol taluka, Nagpur district, has been selected as a study area for implementing Smart Village concept. Figure 3.2 below shows the spatial location of Gondidigras village in the district. The data required for conceptualising smart villages is being collected from different sources such as WMS services, interacting with various departments, etc. For collecting the water conservation structures in the village, a mobile app with GPS data capture facility is being used. The captured structures are depicted in the Figure 3.3 below. The data collected is being standardised and sanitised to improve its quality and data accuracy.

4. Conclusion

The proposed approach for developing a decision support system for Smart Villages focusing on water and soil conservation structures and integrating various natural resources databases is comprehensive and well aligned with the goal of promoting sustainable rural development.

The analysis will help to generate a Spatial Village Development Index (SVDI) which will form the basis of "Smart Village". SDVI can serve as a valuable tool for prioritising resources and interventions in low SVDI villages and thus helps to understand the requirements of each village and

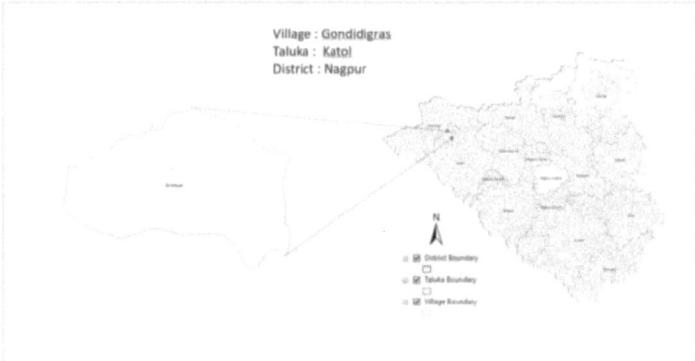

Figure 3.2: Study area.

Source: Author.

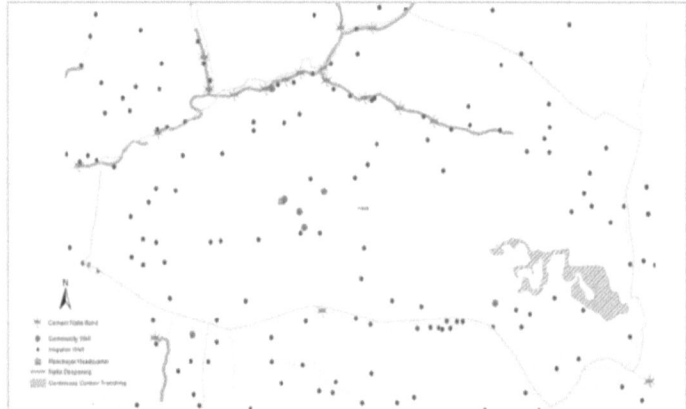

Figure 3.3: Water and soil conservation structures.

Source: Author.

accordingly prepare the action plan for each village.

Overall, the geospatial technology plays a crucial role in conceptualising smart village initiatives worldwide by providing valuable insights into the spatial dynamics of rural communities and supporting evidence-based decision making for sustainable development and inclusive growth. These initiatives demonstrate the potential of geospatial technology to transform rural areas into vibrant, resilient, and digitally enabled smart villages.

References

Aburub, F., and Hadi, W. (2016). Predicting groundwater areas using data mining techniques: Groundwater in Jordan as case study. World Academy of Science, Engineering and Technology, International Journal of Computer and Information Engineering, 10(9), 1621–1624.

Adi, Suroto, Suhartono, Joni, and Janawir. (2017). Smart village geographic information system (GIS) development in Indonesia and its analogous approaches. International Conference on Information Management and Technology (ICIMTech), INSPEC, 65–70. doi: 10.1109/ICIMTech.2017.8273513. https://ieeexplore.ieee.org/document/8273513

Garg, B. S., and Raut, A. V. (2015). Adarsh Gram: A Gandhian dream of Gram Swaraj. Indian Journal of Community Medicine, 40(1), 1–4. doi: 10.4103/0970-0218.149260

Goyal, H., Sharma, C., and Joshi, N. (2017). An integrated approach of GIS and spatial data mining in big data. International Journal of Computer Applications, 169(11), 1–6.

Kamakshaiah, K. (2013). Ground water quality assessment using data mining techniques. International Journal of Computer Applications, 76(15), 39–45.

Khalid, B., and Abdelwahab, N. (2016). A comparative study of various data mining techniques: Statistics, decision trees and neural networks. International Journal of Computer Applications Technology and Research, 5(3), 172–175.

Mazin, A., Jhummarwala, A., and Potdar, M. B. (2019). Multidimensional geospatial data mining in a distributed environment using MapReduce. Journal of Big Data, 6, 82. doi: 10.1186/s40537-019-0245-9

Ravikumar, G., and Sivareddy, M. (2012). An effective analysis of spatial data mining methods using range queries. Journal of Global Research in Computer Science, 3(6), 7–12.

Shuliang Wang. (2014). Spatial data mining: A perspective of big data. International Journal of Data Warehousing and Mining, 10(4), 50–70.

Sinha, S. A. (2024). Secretary, Department of Rural Development, Smart Village India, (saanjhi.gov.in.) Managed by the Ministry of Rural Development, Government of India. https://en.wikipedia.org/wiki/Smart_Village_India (accessed February 12, 2024)

Sivarajah, U., Kamal, M. M., Irani, Z., and Weerakkody, V. (2017). Critical analysis of big data challenges and analytical methods. Journal of Business Research, 70, 263–286.

Tamilarasi, P., and Akila, D. (2019). Ground water data analysis using data mining: A literature review. International Journal of Recent Technology and Engineering (IJRTE), 7(5C), 202–205.

Tomar, D., and Agarwal, S. (2013). Survey on data mining approaches for healthcare. International Journal of Bio-Science and Bio-Technology, 5(5), 241–266. doi: 10.14257/ijbsbt.2013.5.5.25

4 Green hydrogen future fuel to preserve environment

Harikumar Naidu[a] and Swati Kodape[b]

Department of Electrical Engineering, G H Raisoni College of Engineering, Nagpur, India

Abstract

This study shows the importance of the green hydrogen nationally as well as globally. The paper takes a comprehensive approach to India's National Green hydrogen Mission which was launched by Indian government to produce and trade across the globe green hydrogen. This initiative will provide so many opportunities for the employment across the nation. The green hydrogen will be helpful for moving to low carbon economy. There are so many other scheme launched by government for the production of green hydrogen like SIGHT (Strategic Interventions for Green Hydrogen Transition Programme), National Wind-Solar Hybrid Policy, etc. Green hydrogen will be big contribution towards the renewable energy sources for the future aspect.

As a result, the Indian government has formed a ministry group to interrelate and establish to draft the strategy for national hydrogen mission, since India as part of its efforts to decarbonize the economy. Additionally, a number of foreign businesses have showed interest in low carbon hydrogen programs or projects in India.

Keywords: Carbon economy, electrolysis, green hydrogen, renewable energy, solar, wind

1. Introduction

Hydrogen is an element which is not available in the pure form in atmosphere. The gas hydrogen is a periodic element having character H with an atomic count of 1. It is the element with the lowest mass and with the formula H_2, often known as di-hydrogen, but more frequently known as hydrogen. It is transparent, odourless, and bland, non-hazardous and extremely flammable. It has to be extracted from the hydrocarbon compound using electrolysis process. Based on the sources like solar, wind, nuclear, thermal which is used for electrolysis process where hydrogen is classified in so many forms which is shown in the below Figure 4.1 (Report, 2022).

2. Production of Hydrogen

For the production of hydrogen different sort of methods has been used like electrolysis, steam methane reforming (Lord et al., 2014) in which reformer react with Steam at elevated pressures and temperatures interacts with methane and a catalyst composed of nickel to produce hydrogen (H_2) and carbon monoxide (CO). To produce green hydrogen there is need of electricity which is generated through renewable

[a]harikumar.naidu@raisoni.net, [b]swati.kodape.trs@ghrce.raisoni.net

DOI: 10.1201/9781003567653-4

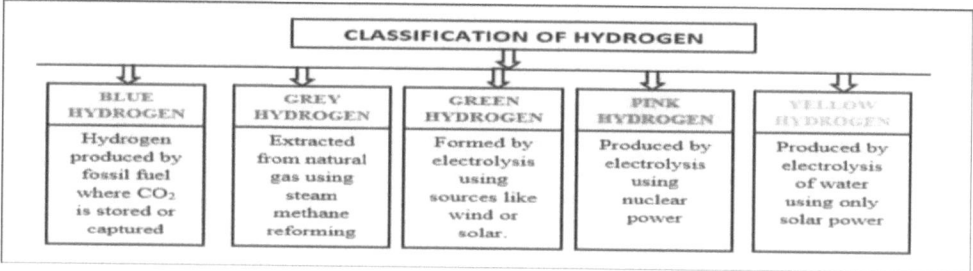

CLASSIFICATION OF HYDROGEN				
BLUE HYDROGEN	**GREY HYDROGEN**	**GREEN HYDROGEN**	**PINK HYDROGEN**	**YELLOW HYDROGEN**
Hydrogen produced by fossil fuel where CO_2 is stored or captured	Extracted from natural gas using steam methane reforming	Formed by electrolysis using sources like wind or solar.	Produced by electrolysis using nuclear power	Produced by electrolysis of water using only solar power

Figure 4.1: Classification of hydrogen.

Source: Report, 2022.

sources like solar, wind, biomass, and hydro for the electrolysis. To produce green hydrogen through electrolysis there are three type of technologies majorly used are Alkaline electrolysis, Proton Exchange Membrane (PEM), and Solid Oxide Electrolysers. By using this process for separation of hydrogen (H_2) and oxygen (O2) from water under the influence of electricity with zero carbon emissions is performed. Process of green hydrogen production shown in Figure 4.2 (Panchenko et al., 2023).

3. Green Hydrogen as Fuel

Fuels which is derived from green hydrogen, such as methanol, ammonia (NH_3), and aviation fuels, can significantly lower greenhouse gas emissions by replacing fossil fuels in the transportation, industrial, and power sectors. They can be used to: (i) heat buildings and industrial operations; (ii) produce chemicals and fuels, such as synthetic hydrocarbon fuels; and (iii) act

as reducing gases in processes that produce goods like computer chips, iron and steel, and glass. They can serve as a solid foundation for CO_2 recycling by being converted into materials, chemicals, and fuels (Vardhan et al., 2022). By enhancing supply stability, green hydrogen can also facilitate a larger share of renewable electricity in the grid (e.g., through storage, fuel cells, and hydrogen turbines). Numerous nations possess abundant renewable energy resources capable of producing hydrogen, and the associated technology.

3.1. Challenges Faced Globally

Currently, worldwide formation of hydrogen is processed through grey hydrogen in which emission of greenhouse gases along with hydrogen occurred which implies harmful for the environment. For the sustainability so many countries like China, Japan adopted the green hydrogen as a fuel for the decarbonisation. But the biggest

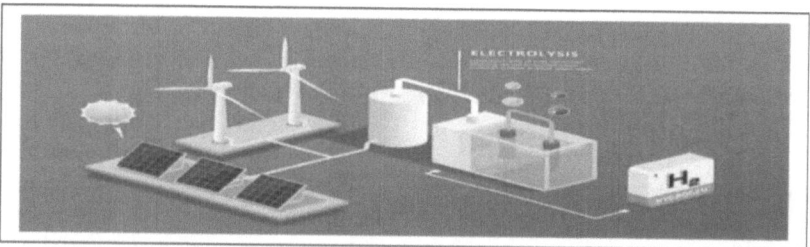

Figure 4.2: Production of Green H_2.

Source: Panchenko et al. (2023).

challenge is the manufacturing of green hydrogen is uneconomical as compare to grey or brown hydrogen. There are several challenges in the form of infrastructure, politics, public acceptance, technology, economics, and public acceptance that restrict green hydrogen products and services from rapidly growing into large markets (Sharma et al., 2023). Several investments and actions are required in order to fully understand the potential of green hydrogen in terms of both economic and climate benefits.

4. Methods and Implementation

The National green Hydrogen Mission (NGHM) aims to make India a "global hub" for the utilization, manufacture, and distributor of green hydrogen, has received approval from the Indian government for a sum of Rs 19,744 crore. Of these, ₹17,490 crore have been set aside by the government for the SIGHT program's implementation ₹1,466 Crore for upcoming pilot programs, ₹400 crore for R&D, and ₹388 cr for extra mission components. The National Green Hydrogen Mission aims to render India an international pioneer in both the manufacturing and distribution of green hydrogen. By 2030, the mission aims to create a minimum of 5 MMT (million metric tons) of annual green hydrogen capability, as well as 125 GW of renewable energy. The incentive's purpose is to mitigate the cost of green hydrogen within a five-year period and increase its accessibility (Report, 2022 NITI Aayog).

To bring the opportunities for the manufacturing of electrolyser government launch the sub scheme named as SIGHT (Strategic Interventions for Green Hydrogen Transition Programme) which provide fund for this. Additionally, it will result in a reduction of nearly 50 MT (metric tons) of yearly greenhouse gas emissions and an overall decrease in consumption of fossil fuels of over Rs 1 lakh crore. It will also entail the generation of 6 lakh jobs in future. Due to the encouragement of government towards the green hydrogen so many industries participating for the production and contribute to mitigate the risk of climate change. The number of companies which are producing green hydrogen by reducing fossil fuel based hydrogen are NTPC, TATA POWER, ONGC, BHEL etc. (Report, 2023a by NITI aayog)

On June 14, 2022, Adani New Industries Ltd. (ANIL), a division of ADANI Group, declared that it has partnered with Total Energies SE of France to invest USD 50 billion in India over the following ten years in order to manufacture green hydrogen and build an ecosystem around it (Hassan et al., 2023). The partnership is expected to focus on developing infrastructure for the manufacture, distribution, and preservation of green hydrogen across India. By the end of the decade, the reliance industries wants to bring down the cost of producing green hydrogen to less than $1/kg. By doing this, the corporation will have taken a big step toward its objective of having zero carbon emanation by 2035. BPCL and the Bhabha atomic research Centre (BARC) have teamed to develop alkaline electrolyser technology in order to progress green hydrogen generation. And many more companies have contributed for the same.

5. Global Scenario of Green Hydrogen

Presently, in world China which now produces the most hydrogen globally, has the ability to employ green hydrogen to lower emissions from important industries and assist the country achieve its climate and energy security goals. With the biggest renewable power capacity in the world, the government wants to "significantly improve" the contribution of clean hydrogen in China's energy consumption by 2035 and build a comprehensive hydrogen economy encompassing the transportation, energy storage, and industrial sectors (Liu & Wan, 2022). The world's biggest utility-scale, commercially operational hydrogen plant powered solely by renewable energy is the NEOM Green Hydrogen Project in Saudi Arabia. And globally so many initiative taken by so many countries towards climate change and zero carbon emission (Report 2023b by WRI).

6. Results and Discussion

1. Creating a Global Market
   o Coordination of policies, pooling of resources for investment and R&D, harmonization of standards for products and procedures, and substantial transition financing.
2. Information of Market
   o To empower the supply-demand energetic to create a cost premium for a higher-performing supplier, a particular low-carbon item is required.
   o Separating resource portfolios to lower the exposure of risk to medium-to long-term showcase development toward a low-carbon future.
   o Better approaches to increase the utilization of intellectual property (IP) beyond the entities of person.
3. Intervention of Policy (IRENA, 2020)
   o Industry pledges to self-regulate and altogether diminish carbon outflows.
   o Carbon taxes or comparable arrangements to reduce the monetary good thing about high-carbon generation.
   o To avoid carbon spillage (i.e., competition from high-carbon imports), purport taxes based on carbon substance ought to be in put.
   o Criteria for carbon execution in open and/or private obtainment.
4. Intervention of Finance
   o To eliminate uncertainty for investors in developing technologies, the government or the private sector could support locking in the value premium for the manufacture of low-carbon steel.
   o Support for late-stage R&D to commercialize inventions currently in pilot stage.

o To enhance the performance of carbon, companies are under pressure of investor.
o Programs for securitization or other money related rebellious to control the conceivable esteem write-down of high-carbon generation resources.

7. Conclusion

By boosting India's industrial competitiveness in a progressively decarbonizing world, cultivating financial development, bringing down CO_2 emanations, and upgrading public wellbeing and quality of life, hydrogen can play a significant part within the country's vitality move. India can take the lead all inclusive in progressing the hydrogen economy since major countries over the world are making noteworthy bets and ventures in hydrogen-based innovation.

References

Hassan, Q., Sameen, A. Z., Salman, H. M., Jaszczur, M., and Al-Jiboory, A. K. (2023). Hydrogen energy future: Advancements in storage technologies and implications for sustainability. Journal of Energy Storage, 72, 108404. doi:10.29119/1641-3466.2023.179.37

IRENA (2020), Green hydrogen: A guide to policy making, International Renewable Energy Agency, Abu Dhabi. www.irena.org/Publications

Liu, W., Wan, Y., Xiong, Y., and Gao, P. (2022). Green hydrogen standard in China: Standard and evaluation of low-carbon hydrogen, clean hydrogen, and renewable hydrogen. International Journal of Hydrogen Energy, 47(58), 24584–24591. doi:10.1016/j.ijhydene.2021.10.193

Lord, A. S., Kobos, P. H., and Borns, D. J. (2014) Geologic storage of hydrogen: Scaling up to meet city transportation demands. International Journal of Hydrogen Energy, 39(28), 15570–15582. doi:10.1016/j.ijhydene.2014.07.121

Panchenko, V. A., Daus, Y. V., Kovalev, A. A., Yudaev, I. V., and Litti, Y. V. (2023). Prospects for the production of green hydrogen: Review of countries with high potential.

International Journal of Hydrogen Energy, 48(12), 4551–4571. doi:10.1016/j.ijhydene.2022.10.084

Report (2022). "Harnessing Green Hydrogen" opportunities for deep carbonization in India, NITI Aayog, RMI.

Report (2023a). National green hyrdogen mission. International Conference on Green Hydrogen. https://www.iea.org/reports/india-energy-outlook-2021 (NITI Aayog and International Energy Agency, India Vision Scenario)

Report (2023b). Accelerating the production and use of green hydrogen. WRI India.

Sharma, S., Kumar, A., and Jain, P. (2023). Green hydrogen in India – Applications, opportunities and challenges. 2023 1st International Conference on Intelligent Computing and Research Trends (ICRT), pp. 1–6. IEEE. doi:10.1109/ICRT57042.2023.10146660

Vardhan, R. V., Mahalakshmi, R., Anand, R. and Mohanty, A. (2022). A review on green hydrogen: Future of green hydrogen in India. 2022 6th International Conference on Devices, Circuits and Systems (ICDCS), Coimbatore, India, pp. 303–330. doi: 10.1109/ICDCS54290.2022.97808

5 Cold storage solutions for sustainable agriculture 4.0: A design outlook for Indian agri-product

Milind Sadashiv Swami[1,a], Manish Sheshrao Deshmukh[2], and Manoj Ramesh Dahake [3]

[1]Research Scholar at Mechanical Engineering Department, AISSMS'S College of Engineering, Pune, India
[2]Associate Professor at Mechanical Engineering Department, AISSMS'S College of Engineering, Pune, India
[3]Assistant Professer at Mechanical Engineering Department, AISSMS'S College of Engineering, Pune, India

Abstract

Applying Industry 4.0 technology to Indian agriculture product design can be expensive. Small-scale farmers face challenges in adopting these advanced systems due to limited resources. Industry 4.0's flexibility approach can benefit agriculture by adapting to its unique requirements. Detailed design consideration with diverse parameters is applied for cold storage system. Addressing the critical need for improved shelf-life preservation of agricultural products, the current study investigates creative design ideas with case studies.

This method is based on the utilization of Agriculture 4.0 advancements like the Internet of Things (IoT) and artificial intelligence and automation to monitor and control cold storage settings in real time. These systems use data analytics with localized actions to precisely regulate temperature, humidity and other characteristics that are crucial for maintaining product quality. Industry 4.0 practices should benefit both large and small farmers balancing costs and benefits are essential for sustainable adoption.

Keywords: Cold storage, temperature, humidity, shelf-life, agriculture 4.0.

1. Introduction

India has the world's second highest production of fruits and vegetables. India's agricultural prowess yields over 100 million metric tons (MT) of fruit and a staggering 200 million MT of vegetables. Despite having a substantial agricultural sector, India positions 101st out of 116 nations in the Global Hunger Index 2021 [GHI]. India, fruits, and vegetables are highly fragile, and their loss amounts to a staggering 35%

[a]msswami@aissmscoe.com, swami.milind@gmail.com

DOI: 10.1201/9781003567653-5

to 40%. Which translates to a substantial monetary loss of 1,160.1 billion INR (Arunendra V Tiwari and Rane Milind, 2022). The main reason for post-harvest losses in poor nations is the lack of suitable storage facilities. [FAO, 2019] (Aysel Elik et al., 2019). Table 5.1, give losses compassion between developed and developing countries. The study estimated annual losses at approximately INR 9,92,717 million, while the nation's overall food security expenditure amounts to INR 15,00,000 million. Prioritizing fruits and vegetables can lead to cost savings through value addition (India's agriculture and food exports: opportunities and challenges, 2022).

In rural communities, refrigeration is costly due to epileptic power supplies and low income (Lal Basediya et al, 2013). Considering global need of agrifood the demand for affordable cooling systems is growing in cold storage systems for agricultural products (Chattopadhyay & Ghosh, 2020). Cold storage procedures can postpone ripening and senescence by controlling respiration rates (Brizzolara et al., 2020).

2. Current Landscape of Cold Storage in India

2.1. Cold Chain System in India

Cold chain infrastructures are an important instrument for reducing post-harvest losses and ensuring India's overall food security. Figure 5.1 illustrate supply chain of agriculture products after harvest. The perfect supply chain for agricultural products requires temperature and humidity management at all stages. Cold storage refrigerated vehicles and refrigerated containers enhance the cold chain by extending storage duration and minimizing damage rates (Zhao et al, 2020). In the supply chain, the absence of a recognized pricing system that prioritizes harvest quality appears to be a discouraging truth at the grassroots level in terms of reducing produce waste (Rajapaksha et al., 2021). Small-scale farmers in underdeveloped nations frequently lack the requisite start-up capital, operating costs, and dependable electrical supply for refrigeration and cold storing. As a result, farmers have restricted access to intact refrigeration systems and sustainable storage facilities(Makule et al, 2022).

Cold storage is important stage in Cold Chain. The development of cold storage facilities in the nation has likewise been

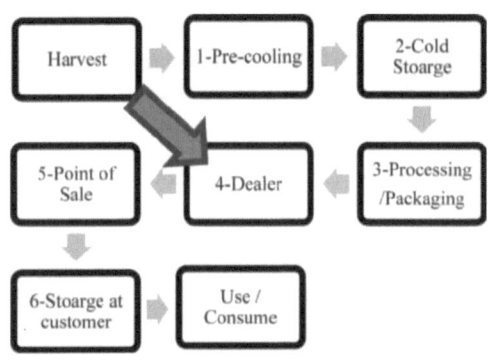

Figure 5.1: Supply Chain of Agriculture Product in India.

Source: Author.

Table 5.1: Losses of perishable foods through lack of cold chain

Losses of perishable harvests through a lack of refrigeration (% out of total production)	Global	Developed Countries	Developing Countries
	20%	9%	23%

Source: The Role of Refrigeration in Worldwide Nutrition, International Institute of Refrigeration, 2014.

severely skewed. The state Uttar Pradesh accounts for around 33% of the country's cold storage, which is primarily utilized to store potatoes (Chakraborty Mahindra, 2020). A huge regional imbalance is observed in cold storage facilities in India. As of March 2014, ten states accounted for more than 90% of the total capacity that was installed in India (Singhal & Saksena, 2017). The Present situation highlights need of cold storage and automated cold chain for more internal and external supply of agrifoods in India.

2.2. Cold Storage Technologies in India

The main technologies used in Cold Storage are Vapor Compression Refrigeration (VCR) and Evaporative Cooling System (ECS). Vapor Compression Refrigeration (VCR) System operations require high-grade electric power to generate a cooling effect, whereas Vapor Absorption Refrigeration (VAR) systems require inferior thermal power to operate like Biomass, solar etc. (Chattopadhyay & Ghosh, 2020). The greatest issue concerning VCRS based Cold Storage is its high upfront and operating costs. Phase Change Energy Storage (PCES) technology in cold chain transport ensures high internal humidity, stable temperatures, energy conservation,

and efficient refrigerated transportation for fruits and vegetables (Qi et al., 2022). The zero-energy cool chamber (ZECC) is a type of evaporative cooling system (ECS) used on farms to store fruits and vegetables. ZECC does not require electricity; it relies on evaporative cooling to prevent produce from drying out, thus extending its shelf life. Farmers can store their harvest without energy cost concerns (Ghosal M. K., 2019). The amalgamation of energy from the sun, wind energy and thermal power storage can cut operating costs and improve system reliability. On-farm cold storage spaces can dramatically lessen losses, and integrating them with solar power would lower operational costs (Munir et al., 2021).

3. Cold Storage-Sustainable Agriculture 4.0

Technological upgradation and appropriate management in field of post-harvest with connecting value addition cold chain system is required to tackle life-threatening encounters of farmers. Industry 4.0, introduces smart automation, data analytics, and networking. It is like giving tiny farmers a digital boost. Figure 5.2 describe IoT based Cold Storage Administration. Cold storages require monitoring of ecofriendly parameters such as humidity, temperature,

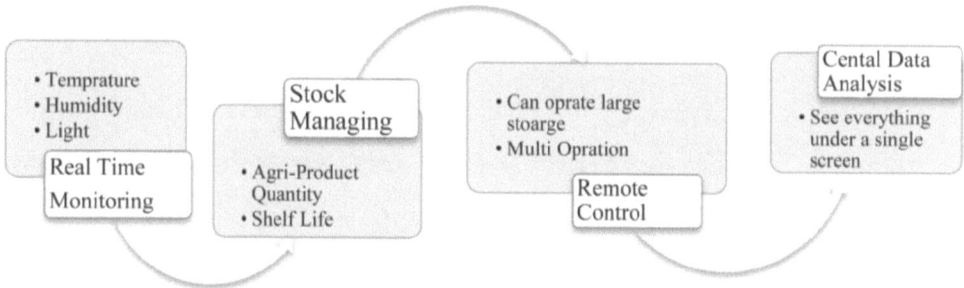

Figure 5.2: IoT Based Cold Storage Administration.

Source: Author.

hazardous gases, and light concentration to ensure commodity quality (Nagpal & Sudha, 2015). The sustainability performance helps determine how efficiently the cold chain preserves freshness while minimizing energy usage (Gallo et al., 2017). The REAMIT- LCA tool enabled quantifying of environmental impacts such as global warming, source use, and extra types of impact (Costa et al., 2023).

The capacity to design and self-adapts can significantly improve the operational effectiveness of cold-storing systems and reduce wasteful use of resources (Appasani et al., 2021). Strategically embedding sensor nodes in storage systems improves product quality and extends shelf life, benefiting everyday consumers with greater access to fresh produce (Nagpal & Sudha, 2015).

Using a basic neural network and reinforcing learning achieved exceptional savings in energy of 27.33% while preserving stable temperatures, even throughout scenarios such as entry openings (Park et al., 2023). To monitor fresh food during transportation in the cold chain, employ temperature-controlled vehicles and cloud connected sensors. Mobile alerts are sent to trucks traveling over 50 kilometers, ensuring swift action in case of temperature irregularities (Maiyar et al., 2023).

4. Case Studies and Examples

4.1. Tomato as Product for Designing Cold Storage of Next Generation

India is the world's second-largest tomato grower, with 21.18 million tons of tomatoes in Year 2021 [APEDA Report]. It is a key ingredient in Indian cuisine. Due to restricted processing of tomato products, approximately 31 percent of the tomatoes grown are wasted. When tomato prices plummet due to oversupply, farmers struggle to cover their harvest costs. Consequently, they often abandon tomatoes on the farm, leading to wastage (Mohan et al., 2023). Optimal post-harvest procedures for freshly picked tomato fruit require low temperatures (10°C–15°C) and high moisture content (85–95%) during harvesting, preservation, and transport to the market (Alenazi et al., 2020). Tomatoes have a limited life span and small economic worth

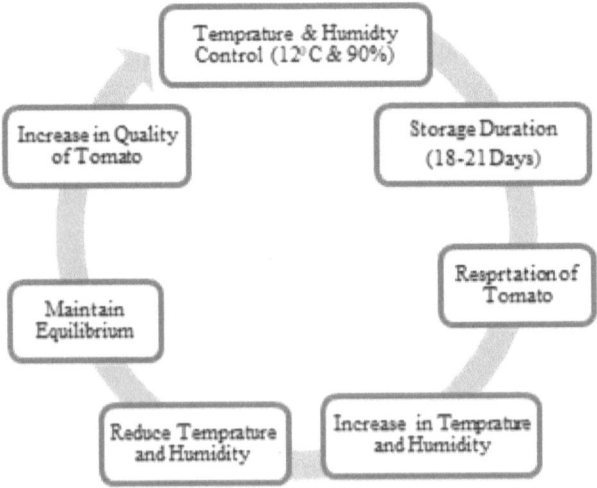

Figure 5.3: Data cycle of Tomato cold storage.

Source: Author.

(Maiyar et al., 2023). Shelf-life is an significant issue for tomato, and it differs from variety to variety (Sinha et al., 2019).

These inputs are essential for the design of cold storage. Since tomato storage requires temperatures within the range of 12°C, the ZECC, equipped with monitoring and control devices, becomes an ideal solution. Evaporative cooling devices can be made locally with low-cost materials such as jute sacks, planks of wood, and basins (Arah et al., 2016).

Additionally, considering storage duration and database choices for market distribution or food processing is crucial. Notably, funding from the National Horticulture Board (NHB) is available for the modernization of cold storage facilities for horticultural products. In combination with traditional methods IoT based equipment are used to collect data of Cold Storage as illustrated in Figure 5.3. Tomato handlers in nations with limited resources may lack efficient postharvest techniques to minimize losses. Effective communication and continual information updates would assist farmers in planning their harvest schedule to prevent post-harvest losses (Mohan et al., 2023). Knowing the best postharvest practices for tomatoes has been useful in combination with IoT-enabled equipment for greater precision. (Arah et al., 2016).

4.2. Grapes Production in Nasik Maharashtra

Maharashtra is India's biggest grape producer, which produces approximately 81 percent of the total grapes production in the country. Grape producers in Maharashtra employ modern agricultural methods to achieve and uphold world-class grape quality standards (Sadi & Arabkoohsar, 2020). During the COVID-19 epidemic, Maharashtra's grape value chain encountered numerous obstacles. In response, grape producers and cooperative organizations adopted techniques such as digital advertising, online platforms, sales direct, contracted farming, and value addition. To retain the postharvest taste of table grapes, they are typically stored at a low temperature of roughly 0°C with elevated relative humidity. Emerging countries has significantly increased table grapes exports as an indicator of economic prosperity (Romero et al., 2020). Figure 5.4 describe the flow of grapes supply after harvest. Every stage decision required more accuracy and reference of National or International norms to be implemented which are achieved with help of IoT based equipment and Dada analysis. Efficient logistics that prioritize product quality can help reduce losses and enhance shelf life (de Oliveira et al., 2021). Image analysis can help analyze and assure the nutritional value and freshness of agricultural products. Innovative models based on selected texture and predictive algorithms can objectively assess changes in fruits and vegetables during storage (Ropelewska & Noutfia, 2024).

The support of advance technology, contributes to high income, employment, foreign exchange earnings and address food security challenges (Study of Value Chain for Grapes Nasik, Maharashtra National

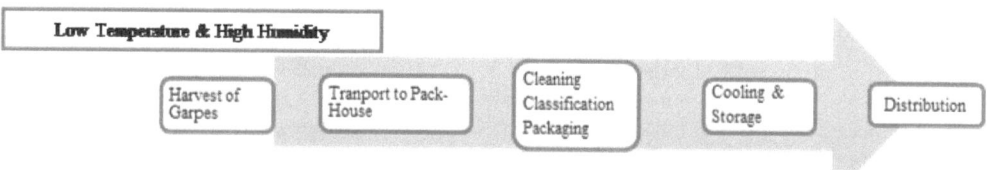

Figure 5.4: Post harvest Grapes Supply Flow Charn.

Source: Author.

Committee for Plasticulture Applications in Horticulture).

Conclusion

Cold storage facilities contribute to reduce improve food quality, post-harvest losses, ensure safety, and increase farmers' revenue and market reach. However, in rural areas of developing nations such as India, the availability and affordability of cold storage facilities are severely limited due to a lack of electricity, infrastructure, financial resources. To overcome challenge a collaborative approach for technology application including several stakeholders' farmers, cooperatives and the Government is required. Modern systems use data analytics to precisely control temperature, humidity, and other essential conditions, resulting in optimal product quality. Working together with stakeholders, remove constraints and support the development and acceptance of cold storage facilities in rural areas, in line with Agriculture 4.0 principles. Finally, the enduring growth of agricultural profits is dependent on developing a healthy agrifood processing, which is backed by a well-connected network of cold storage facilities and efficient cold chain infrastructure.

References

Alenazi, M. M., Shafiq, M., Alsadon, A. A., Alhelal, I. M., Alhamdan, A. M., Solieman, T. H., ... & Al-Selwey, W. A. (2020). Improved functional and nutritional properties of tomato fruit during cold storage. *Saudi Journal of Biological Sciences*, 27(6): 1467–1474.

Appasani, B., Jha, A. V., Ghazali, A. N., & Gupta, D. K. (2021). Analytical modeling and optimal control of cold storage system with large-scale implementation using IoT. In *Advances in Smart Grid Automation and Industry 4.0: Select Proceedings of ICETS-GAI4. 0* (pp. 51–59). Springer Singapore.

Arah, I. K., Ahorbo, G. K., Anku, E. K., Kumah, E. K., & Amaglo, H. (2016). Postharvest handling practices and treatment methods for tomato handlers in developing countries: A mini review. *Advances in Agriculture*, 2016(1): 6436945.

Brizzolara, S., Manganaris, G. A., Fotopoulos, V., Watkins, C. B., & Tonutti, P. (2020). Primary metabolism in fresh fruits during storage. *Frontiers in Plant Science*, 11: 80.

Chakraborty Mahindra, M. (2020). Cold storage in India: Challenges and prospects. *Agriculture & Food E-Newsletter*, 2(10): 458–460.

Chattopadhyay, S., & Ghosh, S. (2020). Comparative energetic and exergetic assessment of different cooling systems in vegetable cold storage applications. *Journal of the Institution of Engineers (India): Series C* 101 (4): 643–50. https://doi.org/10.1007/s40032-020-00579-2

Costa, T. P. D., Gillespie, J., Pelc, K., Adefisan, A., Adefisan, M., Ramanathan, R., & Murphy, F. (2023). Life cycle assessment tool for food supply chain environmental evaluation. *Sustainability*, 15(1), 718. https://doi.org/10.3390/su15010718

Elik. A. (2019). Strategies to reduce post-harvest losses for fruits and vegetables. *International Journal of Scientific and Technological Research*, March. https://doi.org/10.7176/jstr/5-3-04

Gallo, A., Accorsi, R., Baruffaldi, G., & Manzini, R. (2017). Designing sustainable cold chains for long-range food distribution: Energy-effective corridors on the Silk Road Belt. *Sustainability*, 9(11): 2044.

Ghosal, M. K. (2019). Storage of vegetables in zero energy cool chambers: A review. *EAS Journal of Nutrition and Food Sciences*, 1(4): 68–73. https://doi.org/10.36349/easjnfs.2019.v01i04.002

Lal Basediya, A., Samuel, D. V. K., & Beera, V. (2013). Evaporative cooling system for storage of fruits and vegetables: A review. *Journal of Food Science and Technology*, 50: 429–442.

Maiyar, L. M., Ramanathan, R., Roy, I., & Ramanathan, U. (2023). A decision support model for cost-effective choice of temperature-controlled transport of fresh food. *Sustainability*, 15(8): 6821.

Makule, E., Dimoso, N., & Tassou, S. A. (2022). Precooling and cold storage methods for fruits and vegetables in Sub-Saharan Africa—A review. *Horticulturae*, 8(9): 776.

Mohan, A., Krishnan, R., Arshinder, K., Vandore, J., & Ramanathan, U. (2023). Management of postharvest losses and wastages in the Indian tomato supply chain—a temperature-controlled storage perspective. *Sustainability*, 15(2): 1331.

Munir, A., Ashraf, T., Amjad, W., Ghafoor, A., Rehman, S., Malik, A. U., ... & Morosuk, T. (2021). Solar-hybrid cold energy storage system coupled with cooling pads backup: A step towards decentralized storage of perishables. *Energies*, 14(22): 7633.

Nagpal, S. K., & Sudha, S. (2015, April). Design and development of a sensor node to monitor and detect change in internal parameters of a cold storage system. In *2015 Global Conference on Communication Technologies (GCCT)* (pp. 745–748). IEEE.

Oliveira, C. C. M. D., Oliveira, D. R. B. D., & Silveira, V. (2020). Variability in the shelf life of table grapes from same batch when exposed under different ambient air conditions. *Food Science and Technology*, 41(suppl 1): 290–300.

Park, J. W., Ju, Y. M., Kim, Y. G., & Kim, H. S. (2023). 50% reduction in energy consumption in an actual cold storage facility using a deep reinforcement learning-based control algorithm. *Applied Energy*, 352: 121996.

Qi, T., Ji, J., Zhang, X., Liu, L., Xu, X., Ma, K., & Gao, Y. (2022). Research progress of cold chain transport technology for storage fruits and vegetables. *Journal of Energy Storage*, 56: 105958.

Rajapaksha, L., Gunathilake, D. M. C. C., Pathirana, S. M., & Fernando, T. (2021). Reducing post-harvest losses in fruits and vegetables for ensuring food security—Case of Sri Lanka. *MOJ Food Process Technols*, 9(1): 7–16.

Romero, I., Vazquez-Hernandez, M., Maestro-Gaitan, I., Escribano, M. I., Merodio, C., & Sanchez-Ballesta, M. T. (2020). Table grapes during postharvest storage: A review of the mechanisms implicated in the beneficial effects of treatments applied for quality retention. *International Journal of Molecular Sciences*, 21(23): 9320.

Ropelewska, E., & Noutfia, Y. (2024). Application of image analysis and machine learning for the assessment of grape (Vitis L.) berry behavior under different storage conditions. *European Food Research and Technology*, 250(3): 935–944.

Sadi, M., & Arabkoohsar, A. (2020). Techno-economic analysis of off-grid solar-driven cold storage systems for preventing the waste of agricultural products in hot and humid climates. *Journal of Cleaner Production*, 275: 124143.

Singhal, R., & Saksena, S. (2017). Performance assessment of the storage and warehousing industry in India. *The Journal of Industrial Statistics*, 6(1): 15–40.

Sinha, S. R., Singha, A., Faruquee, M., Jiku, M. A. S., Rahaman, M. A., Alam, M. A., & Kader, M. A. (2019). Post-harvest assessment of fruit quality and shelf life of two elite tomato varieties cultivated in Bangladesh. *Bulletin of the National Research Centre*, 43: 1–12.

Study of Value Chain for Grapes Nasik, Maharashtra National Committee for Plasticulture Applications in Horticulture. (n.d.). Study Report by National Committee for Plasticulture Applications in Horticulture (NCPAH). Ministry of Agriculture & Farmers Welfare, Govt. of India, New Delhi.

Tiwari, A., Harischander, H., & Rane, M. V. (2022). Cold Storage in India for Small Farmers - Current Status And. In *International Refrigeration and Air Conditioning Conference*. Paper 2472. https://docs.lib.purdue.edu/iracc

Zhao, Y., Zhang, X., & Xu, X. (2020). Application and research progress of cold storage technology in cold chain transportation and distribution. *Journal of Thermal Analysis and Calorimetry*, 139(2): 1419–1434.

6 Design and fabrication of automated commode cleaning machine

Rakesh Adakane[a], Suvarsh Rajankar[b], Abhishek Dharme[c], Utkarsh Patil[d], and Ankit Kumar[e]

Department of Mechanical Engineering, Yeshwantrao Chavan College of Engineering, Nagpur, India

Abstract

Maintaining proper hygiene in public and private restrooms is crucial for public health. Manual commode cleaning is a laborious and time-consuming operation. By creating an automated toilet cleaning system that efficiently cleans commodes without the need for human interaction, our research seeks to address this problem. The suggested automated commode cleaning system comprises of a rotating horizontal shaft driven by a DC motor that extends above the commode. A lead screw mechanism for vertical movement and a brush at one end are housed in a vertical shaft that is attached to a horizontal shaft. Another DC motor turns on the brush when it comes into contact with the commode's surface, cleaning the appliance thoroughly. This mechanism is automated using Arduino. Compared to human cleaning techniques, this automated commode cleaning device has benefits as it lowers the danger of infection exposure and raises standards for general hygiene.

Keywords: Automated, cleaning, commode, lead screw mechanism

1. Introduction

Ensuring proper hygiene in both public and private restrooms is not just a matter of convenience but a critical aspect of safeguarding public health. Unfortunately, the manual task of commode cleaning, although essential, often falls short due to its labor-intensive and unsanitary nature. Recognizing this challenge, our research seeks to revolutionize restroom maintenance by introducing an innovative automated toilet cleaning system. This solution aims to efficiently clean and sanitize commodes without the need for human interaction, marking a paradigm shift in restroom cleanliness.

The proposed automated commode cleaning system serves as a testament to technological ingenuity. At its core is a 90° rotating horizontal shaft, meticulously powered by a 10 RPM motor, extending gracefully above the commode. This design lays the foundation for a transformative approach to toilet hygiene. A vertical shaft, housing a lead screw

[a]rvadakane@ycce.edu, [b]suvarsh29@gmail.com, [c]abhishekdharme687@gmail.com, [d]utkarshpat27@gmail.com, [e]ankitkumarroy82@gmail.com

DOI: 10.1201/9781003567653-6

mechanism for precise vertical movement, intricately connects to the horizontal shaft. At one end of this vertical shaft sits a cleaning brush, poised to make contact with the commode's surface and execute a thorough cleaning operation.

The choreography of this automated cleaning ballet is orchestrated by the synchronized movements of motors and mechanisms. When activated, the cleaning brush descends gracefully with the precision afforded by the lead screw mechanism, ensuring optimal coverage of the commode's surface. The 90° rotation of the horizontal shaft allows the brush to reach all corners and contours. The result in a cleaned commode, free from the limitations inherent in manual cleaning methods.

The advantages of this automated commode cleaning device over traditional human-centric techniques are multifaceted. Foremost is the elimination of direct human touch with contaminated surfaces, thereby mitigating the risk of infection exposure. This not only prioritizes the well-being of restroom maintenance personnel but also sets a new standard for general hygiene in public and private spaces.

To ensure widespread adoption and user acceptance, a user-friendly interface is paramount. Designing an intuitive control panel or app that allows users to schedule cleaning times, adjust settings, and monitor the system's performance can enhance the overall user experience.

Although the suggested automated commode cleaning system has promise, it is imperative to address any possible issues and difficulties that may arise during implementation. One such factor is the requirement for routine upkeep and observation to guarantee the system's durability and continuous operation. Compatibility with different bathroom designs and fixtures could also be problematic.

Looking ahead, the success of the automated commode cleaning system could serve as a catalyst for further innovations in restroom hygiene.

2. Literature Review

2.1. *Bathroom Floor Cleaning Robot System*

The research paper titled "Design and Modelling of Self-sustainable Bathroom Floor Cleaning Robot System" by Azamat Yeshmukhametov, Ainur Baratova, Ainur Salem khan, Zholdas Buribayev, and Kassymbek Ozhikenov, presented at the 21st International Conference on Control, Automation and Systems (ICCAS) in 2021, proposes a novel self-sustainable bathroom floor cleaning robot system to address the challenges of maintaining cleanliness in public restrooms. The proposed system utilizes a robot arm with an extended long reach to effectively clean the entire floor area of a standard bathroom cubicle. From this paper we got an idea for developing a simple robotic arm like mechanism to clean the commode.

2.2. *Automatic Restroom Cleaner*

Authored by T. Kannan, V. R. Gandhiram, V. Hariharan, and S. Dinesh Kumar, and published in the International Journal of Research in Engineering, Science, and Management in June 2020, this paper's fundamental objective revolves around designing a novel system capable of significantly reducing human intervention while ensuring thorough and timely cleaning routines. In essence, the proposed solution aims to alleviate the challenges associated with maintaining restroom hygiene by leveraging cutting-edge technology. This study inspired us to automate the mechanism to reduce the human efforts.

2.3. *Review of Lavatory Cleaning Devices*

The research paper titled "Lavatory Cleaning Devices and Their Feasibility in Public Toilets in Developing Countries" by R. Hari Krishnan, published in the International

Journal of Intelligent Robotics and Applications in 2020, this study surveys various types of intelligent robotic systems equipped with capabilities to clean and sanitize public toilets autonomously. These devices incorporate advanced features such as sensors, AI algorithms, and adaptability to different restroom layouts. Moreover the paper examines the challenges and barriers associated with the adoption of intelligent robotic systems in developing nations. Issues such as initial investment costs, technological infrastructure, user acceptance, and operational sustainability are thoroughly evaluated to provide a holistic view of the implementation challenges. The paper examines the challenges associated to high cost thus there is need to design a mechanism which will be cost effective.

2.4. Automatic Toilet Bowl Cleaner

The research paper titled "Automatic Toilet Bowl Cleaner" in the International Journal of Advanced Research in Science, Communication, and Technology (IJARSCT) authored by Ashwini Patil, Rutuja Patil, Snehal Phalke, and Prof. S.T. Khot discusses the development of an automatic system aimed at revolutionizing toilet hygiene as part of India's "Swachh Bharat Abhiyan" initiative. From this study we got tentative list of components to make the mechanism automated like Arduino board, motor drivers, voltage regulators, etc.

2.5. Lead Screw Mechanism

The research paper titled "Design and Refabrication of Advanced Mechanism for Indian Toilet Dome Cleaning" presents a comprehensive study detailing the design, development, and implementation of an advanced cleaning mechanism specifically targeted at Indian-style toilet domes. This research paper gave us idea of using lead-screw mechanism to give vertical motion to brush.

3. Patent

Zvi Elster Ramla (2005), Cleaning Brush for Sanitary Appliance, United States Patent Application Publication.

A cleaning brush with a housing reservoir that holds cleaning fluid and is in fluid communication with a pump; a hollow spindle that is rotatably mounted in the housing and powered by a motor; a brush with bristles in fluid communication with the hollow spindle so that fluid passing through the hollow spindle enters the brush's bristles directly [VI].

Chin Chuan Chang, Zhen Chyan, (2001), Automatic Cleaner for Toilet Seat, United States Patent.

The automatic toilet seat cleaner that comes with both a cleaner and a seat is the subject of the current invention. The toilet seat has a turning seat and a bottom seat, while the cleaner has a washing, flushing, drying, and transmission unit. The flushing mechanism of the cleaner sprays a specific quantity of cleaning liquid while the gearbox unit within the cleaner rotates the toilet seat. The washing machine is then turned on to begin washing. Furthermore, the turning seat's surface is dried by the drier to produce a sterile, hygienic, and warming effect [VII].

4. List of Components

Table 6.1: List of Components

COMPONENTS
Control Unit - Arduino Nano board
Voltage Regulator (5.2 Volt) – 7805
Motor Drivers (2) - L293D or L298N
DC Motors (Square gear box) - 12 Volt 10RPM, 60RPM
Power Supply - 12 Volt 2 Amp
Display unit – Alphanumeric
Buzzer – 5 Volt
Hollow Pipe – Mild Steel
Lead Screw
Bearings
Chain Sprocket

Source: Author.

4.1. CAD Model

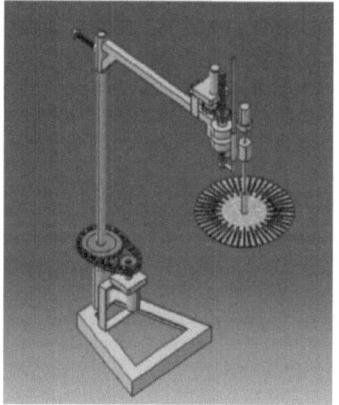

Figure 6.1: 3D CAD model.

Source: Author.

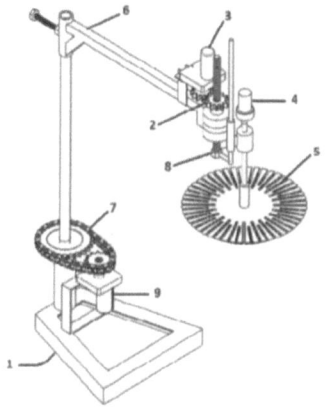

Figure 6.2: Position.

Source: Author.

Table 6.2: Position and description

Pos.No.	Description
1	Base (foundation/ support)
2	Gears & bearing assembly
3	Motor (driving gear)
4	Motor (brush)
5	Brush assembly
6	Horizontal shaft
7	Chain-sprocket assembly
8	Leadscrew
9	Motor for chain sprocket

Source: Author.

Figure 6.3: Fabricated model.

Source: Author.

5. Working Principle

The commode cleaning system consists of a Vertical rotating horizontal shaft driven by a 10 RPM DC motor that extends the horizontal shaft above the commode by 90°. A lead screw mechanism for vertical movement is controlled by a 60 RPM DC motor, and a brush at one end is housed on a vertical shaft. Another DC motor of 60 RPM turns on and rotates the brush, and when it comes into contact with the commode's surface, it cleans the commode thoroughly by friction. All of this is automated using Arduino Nano board, Motor drivers.

6. Maintenance

This automated commode cleaning system will face challenges related to routine upkeep and maintenance. To ensure its continuous operability, the brush needs to be changed after a certain period of time. The gears & bearings should be taken care of since they are in continuous friction conditions, so they are to be greased at regular intervals. When it comes to electronic components, if they malfunction, they eventually need to be replaced.

7. Compatibility

The proposed model is designed as a portable system, whose portability is ensured by its base/foundation structure, such that it can be placed wherever we want at the required location (lavatory positions). In terms of its compatibility with various changing commode designs and positions, a minor change in its programming can make it suitable for those conditions.

8. Conclusion

The automated commode cleaning system is a game-changer for restroom hygiene. It uses synchronized rotating movements to clean every part thoroughly without human touch. This not only reduces infection risks but also sets a new cleanliness standard. Plus, it saves time, money, and is eco-friendly. Overall, it's a big step forward for public health and keeping restrooms clean.

Limitations of this device include that it needs a required space near the area of the commode to get installed and work properly, and it is unable to clean the commode properly, which is extremely stained and dirty.

As of now, only the working operations of this model are automated using Arduino, but it can be integrated with other automation variants, which include various electronics sensors, to change its way of operating accordingly.

Note: Since the model is still under fabrication and R & D the results are yet to be obtained.

References

Chin Chuan Chang, and Zhen Chyan (2001). Automatic Cleaner for Toilet Seat. United States Patent WO 2016/203475.

Hari Krishnan, R. (2020). A brief review on lavatory cleaning devices and their feasibility in public toilets in developing countries. International Journal of Intelligent Robotics and Applications, 4(3), 354–369. doi: 10.1007/s41315-020-00143-2

Kale, S., Hedaoo, P., Sakharkar, M., Kewate, A., Lohakare, G., and Mokhare, P. (2022). Design and refabrication of advanced mechanism for indian toilet dome cleaning. International Journal of Advanced Research in Science, Communication and Technology (IJARSCT), 2(2), 771–775.

Kannan, T., Gandhiram, V. R., Hariharan, V., and Dinesh Kumar, S. (2020). Automatic restroom cleaner. International Journal of Research in Engineering, Science and Management, 3(6), 261–262.

Patil, A., Patil, R., Phalke, S., and Khot, S. T. (2022). Automatic toilet bowl cleaner. International Journal of Research in Electronics and Computer Science (IJRECE), 7(2), 2764–2767.

Ramla, Z. E. (2005). Cleaning brush for sanitary appliance. United States Patent Application Publication US2005/0022324A1.

Yeshmukhametov, A, Baratova, A., khan, A. S., Buribayev, Z. D., and Ozhike, K. (2021). Design and modelling of self-sustainable bathroom floor cleaning robot system. 2021 21st International Conference on Control, Automation and Systems (ICCAS), Jeju, Korea. doi: 10.23919/ICCAS52745.2021.9649969.

7 A novel study of pesticide sensors to preserve the global environment

Manav Wasekar[a] and Harikumar Naidu[b]

Department of Electrical Engineering, G H Raisoni College of Engineering, Nagpur, India

Abstract

Pesticides exercise a fundamental influence in the degradation of soil and water. An array of techniques is employed for the assessment and identification of pesticides. Pesticide sensors could be broadly classified into the electrochemical, optical and mechanical domains. An extensive study was conducted using field-effect transistor biosensors, which is an innovative method for chemical and pesticide detection. In addition, FET biosensors based on nanomaterials provide Immediate and speedy responses with heightened sensitivity and minimal voltage inputs. A comparative study of biosensors is reviewed with the objective of examining the use of recognition elements such as enzymes, antibodies, and glucose oxidase, as well as polymers in FET biosensors for chemical and pesticide detection which will be helpful to R&D experts in their experimental domain to preserve the environment.

Keyword: FET, biosensor, pesticide, nonmaterial, bio receptor

1. Introduction

Pesticide overuse pollutes the environment. Pesticides are dangerous chemicals that are commonly found in soils, waterways, and agricultural goods. Pesticide residues can be highly contaminated, which can be dangerous for human health and severely detrimental to the environment. To ensure their quality and protect people from any risks, it is crucial to regularly do on-site and real-time analysis and monitoring traces of pesticide in food, soil, air and water.

There are several approaches to analyse pesticide residues. Traditional methods for the analysis include mass spectrometry (MS), chromatography. GC-MS and LC-MS chromatographic methods are frequently used in conjunction with mass spectrometry to detect and identify pesticides. Another technique that is frequently used to analyses pesticides in complicated matrices, such food samples, is high-performance liquid chromatography (HPLC). One method that is frequently used to detect pesticides is immunoassay. To find and quantify the concentration of pesticides in a sample, it uses antibodies. These methods have shown effective analysis of traces with good reproducibility and high sensitivity, but they are limited in their applicability for on-site and

[a]manav.wasekar.mtechips@ghrce.raisoni.net, [b]harikumar.naidu@raisoni.net

DOI: 10.1201/9781003567653-7

real-time applications by issues like laborious procedures, reliance on expensive equipment, and protracted sample preparation processes. As a result, new and different ways for detecting pesticides have surfaced in recent years. These methods include mechanical, optical, and electrochemical examination. The relevance of the electrochemical approach for on-site sensor detection has grown. The pesticide sensor categorization, as documented by Hashwan et al., 2020, is displayed in Figure 7.1.

The focus on FET-based transducers arises from their capability to directly interpret interactions between analytes and FET surfaces. This research investigates the application of enzymes, antibodies, glucose oxidase, and polymers within FET sensors for detecting chemicals and pesticides.

2. Field Effect Transistor

FET based biosensors are state-of-the-art devices for chemical and pesticide detection. These biosensors utilize the FET principle, wherein the conductance between the terminals of source and drain is regulated by the voltage applied to the gate terminal. Three terminals make up the FET. Source, which allows carriers to enter the channel; the drain, where carriers leave the channel; and the Gate Terminal, responsible for regulating the conductivity of the channel.

Based on the kind of semiconductor material employed and the major charge carriers, FET biosensors may be divided into two primary categories n-type and p-type FET Biosensors.

Both n-type and p-type FET biosensors use changes in conductivity near the gate terminal to detect specific substances. By attaching bioreceptor molecules to the gate electrode, which selectively bind to target substances, FET biosensors achieve high sensitivity and specificity, making them essential in healthcare, environmental monitoring, and food safety. In n-type FET biosensors, detecting positively charged molecules increases conductivity due to electron accumulation, while in p-type FET biosensors, detecting positively charged molecules decreases conductivity but increases conductivity with negatively charged molecules due to changes in hole concentration. These basic principles underlie the functionality of FET biosensors in various substance detection scenarios.

The Key Principles of How FETs Operate

1. FETs are equipped with three terminals they are the source, the drain, and the gate. The source and drain terminals establish connections with the semiconductor material through which current travels, while the gate terminal regulates the conductivity of semiconductor

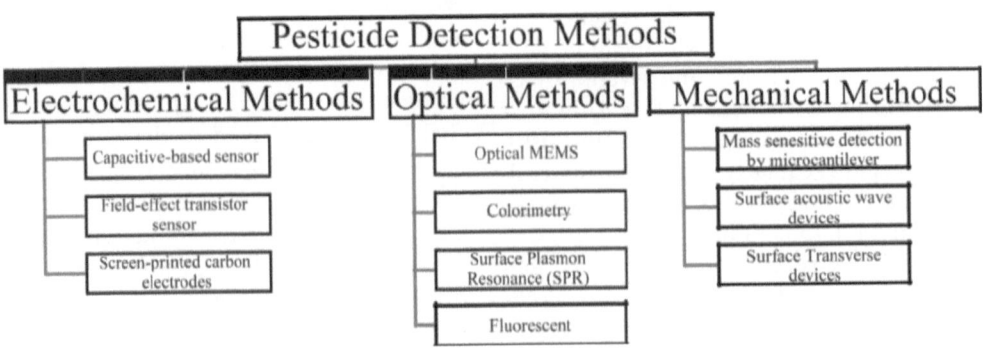

Figure 7.1: Pesticide sensor classification.

Source: Hashwan et al. (2020).

channel that links terminals source and drain.

2. FETs manage the movement of current from the source and drain terminals via manipulating the electric field within the semiconductor channel. This electric field is created by the voltage administered to the gate terminal. By adjusting this voltage, the FET can amplify or diminish the conductivity of the channel, thereby controlling the flow of current. Figure 7.2 represents the self-bias circuit, which helps in achieving a stable operating point over a range of operating conditions.

3. The region between the source and drain terminals, composed of semiconductor material, serves as a conduit for current transmission. The resistance of this channel is adjusted by the voltage administered to the gate terminal. Upon

application of a voltage, an electric field is generated, influencing the density of charge carriers (either electrons or holes) within the channel, thereby modifying its resistance. A biosensor employing a field-effect transistor, also referred to as a Bio-FET, field-effect biosensor (FEB), or biosensor MOSFET, operates by utilizing changes in surface potential induced by molecular binding to modulate the transistor's gating. When charged molecules, such as biomolecules, adhere to the FET gate, typically composed of a dielectric material, they alter the charge distribution of the semiconductor beneath, leading to a shift in the FET channel's conductance. A Bio-FET comprises two primary components: a biological recognition element and the field-effect transistor itself as shown in Figure 7.3 reported by Fenoy et al., and Figure 7.4.

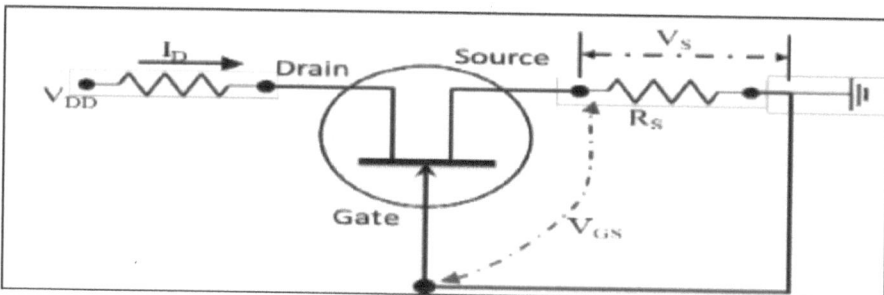

Figure 7.2: Self-bias circuit represents of FET for DC analysis.

Source: Author.

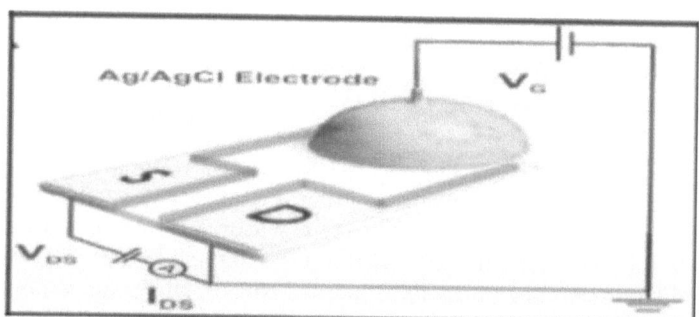

Figure 7.3: Diagrammatic representation of the electrolyte-gated setup.

Source: Fenoy et al. (2020).

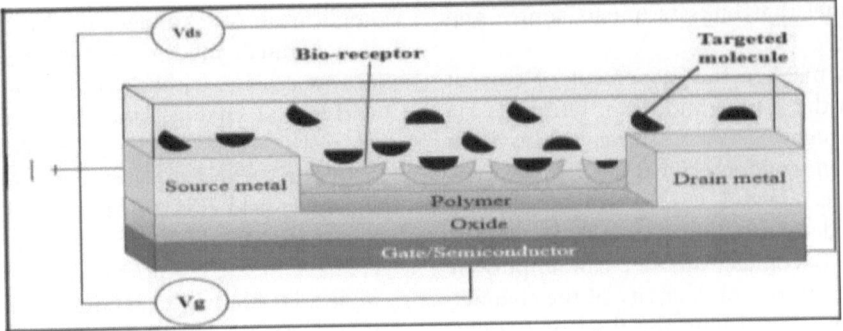

Figure 7.4: Diagrammatic representation of the electrolyte-gated setup.

Source: Author.

3. Nano-material for Sensor

Incorporating nanomaterials like graphene, carbon nanotubes, and metal oxides has greatly improved the performance of field-effect transistor (FET) biosensors. These materials offer distinct qualities suitable for biosensing applications. Graphene, with its large surface area and excellent electrical conductivity, is particularly well-suited for FET biosensors, enabling precise detection of analytes. For instance, a FET biosensor based on graphene exhibited high sensitivity in detecting chlorpyrifos (Islam et al., 2019).

Carbon nanotubes, valued for their mechanical strength and conductivity, can be modified to selectively capture specific analytes, thereby enhancing the sensitivity and selectivity of FET biosensors. Metal oxides like zinc oxide, titanium dioxide, and tungsten oxide possess advantageous electrical and catalytic properties for biosensing, improving the effectiveness of FET biosensors in aqueous environments and allowing for real-time responsiveness. Furthermore, these metal oxides can be engineered to exhibit selectivity towards particular analytes, thus expanding the detection capabilities of FET biosensors. Table 7.1 provides an overview of the FET-based electrochemical biosensor device for detecting chemicals and pesticide residues, which gives the gaps found in the various method detected vividly.

4. Result and Analysis

The method of bio receptor enzyme with GFET and ISFET method gives a wide spectrum of gap which was found in the analysis. The recovery and response time of the IFIT-based biosensor using various methods such as GFET, OFET, ISFET and FET with targeted compounds like Acetylcholine, NO_2, Atrazine and Chlropyrifos were assessed using current change and resistance change detection methods were employed in building LOD and LR. Figure 7.5 shows linearity range of sensor detect. Sensors limit of detection (LOD) has shown in Figure 7.6 compound v/s concentration. All the values in the graph shown in Figures 7.5 and 7.6 was converted to ppm (parts per million) and used.

5. Conclusion

The various method adopted for FET shows that the response time for graphene-FET type sensor is 130 seconds whereas response time of organic-FET is 171 sec. Also, bimetallic Ni/Cu FET sensor response time found 5sec. The recovery for Carbaryl was formed to 86% using graphene material as ISFET.

Table 7.1: FET based biosensor

Method (Reference)	Targeted Compound	Material	Detection principle	Bio receptor	Sample	LOD and LOR	Recovery and Response Time
G-FET (Fenoy et al., 2020)	Acetylcholine	Graphene	Current Change	Enzyme	Urine	LOD = 2.3 μM LR = 5 to 10³μM	Rec. = 97.5% RT = 130 sec
OFET (Kumar et al., 2018)	NO2 gas	Organic	Current Change (density of holes)	PCDTBT	Air	LR = 1- 60 ppm	RT= 171 sec.
ISFET (Thanh et al., 2018)	Carbaryl	Graphene	Current Change	Enzyme (Urease)	Spiked	LOD= 10⁻⁸ μg/mL LR= 2.58 × 10⁻⁷ to 2.58 × 10⁻² μg/mL	Rec. = 86%
FET (Belkhamssa et al., 2016)	Atrazine	SW-CNT	Current Change	Antibodies	Spiked Water	LOD = 0.001 ng/mL LR = 0.001 to 10 ng/mL	Rec.=87.3 to 108.0%
FET (Wang et al., 2021)	Glucose, Sucrose and Fructose	Bimetallic Ni/Cu	Current Change	Glucose oxidase	Spiked	LOD = 0.51μM LR = 0.001 to 20 mM	RT = 5 sec

Source: Author.

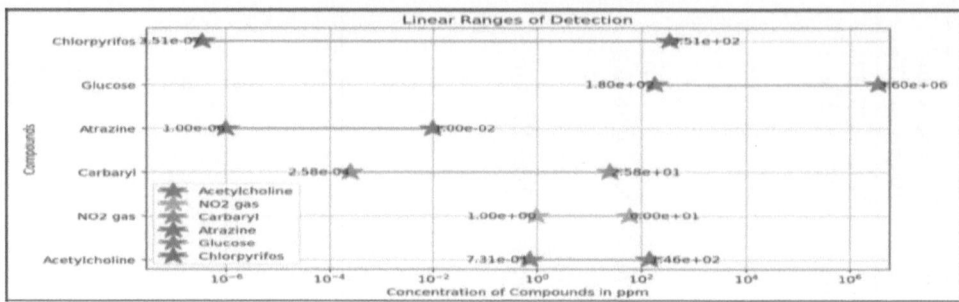

Figure 7.5: Linear range of detection.

Source: Author.

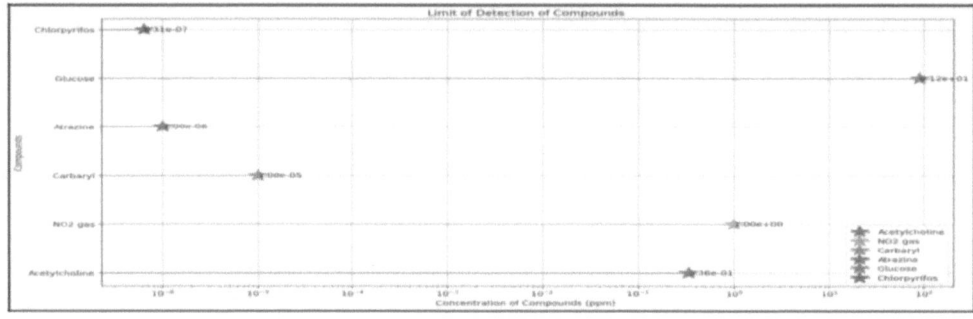

Figure 7.6: Limit of detection of the sensors.

Source: Author.

While got 97% with GFET for acetylcholine. The objective and its subsequent findings were found satisfactory. The endeavour was to investigate and review various methods to help the R&D professional in keeping the soil and water environment clean.

References

Belkhamssa, N., Justino, C. I. L., Santos, P. S. M., Cardoso, S., Lopes, I., Duarte, A. C., Rocha-Santos, T., and Ksibi, M. (2016). Label-free disposable immunosensor for detection of atrazine. Talanta, 146, 430-434. doi: 10.1016/j.talanta.2015.09.015.

Fenoy, G. E., Marmisollé, W. A., Azzaroni, O., and Knoll, W. (2020). Acetylcholine biosensor based on the electrochemical functionalization of graphene field-effect transistors. Biosensors and Bioelectronics, 148, 111796. doi: 10.1016/j.bios.2019.111796

Hashwan, S. S. B., Khir, M. H. B. M., Al-Douri, Y., and Ahmed, A. Y. (2020). Recent progress in the development of biosensors for chemicals and pesticides detection. *IEEE Access*, 8, 82514-82527.

Islam, S., Shukla, S, Bajpai, V. K., Young-Kyu Han, Yun Suk Huh, Ghosh, A., and Gandhi, S. (2019). Microfluidic-based graphene field effect transistor for femtomolar detection of chlorpyrifos. Scientific Reports, 9(1), 276. doi: 10.1038/s41598-018-36746-w

Kumar, A, Jha, P., Singh, A, Chauhan, A. K., Gupta, S. K., Aswal, D. K., Muthe, K. P., and Gadkari, S. C. (2018). Modeling of gate bias controlled NO_2 response of the PCDTBT based organic field effect transistor. Chemical Physics Letters, 698, 7-10. doi: 10.1016/j.cplett.2018.02.043

Thanh, Cao Thi, Nguyen Hai Binh, Nguyen Van Tu, Vu Thi Thu, Bayle, M, Paillet, M., Sauvajol, J. L., et al. (2018). An interdigitated ISFET-type sensor based on LPCVD grown graphene for ultrasensitive detection of carbaryl. Sensors and Actuators B: Chemical, 260, 78-85. doi: 10.1016/j.snb.2017.12.191

Wang, B., Yuanyuan Luo, Lei Gao, Bo Liu, and Guotao Duan. (2021). High-performance field-effect transistor glucose biosensors based on bimetallic Ni/Cu metal-organic frameworks. Biosensors and Bioelectronics, 171, 112736. doi: 10.1016/j.bios.2020.112736

8   Reduction in production lead time of Pneumafil suction pipe in textile equipment industry: a case study

Hari Chealvan S.[1,a], Amirtha K. M.[1,b], Barath M.[1,c], Venkatesh E.[1,d], Aathi R.[1,e], and Brajesh Kumar Kanchan[2,f]

[1]Department of Production Engineering, PSG College of Technology, Coimbatore, India
[2]Cell Design and Simulation, Amararaja Advanced Cell Technologies, Hyderabad, India

Abstract

This research delves into the underexplored realm of process efficiency within textile equipment production in the Indian subcontinent, with a specific focus on the Pneumafil suction pipe—an integral component crucial for the threading of raw fabric. Through a comprehensive examination involving meticulous time studies and current value stream mapping, this study successfully pinpointed significant bottlenecks hindering the seamless flow of production processes. Subsequently, the implementation of kaizen principles and their visualization through future value stream mapping emerged as the strategic intervention to address these identified challenges. The outcomes of this research are marked by a substantial 30% and 37% reduction in both production lead time and processing time respectively following the implementation of future value stream mapping. This not only underscores the effectiveness of the applied methodologies but also serves as a practical guide for textile industry stakeholders seeking to optimize critical components and streamline overall lead time. The findings underscore the significance of operational enhancements, offering actionable insights that have the potential to revolutionize production processes, increase efficiency, and ultimately elevate the value of textile products for discerning customers.

Keywords: Kaizen, lean manufacturing, production lead time, textile industry, value stream mapping

1. Introduction

In the contemporary business landscape characterized by competition and rapidity, enterprises are perpetually exploring methods to optimize their operational effectiveness, avoid unnecessary expenses, and increase customer value. In textile equipment business also, the requirement for operational excellence is increasing to

[a]harichealvan18@gmail.com, [b]amirtha221@gmail.com, [c]studentbarath2002@gmail.com, [d]venkateshyadav094@gmail.com, [e]aathirajr76@gmail.com, [f]brajeshlean@gmail.com

DOI: 10.1201/9781003567653-8

provide higher valuable product to the customers. Some of the common equipment used in textile business are Woollen mill machines, thread winding machines, spinning machines and yarn gassing machines for Indian context. Pneumafil Suction Pipe is one of the critical equipment used in the textile industry owing to its use of conversion of raw fabric to threads.

Several researchers worked on the lean implementation in textile industries (Sahoo & Yadav, 2018; Rother & Harris, 2001; Chiarini, Baccarani, & Mascherpa, 2018; Stapleton et al., 2009; Lu, Liu, & Min, 2021). According to (Hines & Taylor, 2000), VSM can assist businesses in identifying and removing non-value-added tasks, creating a more productive and efficient workflow. VSM often involves employees mapping and improving their work processes at all levels. This empowerment of employees fosters a sense of ownership and responsibility, leading to increased engagement and continuous improvement (Rother & Harris, 2001) show how VSM can eliminate processes that do not directly add value for the customer in a case study. VSM can boost productivity in small-scale industries by cutting waste and simplifying procedures. Achieving and sustaining high levels of customer satisfaction requires this customer-centric strategy. These principles are highlighted in the works of (Liker, 2004). According to a study by (Rother, 1999), eliminating non-value-added activities can result in significant cost savings for small-scale industries. A common understanding of how value is created and delivered to the customer is facilitated by involving staff members from different departments in the mapping process. Better problem-solving and decision-making follow from enhanced communication (Dennis, 2017). Additionally, VSM can support initiatives for sustainability in small-scale industries. VSM aligns with the objectives of environmentally and socially conscious business practices by cutting waste and minimizing the adverse effects of processes on the environment.

2. Problem Statement

ABC Industries established in 2006 specialized in manufacturing and supplying Pneumafil products as retrofit to Long Ring Spinning Frames in the Textile Industry. While there is a lot of research on cutting production lead times across industries, there isn't much on the application of lean principles to textile equipment or the importance of Pneumafil in the Indian textile industry. In order to close this gap, the authors examine how lean methods and the manufacturing of textile equipment interact, highlighting the significance of Pneumafil in the context of Indian manufacturing.

3. Sequence of Operations in Manufacturing of the Product

The flowchart depicted below in Figure 8.1 shows the sequence of operations in manufacturing the common Pneumafil suction pipe. The flowchart focuses on the manufacturing sequence only and other material and information flow such as the raw material procurement and finished goods delivery are mapped in the current state VSM.

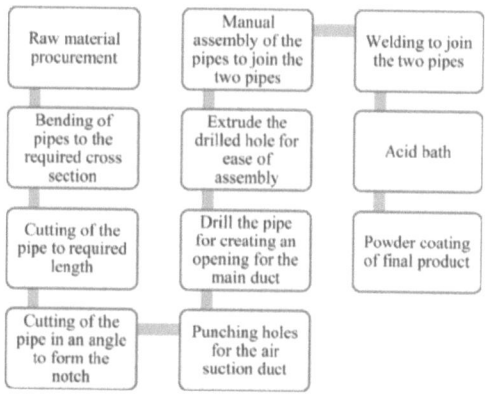

Figure 8.1: Sequence of operations of manufacturing the Pneumafil product.

Source: Author.

4. Current State Value Stream Map

The Current State VSM, is the first stage of VSM implementation wherein organizations create a comprehensive visual depiction of their current processes. This covers the processes, distribution of resources, information flow, and amount of time needed to provide a good or service. This mapping aids in locating wasteful, inefficient, and bottleneck areas in the current operations. It acts as a visual guide and diagnostic tool, pointing out inefficiencies, redundancies, and waste that may have gotten ingrained in the system over time.

4.1. Time Study

The team visited the industry during a regular production shift and the time study was performed on all operations using the conventional stopwatch method. All data were recorded and tabulated in the current state value stream map as shown in Figure 8.2. All details concerning the information flow and material flow were collected by enquiring the same with the Administrative Manager of the company. The current state VSM results in a Processing Time (PT) of 5.8 mins per product and a Production Lead Time (PLT) of 233 mins. This can be further reduced in the future state VSM by various optimizing methods discussed in this paper.

5. Inference of the Current State VSM

Our value stream mapping (VSM) analyses 12 manufacturing processes in a pull-type production system that is controlled centrally in the production department and is motivated by direct consumer demand. Strategic inventory management and batchwise material acquisition provide seamless transitions between operations. Cycle time, changeover time, and batch size are important variables that help with analysis and are included in the VSM data box (Ng et al., 2010). Calculations of production lead time and processing time are informed by timeline charts, which record temporal aspects. With references to Salwin et al. (2021), Hidalgo Martins and Cleto (2013), and Lacerda et al. (2016), this study establishes a foundation for strategic decision-making by highlighting areas that require enhancement and optimisation. From the current state VSM, certain areas are identified as bottlenecking or time-consuming areas that need to be focused on to reduce or optimize the cycle time of a particular operation so that we can obtain an overall reduced production lead time in the future state VSM (Wang et al., 2022). This is necessary to achieve the goal of value stream mapping of reduction in unwanted leakage of time in an industry. The areas to be considered are: 150 seconds for every joint to be welded and stored in the inventory.

5.1. Punching Operation

In punching operation, the product requires to have 8 parallel small holes in a hollow tube. This is achieved by using a punching machine with a single tool of the required dimension and all the holes are punched manually one by one. The changeover to the next hole is done using a position gauge that locates the holes in a required manner according to the production drawing. This manual changeover of gauges one by one results in an increased cycle of 64 seconds for that particular operation.

5.2. Welding Operation

As it is clear from the current state of VSM the welding operation here has 2 operators at present, one for holding the joints and one for holding the welding torch. Due to the manual holding of tubes, this welding operation takes

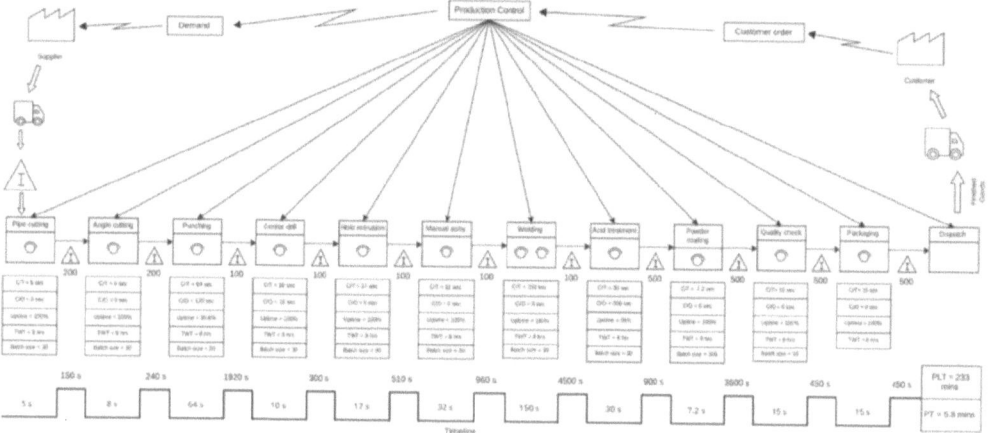

Figure 8.2: Current state value stream map drawn for the product of interest – Pneumafil.
Source: Author.

6. Future State Value Stream Map

Twelve manufacturing processes in a pull-type production system, controlled centrally in the production department and motivated by direct customer demand, are examined in our value stream mapping (VSM). Smooth transitions between processes are ensured by smart inventory management and batch-wise material acquisition. To facilitate analysis, the VSM's data box contains important parameters such batch size, cycle time, and changeover time. Calculations of processing time and production lead time are aided by timeline charts, which record temporal features. This research establishes a baseline for strategic decision-making by highlighting areas for optimisation and improvement, and is backed by citations (Salwin et al., 2021; Hidalgo Martins & Cleto, 2013; Lacerda et al., 2016).

6.1. Implementation of Techniques for Future State VSM

Once the bottleneck stations were identified, several ideas were generated and brainstormed in the form of kaizen opportunities to reduce the operation cycle time. A proposal was given for a special purpose machine (SPM) in the punching station, replacing the existing one. The new design consists of 8 punching tools in a single tool head instead of having a single tool that does the punching operation. This new design has a tremendous advantage over the existing ones since the change over time of every gauge is nullified. Now the improved cycle time of the punching operation is 8 seconds instead of 64 seconds, resulting in reduced production lead time for the overall manufacturing operation.

Another kaizen opportunity came in process reduction by removing certain existing operations in the current state VSM. The usage of welding fixtures or jigs helps us significantly optimize the entire process flow. The prerequisite of welding operation is the previous two operations i.e.: hole extrusion and manual assembly. The operator count is reduced to one instead of using two operators using a welding fixture. A welding fixture is a specialized tool or device used in welding processes to securely hold and position workpieces during the

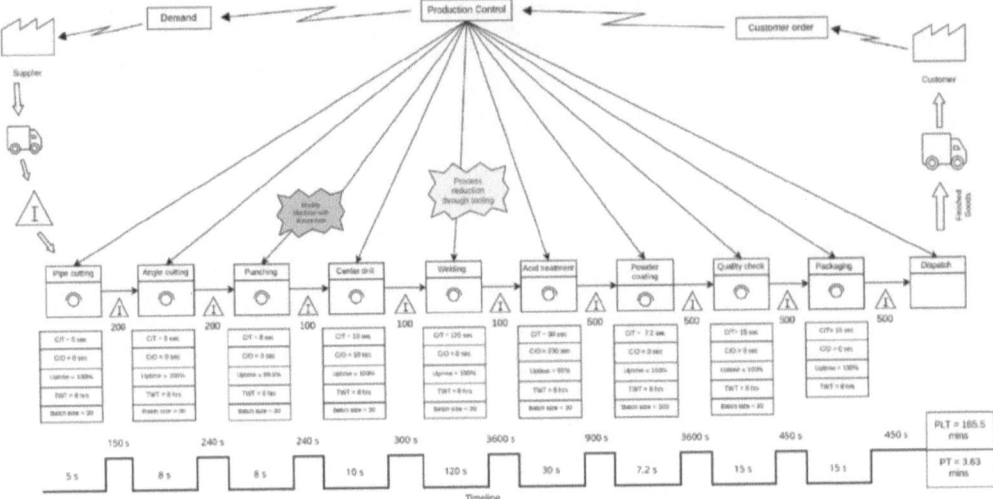

Figure 8.3: Future State Value Stream Map drawn for the product of interest – Pneumafil.

Source: Author.

welding operation. Its primary purpose is to ensure precision, consistency, and repeatability in the welding process by providing a stable and accurate environment for the welder. This facilitates both the holding of the workpiece and the guiding of the welding torch. As a result of this implementation, two operations namely hole extrusion and manual assembly can be comfortably removed from the manufacturing operation sequence reducing the cycle time from 150 seconds to 120 seconds.

This resulted in the reduction of processing time from 5.8 minutes to 3.63 minutes and PLT from 233 minutes to 165.5 minutes. This significant reduction in PT and PLT can result in an increased production rate and satisfy the customer needs in terms of delivery time. The future state value stream map was presented in Figure 8.3.

7. Conclusion

The present study discusses lean implementation in the textile equipment industry by considering a critical product i.e. Pneumafil Suction Pipe. Time Study, Value Stream Mapping, and Kaizen are some common lean tools utilized in the investigation. The following conclusion can be derived from the present study:

- 30% reduction in the production lead time is reported owing to Welding Jig.
- 88% reduction in Punching operation is concluded because of the proposal of using a special multi holes punch machine
- Overall, 37% reduction in the processing time for the Pneumafil suction pipe

The present study motivates industrial engineers, and managers to identify the bottleneck and improve the process by utilizing common lean tools like time study, VSM, and Kaizen, particularly in the textile equipment industry.

References

Chiarini, A., Baccarani, C., and Mascherpa, V. (2018). Lean production, Toyota production system and Kaizen philosophy: A conceptual analysis from the perspective of Zen Buddhism. *The TQM Journal* 30 (4): 425–38.

Dennis, P. (2017). *Lean production simplified: A plain-language guide to the world's most powerful production system* (3rd ed). Productivity Press.

Hidalgo Martins, G., and Cleto, M. (2013). Value stream mapping and earned value analysis: A case study in the paper packaging industry in Brazil. 22nd International Conference on Production Research, ICPR 2013, p. 13.

Hines, P., & Taylor, D. (2000). Going lean. *Cardiff, UK: Lean Enterprise Research Centre Cardiff Business School*, 1(528–534), 43–44.

Lacerda, A. P., Xambre, A. R., and Alvelos, H. M. (2016). Applying value stream mapping to eliminate waste: A case study of an original equipment manufacturer for the automotive industry. International Journal of Production Research, 54(6), 1708–1720.

Liker, J. K. (2004). Toyota way: 14 management principles from the world's greatest manufacturer (1st ed.). McGraw-Hill Education.

Lu, Yangguang, Zhiyong Liu, and Qingfei Min. (2021). A digital twin-enabled value stream mapping approach for production process reengineering in SMEs. International Journal of Computer Integrated Manufacturing, 34, 1–19.

Ng, D., Vail, G., Thomas, S., and Schmidt, N. (2010). Applying the lean principles of the toyota production system to reduce wait times in the emergency department. *CJEM*, 12(1), 50–57.

Rother, M., & Shook, J. (2003). Learning to see: value stream mapping to add value and eliminate muda. Lean enterprise institute.

Rother, M., and Harris, R. (2001). Creating continuous flow: An action guide for managers, engineers & production associates. Lean Enterprise Institute.

Sahoo, S., and Yadav, S. (2018). Lean implementation in small and medium-sized enterprises: an empirical study of Indian manufacturing firms. Benchmarking: An International Journal, 25 (4), 1121–1147.

Salwin, M., Jacyna-Gołda, I., Bańka, M., Varanchuk, D., and Gavina, A. (2021). Using value stream mapping to eliminate waste: A case study of a steel pipe manufacturer. Energies, 14(12), 3527.

Stapleton, F. B., Hendricks, J., Hagan, P., and Del-Beccaro, M. (2009). Modifying the Toyota production system for continuous performance improvement in an academic children's hospital. Pediatric Clinics of North America, 56(4), 799–813.

Wang, Fu-Kwun, Rahardjo, B., and Rovira, P. (2022). Lean six sigma with value stream mapping in Industry 4.0 for human-centered workstation design. Sustainability, 14, 11020.

Womack, J. P., & Jones, D. T. (1997). Lean thinking—banish waste and create wealth in your corporation. Journal of the operational research society, 48(11), 1148–1148.

9 Innovations in hybrid dryers for enhanced agricultural product drying: a comprehensive review

Ankita Balpande,[1,a] P. B. Maheshwary,[1,b] and P. N. Belkhode[2,c]

[1]Department of Mechanical Engineering, Rashtrasant Tukdoji Maharaj Nagpur University, Nagpur, India
[2]Department of Mechanical Engineering, Laxminarayan Institute of Technology, Nagpur, LIT University, India

Abstract

This comprehensive review delves into the design and evolution of hybrid solar dryers tailored specifically for diverse agricultural products. Amidst the burgeoning need for effective post-harvest processing methods, the integration of hybrid drying systems stands out as a pivotal solution within the agricultural domain. By amalgamating traditional drying methodologies with cutting-edge technologies, hybrid dryers aspire to amplify drying efficiency, curtail energy consumption, and alleviate product deterioration. This paper meticulously evaluates an array of design configurations, operational parameters, and performance indicators associated with hybrid drying systems, accentuating their adaptability to various agricultural commodities. Through a meticulous survey of literature, the review elucidates significant advancements, hurdles, and prospects in hybrid dryer development, encompassing empirical studies and theoretical frameworks. Furthermore, it scrutinizes the influence of environmental variables such as temperature and ambient humidity on the drying process and delves into strategies for optimizing dryer performance amidst fluctuating conditions. By synthesizing insights gleaned from a multitude of research endeavours, this review fosters a nuanced comprehension of hybrid dryer technology, fostering improvements in agricultural product drying efficiency and sustainability.

Keywords: Hybrid, agricultural products, drying, dryer, solar energy, electricity

1. Introduction

Post-harvest drying of agricultural products is vital for preserving quality, extending shelf-life, and enabling efficient storage and transportation. With the global population expected to surpass 9 billion by 2050, the demand for effective and sustainable drying technologies has intensified to ensure

[a]ankitabalpande88@gmail.com, [b]prashantmaheshwary51@gmail.com, [c]pramodb@rediffmail.com

DOI: 10.1201/9781003567653-9

food security and minimize losses (FAO, 2021). Traditional methods like sun drying and mechanical drying, while widely used, often suffer from drawbacks such as prolonged drying times and inconsistent quality due to environmental factors. In response, researchers have been exploring innovative hybrid drying systems that integrate various techniques to enhance efficiency and product quality. These hybrids leverage synergies between different drying methods and employ diverse technologies including solar, thermal, mechanical, and renewable energy sources to minimize energy consumption and environmental impact (Govindan et al., 2021). The design and optimization of hybrid dryers involve multidisciplinary considerations such as engineering, thermodynamics, and agricultural science, with key parameters like dryer configuration and control strategies crucial for performance. This review aims to summarize advancements in hybrid dryer technology, offering insights into operational principles, design considerations, and future prospects. By synthesizing empirical findings and theoretical insights, it provides valuable guidance for researchers, engineers, and policymakers involved in agricultural post-harvest processing and food preservation (Abdul Razak et al., 2021). Subsequent sections will delve deeper into hybrid dryer technology, examining design principles, performance evaluation, and future directions, aiming to serve as a comprehensive resource for industry professionals and academics alike (Figure 9.1).

Some quantitative metrics and comparisons between hybrid dryers and traditional methods are:

1. Drying Rates: Hybrid dryers remove moisture faster than traditional methods, with an estimated rate of 30 g/hr and crops typically drying within 2–3 days.
2. Energy Efficiency:
   o Top flow configuration: 27.5% efficiency.
   o Bottom flow configuration: 38.21% efficiency.
   o Maximum reported drying efficiency: 35%.
   o Solar dryers reduce energy consumption by 27–80%, with exergy efficiency ranging from 5.6% to 95.13%.
3. Cost-Effectiveness:
   o Although initially costly due to additional devices, hybrid dryers outperform passive and active dryers.
   o Payback time is typically 1–2 years, making them commercially viable.
   o Hybrid solar dryers reduce costs compared to electric dryers and offer better quality control.
   o Despite higher initial costs, hybrid dryers offer significant improvements in drying rates and energy efficiency, making them cost-effective in the long run and environmentally friendly compared to traditional methods.

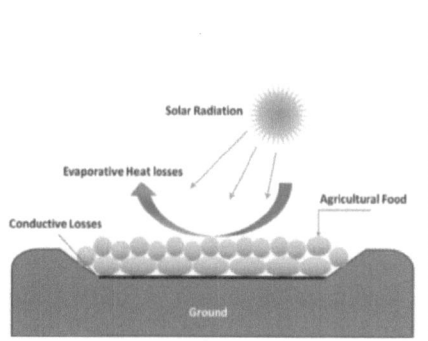 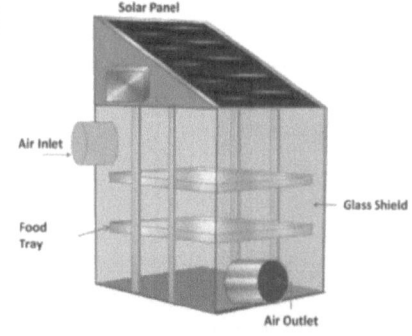

Figure 9.1: Schematic diagram of traditional drying System (left) (Gorjian, Shiva & Hosseingholilou, 2021) and emerging drying system (right) (Hao, W.; Liu, 2020).

2. Literature Review

The literature on agricultural drying technologies covers a wide range of research efforts aimed at improving efficiency, quality, and sustainability. This section evaluates contemporary hybrid drying techniques, drawing from academic and industrial sources.

Hybrid drying systems integrate multiple techniques to improve efficiency and flexibility compared to traditional methods. A key driver is the need to reduce energy consumption and environmental impact, as highlighted by Mujumdar and Devahastin (2007), who emphasized renewable energy use.

Researchers have tailored hybrid dryers for specific agricultural products, like Tunde-Akintunde and Akintunde's (2016) solar-biomass dryer for maize. Other configurations, such as solar-thermal and solar-wind hybrids, have been explored, such as Sharma et al.'s (2009) solar-thermal dryer for mango slices with a phase change material storage system.

Advancements in computational modelling aid in optimizing hybrid dryer designs, though challenges persist in integrating energy sources and control systems. Scalability and adaptability are crucial, considering variations in crops and environmental conditions. Despite the benefits, high costs and technical requirements may limit adoption, especially in resource-constrained areas.

Further research and interdisciplinary collaboration are needed to overcome optimization challenges and facilitate widespread adoption, ultimately contributing to sustainable agriculture.

2.1. Technical Gaps

This paper explores technical gaps and innovations in hybrid dryers for agricultural product drying. It integrates advanced technologies like renewable energy sources and innovative control strategies to enhance efficiency and sustainability. Emphasizing the optimization of energy utilization, it highlights the use of solar energy and biomass to reduce environmental impact. The paper also advances control strategies and scalability considerations for various agricultural products and processing scales. It stresses interdisciplinary collaboration to drive innovation and offers valuable insights for stakeholders in post-harvest processing.

3. Methodology Review

This section reviews the methodologies in hybrid drying system research for agricultural products, including experimental studies, theoretical modelling, computational simulations, and techno-economic analyses. It covers approaches in design, development, and evaluation of these dryers.

3.1. Experimental Studies

Experimental studies validate theoretical models and assess hybrid drying system performance. Researchers construct prototype dryers, measure parameters, and analyse drying kinetics and product quality. They select equipment based on research objectives and consider factors like drying method, heat sources, and operational variables. Data acquisition systems monitor parameters, with analysis using statistical techniques and software like MATLAB or SPSS to derive empirical relationships and predictive models.

3.2. Computational Simulations

CFD modelling optimizes hybrid drying systems by visualizing airflow patterns, temperature distributions, and moisture transport. It involves discretizing chamber geometry, solving fluid flow and heat transfer equations, and specifying boundary conditions. Visualized results aid in design optimization and performance prediction, while sensitivity analysis evaluates design and operating variables' effects. We have created Table 9.1

Table 9.1: A comparison of different research works

Researchers/Authors	Key Techniques
Mujumdar (2014)	• Focused on modular design concepts. • Proposed adaptable drying chambers for specific crops • Explored feasibility of interchangeable components.
Aktaş, Midilli, & Dinçer (2010)	• Evaluated performance indicators. • Convective Heat Transfer Coefficient • Overall, Loss Coefficient • Drying Efficiency. • Collector Efficiency Factor • Suggested novel control strategies for optimization.
Lamidi, El-Wail, & Al-Madani (2016)	• Investigated sensor-based feedback mechanisms • Developed intelligent algorithms for real-time adjustments • Addressed challenges related to varying weather conditions.

Source: Other Researches.

with the researchers and their techniques for solar dryer.

3.3. Modern Tools and Techniques

Modern innovations enhance the efficiency of solar hybrid drying systems. Phase Change Materials (PCMs) improve temperature regulation, while Metal Matrix and Metal Fibre enhance thermal conductivity. Integrating heat pumps offers better temperature control, and Photovoltaic/Thermal (PVT) Panels improve energy efficiency. Thermal storage materials extend drying hours, and advanced simulation tools optimize dryer designs for optimal performance, collectively contributing to more efficient and sustainable agricultural product drying.

4. Overview of Design and Development of Hybrid Dryers

4.1. Overview of Hybrid Dryer Configurations and Components

Hybrid dryers optimize agricultural drying by combining multiple techniques and energy sources. They include key components like drying chambers, heat sources, airflow systems, and controls. Configurations vary based on the product, energy availability, and desired outcomes, with common types including solar-biomass, solar-thermal, solar-wind, and solar-heat pump hybrids (Kaveh, Mohammad, & Karami, 2020) (Figure 9.2).

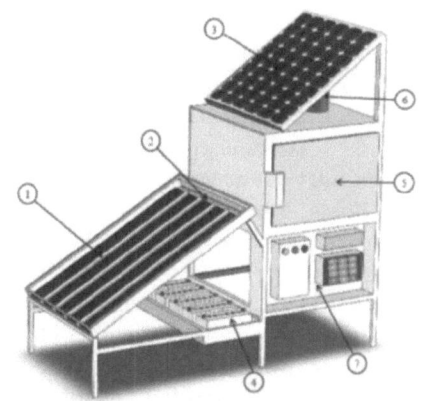

Figure 9.2: Schematic diagram of a hybrid dryer system.

Source: Kaveh, Mohammad, & Karami (2020).

4.2. Design Considerations: Heat Sources, Airflow Dynamics, Control Systems

Hybrid dryer design balances heat sources, airflow, and controls for efficient drying. Sources include solar, biomass, electric, or waste heat, chosen for energy, cost, and environmental factors. The heat transfer equation guides chamber temperature distribution.

$$q = hA(T_s - T_a) \qquad (1)$$

where is the heat transfer rate, A is the surface area for heat transfer, is the surface temperature of the drying material, h is the heat transfer coefficient and is the ambient temperature.

Airflow dynamics play a critical role in facilitating moisture removal and heat transfer within the drying chamber. velocity of airflow distribution can be modelled using computational fluid dynamics (CFD) simulations, considering factors such as duct geometry, fan specifications, and inlet/outlet configurations (Camas-Nafate et al., 2019).

4.3. Critical Issues and Potential Solutions

Hybrid dryers face challenges due to intermittent solar energy, limited control over drying, and quality issues from open sun drying. To overcome these, they combine solar energy with conventional or auxiliary sources, like biomass or fossil fuels. Photovoltaic/Thermal panels provide electric power as a backup. Energy storage materials ensure continuous operation. Thermal recovery units improve efficiency, while Humidification-Dehumidification units control humidity for a controlled drying process. These solutions enhance hybrid dryer performance, though effectiveness varies based on design and usage.

5. Future Directions and Challenges

5.1. Research Gap

The review identifies research gaps in hybrid solar drying technologies:

- Lack of focus on intelligent control systems.
- Challenges in economic viability and scalability.
- Need for localized adaptations.
- Absence of detailed Life Cycle Assessments (LCAs).
- Understanding user acceptance and adoption barriers.
- Enhancing energy storage solutions.
- Bridging the gap between academia and industry stakeholders.

Addressing these gaps will promote sustainable and efficient adoption of hybrid drying technologies in agriculture.

5.2. Challenges

Hybrid drying technology encounters challenges in technology, economics, and policy, hindering widespread adoption. Solutions must address scalability, cost-effectiveness, financing limitations, and regulatory barriers. Collaboration among policymakers, industry, and research communities is essential for innovation and technology transfer.

6. Conclusion

About details study of the literature review concluded that there are various types of hybrid dryers and solar dryers available for removing moisture content from agricultural products but priority is given to the drying effectiveness, dryer selection is based on the dryer effectiveness. A hybrid dryer is efficient and effective with more specifically effective in reducing drying time. Drying time is a most prominent factor in the drying processes of agriculture products as different from other dryers; solar hybrid dryer plays a vital role in drying agricultural product which ultimately decreases losses and increase the profitability and perishability of agricultural products

References

Abdul Razak, A., Tarminzi, M. A., Azmi, M. A. A., Ming, Y., Akramin, M., & Mokhtar, N. (2021). Recent advances in solar drying system: A review. International Journal of Engineering Technology and Sciences, 8, 1–13.

Abhishek, S. R., Gowtham, V., Bishnoi, D. K., Swathi, P., & Manjunath Patel, G. C. (2019). Hybrid solar-electric drier. International Research Journal of Engineering and Technology, 6(3), 2278–0181.

Aktaş, M., Midilli, A., & Dinçer, I. (2010). Performance analysis of solar dryers for drying of onion slices. Renewable Energy, 35(2), 440–449.

Banout, J., Ehl, P., Havlik, J., Lojka, B., Polesny, Z., & Verner, V. (2011). Design and performance evaluation of a double-pass solar drier for drying of red chilli (capsicum annum L.). Solar Energy, 85(4), 506–515.

Bharadwaz, K. (2020). A review on different solar dryers and drying techniques for preservation of agricultural products. International Journal of Scientific Research and Engineering Development, 3(6), 850–859.

Boghali, S., Benmoussa, H., Bouchekima, B., Mennouche, D., Bouguettaia, H., & Bechki, D. (2009). Crop drying by indirect active hybrid solar-electrical dryer in the eastern Algerian Septentrional Sahara. Solar Energy, 83(11), 2223–2232.

Camas-Nafate, M. P., Alvarez-Gutiérrez, P., Valenzuela-Mondaca, E., Castillo-Palomera, R., & Perez-Luna, Y. D. C. (2019). Improved agricultural products drying through a novel double collector solar device. Sustainability, 11(10), 2920.

El-Beltagy, A., Gamea, G. R., & Amer Essa, A. (2007). Solar drying characteristics of strawberry. Journal of Food Engineering, 78(4), 456–464.

Gamea, G. R. (2014). Effect of Drying Conditions on Drying Behavior and Quality of Onion Slices. MISR Journal of Agricultural Engineering, 31(2), 575–598.

Gorjian, S., Hosseingholilou, B., Jathar, L., Samadi, H., Samanta, S., Sagade, A., Kant, K., & Sathyamurthy, R. (2021). Recent Advancements in Technical Design and Thermal Performance Enhancement of Solar Greenhouse Dryers. Sustainability, 13(13), 7025.

Govindan, G. R., Sattanathan, M., Muthiah, M., Ranjitharamasamy, S., & Athikesavan, M. M. (2021). Experimental Study of Solar Dryer with Thermal Energy Storage System for Drying of Agro Products, 03 September 2021, PREPRINT (Version 1) available at Research Square [https://doi.org/10.21203/rs.3.rs-673328/v1].

Hao, W., Liu, S., Mi, B., & Lai, Y. (2020). Mathematical Modeling and Performance Analysis of a New Hybrid Solar Dryer of Lemon Slices for Controlling Drying Temperature. Energies, 13(2), 350. https://doi.org/10.3390/en13020350

Kadam, D. M., & Samuel, D. V. K. (2006). Convective flat-plate solar collector for cauliflower drying. Biosystems Engineering, 93(2), 189–198.

Khawale, V. R., & Kale, B. N. (2020). Analysis of direct, indirect and hybrid solar dryer. International Journal of Mechanical and Production Engineering Research and Development, 10(4), 775–792.

Kaveh, M., Karami, H., & Jahanbakhshi, A. (2020). Investigation of mass transfer, thermodynamics, and greenhouse gases properties in pennyroyal drying. Journal of Food Process Engineering. 43(8), e13446. 10.1111/jfpe.13446.

Lamidi, A., El-Wail, M., & Al-Madani, H. (2016). Design and performance of a solar dryer with intelligent controller. Renewable Energy, 86, 820–832.

Maiti, S., Patel, P., Vyas, K., Eswaran, K., & Ghosh, P. K. (2011). Performance evaluation of a small scale indirect solar dryer with static reflectors during non-summer months in the saturated region of western India. Sol Energy, 85, 2686–2696.

Mujumdar, A. S. (2007). An overview of innovation in industrial drying: current status and R&D needs. Transport in Porous Media, 66, 3–18. https://doi.org/10.1007/s11242-006-9018-y

Mohanraj, M., & Chandrasekar, P. (2008). Drying of copra in a forced convection solar drier. Biosystems Engineering, 99, 604–607.

Nabnean, S., Janjai, S., Thepa. S., Sudaprasert, K., Songprakorp, R., & Bala, B. K. (2016). Experimental performance of new design of solar dryer for drying osmotically dehydrated cherry tomatoes. Renew Energy, 94, 147–156.

Parikh, D., & Agrawal, G. D. (2011). Solar drying in hot and dry climate of jaipur, India. International Journal of Renewable Energy Research, 1, 224–231.

Potdukhe, P. A., & Thombre, S. B. (2016). Development of a new type of solar dryer: its mathematical modeling and experimental evaluation. International Journal of Energy, 94, 147–156.

Sharma, A., Chen, C. R., & Lan, N. V. (2009). Solar-energy drying systems: A review. Renewable and Sustainable Energy Reviews, 13(6–7), 1185–1210.

Tunde-Akintunde, T. Y. (2014). Effect of Pretreatments on Drying Characteristics and Energy Requirements of Plantain (Musa AAB). Journal of Food Processing and Preservation, 38 (4), 1849–1859.

UmayalSundari, A. R.., Neelamegam, P., & Subramanian, C. V. (2013). An experimental study and analysis on solar drying of bitter gourd using an evacuated tube air collector in Thanjavur, Tamil Nadu, India, Conference Papers in Science, 125628, 4. https://doi.org/10.1155/2013/125628.

Weiss, W., & Buchinger, J. (2012). Solar drying. AEE INTEC Publication, A-8200 Gleisdorf, Feldgasse, 19: A-8200.

10 Analysis and design modification of cotton lint opener

Rupesh K Amarghade[1,a], Minhaj Ahemad Rehman,[2,b] and Dinesh Seth[3,c]

[1,2]Department of Mechanical Engineering, St. Vincent Pallotti College of Engineering, Nagpur, India
[3]Department of Mechanical and Industrial Engineering, College of Engineering, Qatar University, Doha, Qatar

Abstract

Opening up and cleaning of cotton lint samples drawn from densely packed bales is essential before samples are used for fibre quality assessment using modern testing equipment. Importance of proper opening of sample introduced into the porosity chamber for micronaire (fibre fineness) measurement is well recognised to ensure not only correct micronaire but also strength values. If lint samples are tested as such without any opening and cleaning, there is every possibility of variations in the micronaire reading from the actual value. To redesign the existing cotton lint opener machine so as to make it more compact, portable and reduce number of drives. First detailed study of existing cotton lint opener was carried out. Completely novel compact and portable design of lint opening machine is proposed. The proposed machine eliminates limitations of existing machines like number of electric motors are reduced to two instead of three, suction blower system is made more compact and duct type, more compact and rigid drive (worm and worm gear) is selected for feeder roller. This paper discusses merits and demerits of two methods of cotton lint opening namely manual hand method and trash separator and some of the related issues to ensure accurate and precise values of fibre fineness.

Keywords: Cotton, fibre, fineness, lint, micronaire, opener, testing

1. Introduction

Quality of cotton fiber depends on many parameters like strength, staple length, micronaire (fiber fitness), trash, spinning consistency index (SCI), uniformity index etc. Market value these parameters. High volume instrument (HVI) is used to determine numerous properties of fiber such as strength, maturity, micronaire and other in bale management system. Cotton sample drawn from densely packed bale for testing and quality check need to be open up without affecting basic quality in order to improve accuracy of measurement. Opening densely packed sample drawn from bale and converting it in to high volume sample with loose and separated fiber is a tedious

[a]rupesh375@gmail.com, [b]mrehman@stvincentngp.edu.in , [c]dseth1966@gmail.com

DOI: 10.1201/9781003567653-10

and time taking task. Country requires large scale and faster fiber testing facilities in order to perform quality checks of cotton fibers produced in various parts of country (Shwetha et al., 2022). High volume instrument used for testing various quality parameters of cotton fiber requires opened up samples of high volume having neatly separated fibers (Ghadge et al., 2016). High volume instrument used for testing various quality parameters of cotton fiber requires opened up samples of high volume having neatly separated fibers. Manual opening of dense cotton in to high volume separated fibers is tedious and time consuming task, also this is not scientific way. During manual opening of fibers there are chances of cotton fibers getting damaged. During manual opening of fibers there are chances of cotton fibers getting damaged (Ghadge et al., 2016; Introduction to HVI). To address this issue cotton lint opener were developed by various researcher, government and non-government organizations. Existing lint openers are larger in size, bulky and not portable. Most of the time it is needed performs on site lint opening and further testing (Chand & Raju, 2009). In such situation existing bulky lint opener machine cannot be used. Hence, need of compact and portable and simplified design of lint opener machine is arise. To address this issue in the present project work small and portable lint opener machine is designed. The main objective of the current study is to design and develop a digital prototype, 2-D detailing of lint opener using CAD software tool (India brand equity foundation). Cotton is most important commercial crop of India (The Classification of Cotton). It is generally regarded as King of Textile Fibers which has made significant contribution to the National economy (Griffin et al., 1970; James St. Clair & Roberts, 1958).

2. Limitations and Design Flaws of Existing Lint Opener

Overall size of machine is very large (920×760×800 mm), each subassembly, Link-in cylinder, feeder cylinder and suction assembly is driven by separate electric motor. Hence, three electric motors are used which lead to more power consumption. Machine is not portable and cannot be carried to different sites. Chain drive power transmission mechanism for feeder roller is complicated and not rigid. Frequent problem of roller-chain run-out occurs. Suction piping system cannot be cleaned without disassembling. Overall selected specifications of machine components are overdesigned.

3. Main Objectives of Present Project Work

On the basis of feedback received from existing users of this lint opener machine and overall limitations it is decided to redesign the complete machine and all mechanisms with following main objectives.

To reduce overall size and weight of lint opener machine so as to make it portable. To reduce number of electric motors so as to reduce power consumption.

To implement modified rigid and compact power transmission system for feeder roller. To implement compact and without piping suction system.

To suggest optimum design specifications of machine components keeping intact all other parameters related to quality of open lint.

4. Scope of Work (Methodology)

To study existing lint opening machine. To study mechanism of power transmission of existing machine. Identify areas of design modifications so as to reduce machine size. To redesign complete machine of smaller size which will be portable. To design and select standard components for power transmission. To design machine frame and blower suction duct. To prepare CAD Models of parts and assembly of complete machine. To perform mechanism simulation, Kinematic and Dynamic analysis of machine assembly. To create 2-D detailing and production drawing of all machine components and assembly. Green cotton balls eventually become brown. Cotton bolls break open at maturation, revealing the white cotton fiber (Good Year Belts).

The data have been taken and tested (Made in Pro/ENGINEER - Creo) Tested result from GTC ICAR-Central Institute for Research on Cotton Technology (CIRCOT) Nagpur. Overall size of existing Lint-Opener is: 790×920×780 mm (H×W×D). No. of motors=3 (One each for Feeder, Licker-In Cylinder and Blower).

Power transmission units: From 0.5 Hp, 3-Ph 1440 rpm motor to Licker-In Cylinder through V-Belt Drive (center dist. 265 mm). Speed of Licker-in cylinder is 1200 rpm. From geared motor to Feeder roller through series of chain drives, to get feeder roller speed about 8 to 10 rpm. Center distance 450 mm. Separate suction unit driven by 0.5 HP, 2800 rpm motor with blower. Other units: Suction piping, lint collection chamber for (20 gm Lint), filters, M.S. Frame. Image processing approach was proposed by Wenzhu Yang et al. [26]. *I have been taken the data and tested (Made in Pro/ENGINEER - Creo) Tested result from*

Table 10.1: Influence of Lint opening on micronaire value of cotton samples

Cotton Variety		H6	Lra5166	H6	H4	G Cot 16
Micronaire Value	Opening	3	3.3	3.5	3.6	3.6
	Un-opening	3.4	3.7	4.2	4	4.3
Deviation		0.4	0.4	0.7	0.4	0.7
Deviation %		13	12	20	11	19

Source: The Classification of Cotton and India Brand Equity Foundation, 2021.

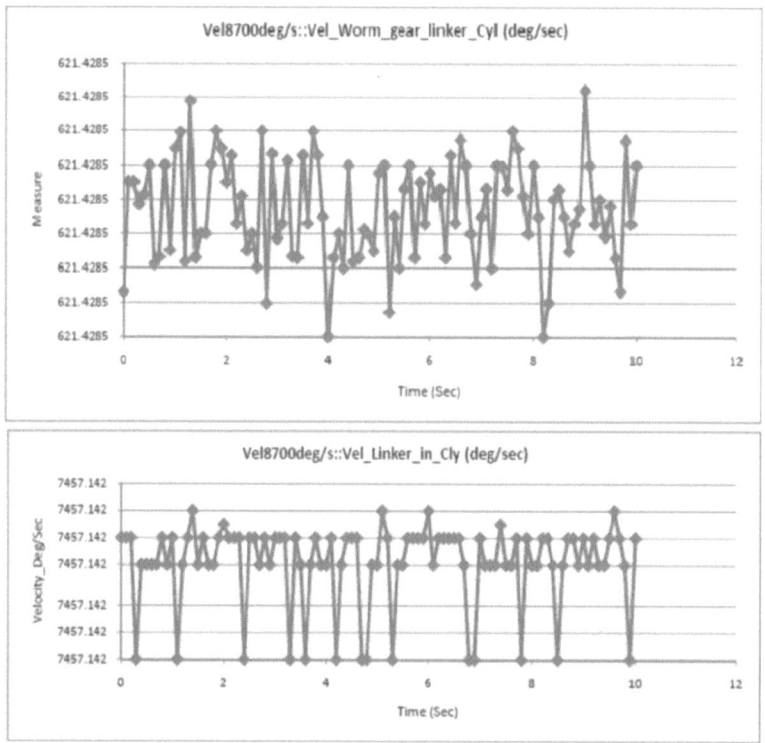

Figure 10.1: Velocity Diagram (worm gear linker cylinder (deg/sec).

Source: Author.

GTC ICAR-*Central Institute for Research on Cotton Technology (CIRCOT) Nagpur* (Ministry of textile).

5. Identified Areas of Design Modifications (Methodology)

Power transmission system to Licker-In Cylinder. Instead of simple V-Belt, timer belt and pulley can be used and center distance can be reduced which will decrease overall size. Power transmission system for feeder roller: Existing chain drive to be replaced with other rigid drive–compact drive which will lead to size reduction. Power can be transmitted from Licker-In cylinder which can eliminate one motor. Suction System: Existing suction piping system if taking too much space. It can be replaced with duct type system with compact design. Existing Suction blower/motor assembly is oversized and bully design. This can be made compact. More particularly the instant invention relates to a lint opener and cleaner for use in cloth and thread mills where there is an increasing need for improved equipment for opening, cleaning and mixing fibrous materials (Eugene, 1961; Worm gear box). Figure 10.2: shows Linter drum.

6. Design of Power Transmission System for Licker-in Cylinder (Methodology)

6.1. *Physical Properties Form Cad Model Material*

Steel VOLUME = 2.5551331e + 06 MM³, SURFACE AREA = 1.1852522e + 05 MM² DENSITY = 7.8270820e – 09 TONNE/MM³, MASS = 1.9999236e – 02 TONNE.

6.2. *Touque Required to Overcome Self Inertia*

Mass =20 Kg (Apporx), Radius of Gyration = 75 mm = 0.75 meter ∴ Mass M.I = I = $M.k^2$ = 11.25 Kg.m², ∴

Figure 10.2: Linter drum.

Soure: Design Data Book, 2022.

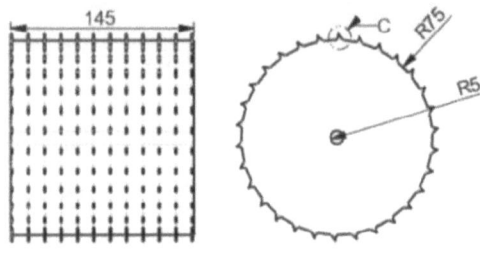

Linter_Drum (Linked-In Cylinder)

Figure 10.3: Licker-in cylinder.

Source: Author.

Initial torque considering angular acceleration (α) as 0.1 m/sec2.

T= I × α = 1.125 N.m, Considering, friction between lint and Licker-In cylinder and other losses.. Design torque, Td = 1.5 × T = 1.68 N.m.

6.3. *Motor and Belt Selection*

Torque developed by 1 HP,1440 RPM E-Motor

P =2πNT/60 = 5 N.m (Approx) >> 1.68 N.m (required 0

Hence, 1 H.P. single Phase, 1440 RPM, electric motor is selected.

Now, Problem Statement: to select suitable Timer (Cogged V-Belt and pulley to transmit max. 1 Hp (0.75 Kw) power with velocity ratio of 1440:1200 rpm.

Solution: From catalogue of Goodyear Company Cogged Belts, Belt section BX with following specifications is selected. Figure 10.4 represents Worm gear and Belt.

To transmit power at minimum center distance with such a high velocity ratio (1:120) only suitable drive is Worm- Worm Gear, also this is rigid. Now. Licker-In

Worm Gear   Belt (BX Series)

Figure 10.4: Worm gear and Belt (Good Year Belts; Good Year Belts).

Source: Author.

Table 10.2: Goodyear_Cogged Belt (BX Series)

Belt Type	Classic
Cord Material:	Polyester
Effective Length (mm):	2609
Item Weight (lbs):	0.9
Outside Circumference (mm):	2624
Rib Angle:	38°
Thickness (mm):	11
Top Width (mm):	17

Source: Good Year Belts; Good Year Belts.

Cylinder and feeder roller shaft axis are parallel. Hence we need to use two worm gear drives. (One of Licker-In Cylinder and other on feeder roller.)

A. **Worm Gear_Licker-In Cylinder:** Torque = 5 N.m , Power = 750 Watt, VR = 10:1, Input RPM =1200.

B. **Worm Gear_Licker-In Cylinder:** Torque = 5 N.m , Power = 750 Watt, VR = 12:1, Input RPM =120

SW-3 & SW-4 Miniature Right Angle Worm Gearbox Speed Reducers (Design Data Book)

SW3-10   70451: Miniature Worm Gear Speed Reducer, 10:1 Ratio, 1/4" Input Shaft, 3/16" Output Shafts0.1873

SW3-20   70452: Miniature Worm Gear Speed Reducer, 20:1 Ratio, 1/4" Input Shaft, 3/16" Output Shafts 20:1 0.1873

Mechanical Features: Machined aluminium housing Bearings - oil impregnated bronze Weight - 0.25 lb.

Technical Features -Torque - 2.5 N.m maximum input torque for all ratios Ratios - 5:1, 10:1, 20:1, Speed to 3000 rpm, Size - 1.5" × 1.5" × 1.08", Max. backlash – 2°

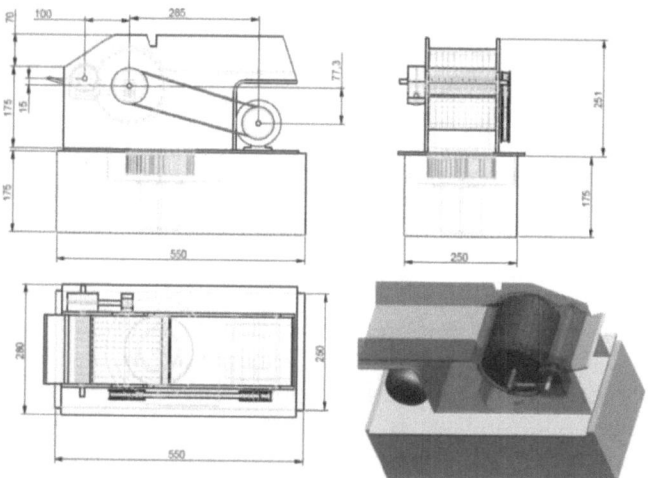

Figure 10.5: Cotton lint opener and isometric views figure, through Pro/ENGINEER - Creo.

Source: Author.

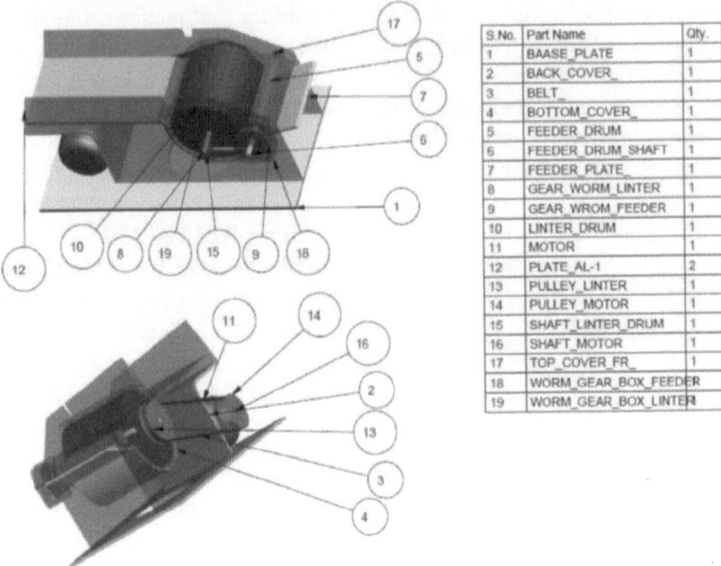

S.No.	Part Name	Qty.
1	BAASE_PLATE	1
2	BACK_COVER	1
3	BELT	1
4	BOTTOM_COVER	1
5	FEEDER_DRUM	1
6	FEEDER_DRUM_SHAFT	1
7	FEEDER_PLATE	1
8	GEAR_WORM_LINTER	1
9	GEAR_WROM_FEEDER	1
10	LINTER_DRUM	1
11	MOTOR	1
12	PLATE_AL-1	2
13	PULLEY_LINTER	1
14	PULLEY_MOTOR	1
15	SHAFT_LINTER_DRUM	1
16	SHAFT_MOTOR	1
17	TOP_COVER_FR	1
18	WORM_GEAR_BOX_FEEDER	
19	WORM_GEAR_BOX_LINTER	

Figure 10.6: Modification of Lint Opener isometric views figure, through Pro/ENGINEER - Creo.

Source: Author.

Cotton lint samples weighing 12–15 g each can be opened manually by hand as shown in the Figure 10.2 where women labourers are employed for the same. This is the simplest method of lint opening in which cotton lint bundles are simply pulled apart by two hands to separate out the entangled fibres and to remove any non-lintimpurities. However, in case of hand opening, the desired optimum extent of lint opening is not achieved and the amount of non-lint content seen is also more as compared to the opening by trash analyser. The presence of non-lint content adversely affects the microanire value of cotton fibres. Besides this opening of samples by hand is very tedious and laborious job that leads to testing of samples without proper opening. The speed of opening of samples is also very slow at only about 35–40 samples per hour. Figure 10.5 represents Cotton lint opener and isometric views figure and Figure 10.6 represent Modification of Lint Opener isometric views figure. Both drawing is prepared through Pro/ENGINEER - Creo.

7. Conclusions

In the current project work portable cotton lint opening machine is designed. Primary purpose of lint opening machine is preparation of opened up lint samples for performing testing on HVI (High Volume Instrument). Existing lint opening machine was designed and developed by ICAR-CIRCOT (Central Institute of Research on Cotton Technology). Detailed study of existing machine with reference to mechanism working, construction, selection of power transmission devices was carried out. Cotton lint samples used for testing fibre quality parameters must be clean and free from any non-lint content. Generally, lint samples received for testing are drawn from densely packed bales and contain some non-lint portions too. Hence, they require cleaning and opening up to ensure accurate measurement of micronaire values. Presently, HVI testing laboratories open cotton lint samples either manually by hand or by using trash analyser. While both these methods have been found to meet the requirement for giving accurate micronaire readings, yet there are some disadvantages of these methods, which need to be looked into while attempting development of new methods of lint opening.

Acknowledgement

The authors gratefully acknowledge the technical assistance from M/s Workshop, SVPCET Nagpur for fabrication of the research prototype of the cotton modification lint opener, as part of the Department of Mechanical Engineering St. Vincent Pallotti college of Engineering & Technology, Nagpur research work. The fabrication of the first prototype involved frequent design changes as per the availability of component parts and the desired performance results. Testing of the research prototype was carried out at scientific, technical GTC, Nagpur for which the timely help received from the Dr. Minhaj A. Rehman Associate Professor, Dr. Amit R. Bhende Assistant Professor, Dr. Ghanshyam R. Boob Assistant Professor (SVPCET) SV Ghadge Principal Scientist of the at Nagpur is also duly acknowledged.

References

Eugene, H. (1961). Brooks, Birmingham, Ala., assignor to Conti mental Gin Company, Birningham, Ala., a corporation of Delaware, Patented, United States Patent Office, 2977461.

"The Classification of Cotton," Agricultural Handbook, 566, The United States Department of Agriculture's Agricultural Marketing Service.

Chand, R., & Raju, S. S. (2009). Instability in Indian Agriculture during different phases of Technology and Policy. Indian Journal of Agricultural Economics, 64(2): 187–207.

Design Data Book: PSG Design Data Handbook - Data Book of Engineers - New Updated 2022 Reprinted Edition, Pub: Kalaikathir Achhangham Coimbtore (15 November 2020); PSG Coimbtore, ISBN-13 978- 8192735504.

Good Year Belts: https://goodyearbelts.com/classic_cogged/

Good Year Belts: https://goodyearbelts.mycar-parts.com/items/BX100

Griffin, A. C., LaFerney, P. E., & Shanklin, H. E. (1970). Effects of lint-cleaner operating parameters on cotton quality (No. 864). Agricultural Research Service and Economic Research Service, United States Department of Agriculture.

India brand equity foundation, Cotton industry and export; 2021, https://www.ibef.org/exports/cotton- industry-india.aspx

Introduction to HVI: https://ordnur.com/spinning/introduce-with-high-volumeinstrument/#Introduce_ with_High_ Volume_Instrument_HVI

James St. Clair and Arthur L. Roberts, "Effects of Lint Cleaning of Cotton—An Economic Analysis". California Gins. Mktg. Res. Rpt. 238. May 1958.

licker-in wire: http://cardindia.com/carding-solutions/card-clothing/cotton-synthetic/lick-erin- wires/#1579261372713-43fd22ef-fef5

Ministry of textile, Annual report. 2020–21, http://texmin.nic.in/documents/annual-report

S.V. Ghadge et al. (2016). Design and development of cotton lint opener for preparation of fiber quality test samples. Cotton Research Journal, 7(2).

Shwetha, M. N., Devi, I. S., Lavanya, T., Suhasini, K., & Meena, A. (2022). Growth in Area, Production and Productivity of Cotton Crop in India: A State Wise Analysis. International Journal of Environment and Climate Change, 12(11), 51–57.

Size-Single-Inlet-Centrifugal-Blower-Fan-for-Air-Suction.html

Suction Blower: https://tzhangda.en.made-in-china.com/product/OFrEBgQugUkL/China-130mm-Small-Timing Pulley: https://www.mak.nl/product_poelies.htm

Yang, W., Li, D., Zhu, L., Kang, Y., & Li, F. (2009). A new approach for image processing in foreign fiber detection. Computers and Electronics in Agriculture, 68(1), 68–77.

Worm gear box: https://catalog.torquetrans.com/catalog3/d/torquetransmission/?c=fsearch&cid=gear-boxes-sw4&filter=ratio_search%3A2%3A6.0#cds-attribute-ratio_search

11 Nanoparticles as effective fuel additive for internal combustion engine: a review

Bhojraj N. Kale[a], Vivek G. Parhate, Ram Wayzode, Kalyani Sengar, Mayuri Wndhare, and Ketan Tonpe

Mechanical Engineering Department, Suryodaya College of Engineering and Technology, Nagpur, India

Abstract

A colloidal dispersion of nanoparticles scattered over a liquid medium is called nanofluid. It enhances the characteristics of heat transport and encourages great energy efficiency in a variety of technical applications. Because of its outstanding thermophysical qualities, adding nanofluid to diesel as well as biodiesel to serve as an additive for internal combustion engines has become a popular strategy in nearby years, particularly in the automobile industry to encourage increased combustion efficiency along with pollution reduction. Numerous studies have already shown that addition of nanoparticles into diesel/biodiesel fuel enhanced the general properties of the engine. The overall aim of this work is to offer an overview of recent research on the impact of nanoparticles on fuel characteristics and engine performance. A summary of the benefits and potential applications of nanofluid as an additional fuel is also provided in this article.

Keywords: Biofuels, IC engine, metal and nonmetal, nanoparticles, properties, pollution

1. Introduction

It is anticipated that the greatest challenge of this century will be global warming. Over the previous ten decades, the Earth's average temperature rose by roughly 0.4 to 0.8°C (Bhatti et al., 2006; Trabalka & Reichle, 2013). According to recent estimates from Intergovernmental Panel on Climate Change, by the year 2100, the avg. global temperature will have increased by an additional 1.4–5.8°C. The primary causes of the greenhouse effects are widely recognised to be carbon dioxide water vapour, sulphur dioxide, nitrogen dioxide, methane and chlorofluorocarbon.

In order to reduce the reliance on petroleum-based fuels as well as regulate the emission of greenhouse gases (GHGs) to environment, researchers are now focusing on exploring a variety of renewable energy sources, such as oxygen-rich fuels, fuel-cell, solar, along with biofuel (n-butanol) techniques (Uyumaz, 2015; He et al., 2015; Kahveci & Taymaz, 2019; Taymaz & Benli, 2014; IPCC, 2014). Cutting the emission of CO_2 by decreasing the utilisation of petroleum fuels is one of the most successful

[a]bnkale.phd2018.me@nitrr.ac.in

DOI: 10.1201/9781003567653-11

procedures to limit the emission of GHGs, as CO_2 is right away linked to the carbon presence of the fuel being used and the quantity of fuel consumed (Shaheen & Lipman, 2007). In addition to CO_2, the production as well combustion of automobile fuels also releases CH_4 plus N_2O, both of which lead to the pollution of GHGs.

In addition to CO_2, the production and burning of transport fuels also releases nitrous oxide and CH_4. These emissions also play a role in the greenhouse gas emissions. To lower greenhouse gas emissions and boost the effectiveness of energy systems, innovative vehicle engines having post-combustion emission monitoring systems should be invented in addition to adopting clean fuel substitutes (Yusof et al., 2019). Over the past few decades, the usage of biodiesel engines in the transportation along with power generating sectors has grown. The most recent scientific and technological trend is to create a new type of internal combustion engine (ICE) with low emissions (Giakoumis's, 2012; Kumar & Saravanan, 2016), energy savings (Mwangi et al., 2015), as well as high efficiencies (Wu et al., 2019; Pedrozo et al., 2018).

Since they possess strong thermal brake efficiency, elevated compression ratio, and abridged air-fuel blend, biodiesel engines possess an outstanding track record for low fuel consumption, excellent reliability, as well as excellent durability (Ghaffarzadeh et al., 2020; Yesilyurt & Aydin, 2020). Nevertheless, there are limits to how much NOx either diesel or biodiesel fuel may produce, which results in inefficient combustion (Devarajan & Madhavan, 2017; Radhakrishnan et al., 2019; Saravanan & Nagarajan, 2014). In order to get around these restrictions, fuel additives are being added more and more to help biodiesel's oxidation properties. Combining diesel or biodiesel using fuel additives has the capacity to enhance performance and efficiently lower greenhouse gas emissions.

The features of nanoparticles as additive for diesel/ biodiesel fuels are presented in this study based on recent research. The current study examines how various diesel/biodiesel-nanoparticle blends and internal combustion engines (ICEs) perform in relation of performance plus emission characteristics. Overall, this review paper will assist industrial engine producers and scientists studying nanotechnology in compiling a brief report on the engine performance along with emission problems associated with various diesel/biodiesel fuel combinations. Although work is carried out using several nanoparticles, but there is scope to investigate the blends of biodiesel (one or more) plus nanoparticles inclusion for single and multicylinder IC engines.

2. Nanoparticles

Distinct results may be obtained when using several types of nanoparticles within distinct diesel/biodiesel blends. It will impact engine performances, such as engine torque, brake power, brake thermal efficiency, brake-specific fuel consumption, and particulate matter (PM) along with other harmful gas emissions. At the moment, the most commonly utilised fuel-additive nanoparticles for diesel as well as biodiesel fuels are mainly classified in two categories like metal oxides and nonmetal oxides-based nanoparticles.

- Metal oxides-based nanoparticles
  - (Cerium oxide (CeO_2), Aluminium oxide (Al_2O_3), silver oxide (Ag_2O), copper oxide (CuO), iron oxide (Fe_2O_3), silicon (Si), zinc oxide (ZnO), titanium oxide (TiO_2), and magnesium (Mg)) and
- Nonmetal oxides-based nanoparticles
  - multiwall CNTs (MWCNTs), carbon nanotubes (CNTs) and graphene oxide (GO), etc.

3. Nanoparticles Influence on Fuel Characteristics

One of the important elements influencing how well fuel is mixed and burned is the fuel's characteristics. Lately, adding nanoparticles has been seen to be a beneficial way to improve the fuel's qualities. By incorporating several nanoparticle kinds into different diesel and biodiesel fuels, multiple investigators have investigated the fuel's qualities (Nema & Singh, 2018). The physio-chemical characteristics of diesel/biodiesel mixture are discussed in Table 11.1.

It has been observed that most of the nanoparticles boost the base fuel thermophysical characteristics with a very few exceptions.

Table 11.1: Characteristics of diesel/biodiesel blends with nanoparticles

Reference	Fuel	Additive	Characteristics of Fuel			
			Viscosity	Flashpoint	Calorific value	Density
Vedagiri et al.	Grapeseed oil biodiesel (GSO)	CeO_2;	Upsurge	Neutral	Upsurge	Upsurge
		ZnO	Upsurge	Neutral	Upsurge	Upsurge
Praveena et al.	GSO	CeO_2;	Upsurge	Neutral	Upsurge	Upsurge
		ZnO	Upsurge	Neutral	Upsurge	Upsurge
Karthikeyan et al.	GSO	CeO_2;	Upsurge	Neutral	Upsurge	Upsurge
		ZnO	Upsurge	Neutral	Upsurge	Upsurge
Chen et al.	Petro-diesel	SiO_2;	Upsurge	Falls	Upsurge	Upsurge
		Al_2O_3;		Upsurge	Upsurge	Upsurge
		CNT	Falls	Neutral	Upsurge	Upsurge
Gumus et al.	Petro-diesel	Al_2O_3;	Falls	Upsurge	Upsurge	Upsurge
		CuO	Falls	Upsurge	Decline	Falls
Sahoo and Jain.	Petro-diesel	CuO	Falls	Upsurge	Decline	Falls
D'Silva et al.	Petro-diesel	CuO ;	Falls	Upsurge	Falls	Falls
		TiO_2	Falls	Upsurge	Falls	
Devarajaan et al.	Palm stearin biodiesel	AgO	Falls	Upsurge	Upsurge	Upsurge
Perumal.	Pongamia biodiesel	CuO	Falls	Upsurge	Falls	Upsurge
Sajin et al.	Mango seed biodiese	ZnO	Falls	Upsurge	Upsurge	Upsurge
Lenin et al.	Petro-diesel	MnO;	Falls	Falls	Upsurge	Upsurge
		CuO	Falls	Upsurge	Falls	Falls

Source: Vedagiri et al., 2020; Praveena et al., 2020; Karthikeyan et al., 2014; Chen et al., 2018; Gumus et al., 2016; Sahoo and Jain, 2019; D'Silva et al., 2015; Devarajan et al., 2019; Perumal and Ilangkumaran, 2018; Lenin et al., 2013.

4. Nanoparticles Influence on Engine Characteristics (Performance and Emission)

In order to boost the ratio of surface area to volume and enhance the total number of reactive surfaces, nanoparticles have primarily been added to diesel and bio-diesel fuel. This makes it possible for the nanoparticles to serve as an efficient chemical catalyst, enhancing both the fuel-air mixing sequence and fuel combustion efficiency, ultimately resulting in an aromatic fuel that burns completely. Together with a few unreported research studies, Table 11.2

Table 11.2: Effect of nanoparticles on performance and emission parameters

Reference	Fuel	Additive	Performance Parameters		Emission Parameters
			BTE	BSFC	
Vedagiri et al.	Grapeseed oil biodiesel (GSO)	CeO_2 (100 ppm); ZnO (100 ppm)	Upsurge Upsurge	Falls Falls	Less-NOx (5%) Less
Praveena et al.	GSO	CeO_2 (50–100 ppm); ZnO (50–100 ppm);	Upsurge Upsurge	Falls Falls	Less- CO; HC & NOx Decrease (3%)
Karthikeyan et al.	GSO	CeO_2 (25 ppm); ZnO (25 ppm)	Upsurge Upsurge	Falls Falls	Less Pollution Less
Chen et al.	Petro-diesel	SiO_2; Al_2O_3 (25, 50, 100 ppm); CNT (25,50,100 ppm);	Upsurge Upsurge Upsurge	Falls Falls Falls Falls	Less Less UBHC (12%) Less- CO; HC & NOx Less - CO; HC & NOx (3–7%)
Gumus et al.	Petro-diesel	Al_2O_3 (12, 20, 30 ppm); CuO (50,100 ppm);	Upsurge Upsurge	Falls Falls	Less -UBHC Less - CO; HC & NOx (5–8%)
Sahoo and Jain.	Petro-diesel	CuO (50,100 ppm);	Upsurge	Falls	Less - CO; HC & NOx (4–7%)
D'Silva et al.	Petro-diesel	CuO (25, 50 ppm); TiO_2 (50 ppm);	Upsurge Upsurge	Falls Falls	Less - CO; HC & NOx (upto 10%) Less -CO, HC (upto 10%); More- NOx
Devarajaan et al.	Palm stearin biodiesel	AgO	Upsurge	Falls	Less - CO; HC & NOx (Max upto 10%)
Perumal.	Pongamia biodiesel	CuO	Upsurge	Falls	Less - CO; HC & NOx
Sajin et al.	Mango seed biodiese	ZnO	Upsurge	Falls	Less - CO; HC & NOx
Lenin et al.	Petro-diesel	MnO; CuO	Upsurge Upsurge	Falls Falls	Less - CO; HC & NOx Less - CO; HC & NOx (upto 10%)

Source: Vedagiri et al., 2020; Praveena et al., 2020; Karthikeyan et al., 2014; Chen et al., 2018; Gumus et al., 2016; Sahoo and Jain, 2019; D'Silva et al., 2015; Devarajan et al., 2019; Perumal and Ilangkumaran, 2018; Lenin et al., 2013.

provides an overview of how diesel and biodiesel fuel's nanoparticle content affect the ICE's performance, along with emission characteristics.

5. Conclusion

An updated analysis of nanofluid's performance as a fuel additive for diesel and biodiesel is provided in this article. Dropping GHG and growing engine efficiency may be achieved by using a nanofluid biodiesel combination in the IC engine. The following are the primary findings of this manuscript.

- The fuel's kinematic viscosity, caloric value, flash point, along with density can all be improved by adding nanoparticles into diesel or biodiesel, which will result in combustion improvement.
- The engine's efficiency as well as its emissions can be significantly improved by increasing the percentage of nanoparticles.
- CNT nanoparticles possessed excellent thermal conductivity and surface deficits so it preferred is as non-metal nanoparticles.
- The presence of excess oxygen and nanoparticles in diesel fuels improved the fuel properties and combustion efficiency, which led to lower BSFC and harmful emissions so, Aluminium is the ideal metal-based nanoparticle choice.

References

Bhatti, J. S., Apps, M. J., Lal, R., and Price, M. A. (2006). Anthropogenic changes and the global carbon cycle. Climate change managed ecosystems (pp. 71–92). CRC Press.

Chen, A. F., Adzmi, M. A., Adam, A., Othman, M. F., Kamaruzzaman, M. K., and Mrwan, A. G. (2018). Combustion characteristics, engine performances and emissions of a diesel engine using nanoparticle-diesel fuel blends with aluminium oxide, carbon nanotubes and silicon oxide. Energy Conversion and Management, 171, 461–477.

D'Silva, R., Binu, K. G., and Bhat, T. (2015). Performance and emission characteristics of a CI engine fuelled with diesel and TiO_2 nanoparticles as fuel additive. Materials Today: Proceedings, 2(4–5), 3728–3735.

Devarajan, Y., and Madhavan, V. R. (2017). Emission analysis on the influence of ferrofluid on rice bran biodiesel. Journal of the Chilean Chemical Society, 62(4), 3703–3707.

Devarajan, Y., Munuswamy, D. B., and Mahalingam, A. (2019). Investigation on behavior of diesel engine performance, emission, and combustion characteristics using nano-additive in neat biodiesel. Heat Mass Transf., 55(6), 1641–1650.

Ghaffarzadeh, S., Toosi, A. N., and Hosseini, V. (2020). An experimental study on low temperature combustion in a light duty engine fueled with diesel/CNG and biodiesel/CNG. Fuel., 262, 116495.

Giakoumis, E. G. (2012). A statistical investigation of biodiesel effects on regulated exhaust emissions during transient cycles. Appl Energ., 98, 273–291.

Gumus, S., Ozcan, H., Ozbey, M., and Topaloglu, B. (2016). Aluminum oxide and copper oxide nanodiesel fuel properties and usage in a compression ignition engine. Fuel, 163, 80–87.

He, B. Q., Liu, M. B., and Zhao, H. (2015). Comparison of combustion characteristics of n-butanol/ethanol–gasoline blends in a HCCI engine. Energ Convers Manage., 95, 101–109.

Kahveci, E. E., and Taymaz, I. (2019). Experimental study on performance evaluation of PEM fuel cell by coating bipolar plate with materials having different contact angle. Fuel, 253, 1274–1281.

Karthikeyan, S., Elengo, A., and Prathima, A. (2014). Performance, combustion and emission characteristics of a marine engine running on grape seed oil biodiesel blends with nano additive. Indian Journal of Geo-Marine Sciences, 43, 12.

Kumar, B. R., and Saravanan, S. (2016). Effects of iso-butanol/diesel and n-pentanol/diesel blends on performance and emissions of a DI diesel engine under premixed LTC (low temperature combustion) mode. Fuel, 170, 49–59.

Lenin, M. A., Swaminathan, M. R., and Kumaresan, G. (2013). Performance and emission characteristics of a DI diesel engine with a nanofuel additive. Fuel, 109, 362–365.

Mwangi, J. K., Lee, W. J., Chang, Y. C., Chen, C. Y., Wang, L. C. (2015). An overview: Energy saving and pollution reduction by using green fuel blends in diesel engines. Applied Energy, 159, 214–236.

Nema, V. K., and Singh, A. (2018). Emission reduction in a dual blend biodiesel fuelled CI engine using nano-fuel additives. Mater Today Proc., 5(9), 20754–20759.

Pedrozo, V. B., May, I., Guan, W., and Zhao, H. (2018) High efficiency ethanoldiesel dual-fuel combustion: A comparison against conventional diesel combustion from low to full engine load. Fuel, 230, 440–-451.

Perumal, V., and Ilangkumaran, M. (2018). The influence of copper oxide nano particle added pongamia methyl ester biodiesel on the performance, combustion and emission of a diesel engine. Fuel, 232, 791–802.

Praveena, V., Martin, M. L., and Geo, V. E. (2020). Experimental characterization of CI engine performance, combustion and emission parameters using various metal oxide nanoemulsion of grapeseed oil methyl ester. Journal of Thermal Analysis and Calorimetry, 139(6), 3441–3456.

Radhakrishnan, S., Munuswamy, D. B., Devarajan, Y., and Mahalingam, A. (2019). Performance, emission and combustion study on neat biodiesel and water blends fuelled research diesel engine. Heat and Mass Transfer, 55(4), 1229–1237.

Sahoo, R. R., and Jain, A. (2019). Experimental analysis of nanofuel additives with magnetic fuel conditioning for diesel engine performance and emissions. Fuel, 236, 365–372.

Sajin, J. B., Pillai, G. O., Kesavapillai, M., and Varghese, S. (2019). Effect of nanoparticle on emission and performance characteristics of biodiesel. International Journal of Ambient Energy, 42(5), 1–16.

Saravanan, S., and Nagarajan, G. (2014). Comparison of influencing factors of diesel with crude rice bran oil methyl ester in multi response optimization of NOx emission. Ain Shams Engineering Journal, 5(4), 1241–1248.

Shaheen, S. A., and Lipman, T. E. (2007). Reducing greenhouse emissions and fuel consumption: Sustainable approaches for surface transportation. IATSS Research, 31(1), 6–20.

Taymaz, I., and Benli, M. (2014). Emissions and fuel economy for a hybrid vehicle. Fuel, 115, 812–817.

Trabalka, J. R., and Reichle, D. E. (Eds.). (2013). The changing carbon cycle: A global analysis. Springer Sci Bus Media.

Uyumaz, A. (2015). An experimental investigation into combustion and performance characteristics of an HCCI gasoline engine fueled with n-heptane, isopropanol and n-butanol fuel blends at different inlet air temperatures. Energy Conversion and Management, 98, 199–207.

Vedagiri, P., Martin, L. J., Varuvel, E. G., and Subramanian, T. (2020). Experimental study on NOx reduction in a grapeseed oil biodiesel-fueled CI engine using nanoemulsions and SCR retrofitment. Environmental Science and Pollution Research, 27, 29703–29718.

Wu, G., Lu, Z., Xu, X., Pan, W., Wu, W., and Li, J. (2019). Numerical investigation of aeroacoustics damping performance of a Helmholtz resonator: Effects of geometry, grazing and bias flow. Aerospace Science and Technology, 86, 191–203.

Yesilyurt, M. K., and Aydin, M. (2020). Experimental investigation on the performance, combustion and exhaust emission characteristics of a compression-ignition engine fueled with cottonseed oil biodiesel/diethyl ether/diesel fuel blends. Energy Conversion and Management, 205, 112355.

Yusof, S. N., Asako, Y., Faghri, M., Tan, L. K., bin Che Sidik, N. A., and bin Aziz Japar, W. M. (2019). Numerical analysis of irreversible processes in a piston-cylinder system using LB1S turbulence model. International Journal of Heat and Mass Transfer, 136, 730–739.

12 Enhancing safety for motorcycle riders with a smart helmet and emergency SOS system using nRF 433 Module

Asfar Siddiqui[1,a], Vinod Waiker[2,b], and Anup Ranade[1,c]

[1]Yeshwantrao Chavan College of Engineering, Nagpur, India
[2]Datta Meghe Institute of Management Studies, Nagpur, India

Abstract

In today's busy world, everything is in fast pace. We use various types of transport systems according to our needs. Road transport is one of the most widely used mode of transport. For personal use cars and two wheelers are the most preferable method of transport all over the world. In developing countries bikes are also in high usage compared to cars. Apart from just mere mode of transport, bikes are a popular preference amongamongst youths used during bike tours or even during daily commute due to its fashion aspect. With the increasing number of bikers, the chances of unwanted accidents are also rising exponentially. Although several laws and regulations are in place to ensure a safe travel, most of the times these rules are not followed or aren't followed up to the standard. Wearing of helmet is one of the **rules** which has shown to save a lot of lives in the case of an accident. **But it is also seen that wearing a helmet is considered only necessary to escape from challan and not considered a mandatory equipment for a safe ride.** In this study we will be discussing and implementing a smart helmet which can provide an optimum level of safety to the rider all around. It will be equipped with features including mandatory helmet wearing and strapping feature prior to starting the ignition of the bike, real time accident detection feature to identify when the rider has got into an accident, emergency SOS system which can send an emergency SOS message to the emergency contact or other authorities in case of an accident along with the exact GPS location of the accident, etc. With all these features along, the system will be able to provide a reliable option for the riders to be safe during their commute and also to get emergency help in case of an accident as soon as possible.

Keywords: Safety, smart helmet, SOS, Internet of Things, GPS

1. Introduction

Modern life is incredibly dependent on transportation. Today, various modes of transport are available, including road, air, water, and rail transit. The choice of mode of transport is influenced by the destination, price, distance of travel and urgency. Each

---

[a]asfar1811@gmail.com, [b]dmimsit@yahoo.com, [c]drarranade@gmail.com

DOI: 10.1201/9781003567653-12

method of transportation has advantages and drawbacks. For long distance travel, air travel is the fastest means of transport. It is a well-preferred option for both business and pleasure travellers as it is effective, safe and enjoyable. Although expensive, not everyone can afford to go by air. Additionally, air travel causes considerable damage to the environment by increasing carbon emissions and climate change. Another preferred means of transport is rail, especially for interstate travel and cargo transit. It is the best choice for long distance travel as it is cost-effective, ecologically friendly and energy-efficient. But it's less adaptable than driving, and the infrastructure and maintenance are expensive. The most traditional and cost-effective way to move large and heavy objects is by water. It is perfect for carrying objects across borders and over large distances. However, it is dull and unfit for things that must be delivered quickly. Additionally, only coastal locations can use water transport, and not every destination is accessible by sea (Buehler & Pucher, 2012; Egset & Nordfjærn, 2019). The most popular means of transportation for carrying people and products on a daily basis is the road. It is a practical choice for short distances because it is adaptable and offers door-to-door services. Moreover, two-wheeled vehicles are the most popular form of personal and commercial transportation in many developing nations, particularly in highly populated areas (Gutierrez & Mohan, 2020). These two-wheelers are well-liked because they are inexpensive, easy to manoeuvre in heavy traffic, and fuel-efficient. Additionally, they can move through alleys and tight streets with ease and occupy less space on the road. In contrast to four-wheelers, riders of two-wheelers are less protected, making them more prone to accidents and frequently involved in deadly ones. As a result, it's critical to prioritise two-wheeler riders' safety by taking steps like mandating the usage of helmets and enforcing traffic laws (Agarwal et al., 2018). Any form of transportation must adhere to safety laws and regulations. The purpose of these laws is to protect the well-being of drivers, passengers, and pedestrians. Nevertheless, despite the fact that these regulations exist, it is frequently seen that they are not fully adhered to. For instance, when it comes to road transportation, people frequently disregarded fundamental safety precautions like utilising seatbelts when operating a four-wheel vehicle or riding a two-wheeler while wearing a helmet (Joewono & Kubota, 2006; Singh et al., 2021). Such actions put at risk not only the lives of those involved but also the safety of other drivers. Therefore, it is critical to raise awareness of the significance of adhering to safety norms and regulations for all forms of transportation and to enforce them (Talbot et al., 2020). Trains, cars, and other frequently used forms of transportation all have a number of safety safeguards in place to protect passengers. For instance, to reduce the danger of accidents and injuries, modern cars come standard with airbags, seat belts, anti-lock braking systems (ABS), electronic stability control (ESC), and blind-spot detection. To protect the safety of passengers while in flight, aeroplanes include sophisticated avionics systems, air traffic control, and emergency protocols. Numerous other safety elements are also included aboard trains, including automatic train control systems, emergency brakes, and track inspections (Singh & Mishra, 2010). However, many of these safety elements are absent from two-wheelers like motorcycles and scooters. They lack the robust structural design of cars and the advanced technology of trains or aeroplanes. But basic safety equipment that can shield passengers in an accident, such airbags and seat belts, are also absent from two-wheelers. Due to their smaller size and fragility in traffic, two-wheelers are also more likely to be in accidents. To ensure the safety of riders on the road, it is essential to design and deploy more two-wheeler safety measures (Kothakonda, 2021). Road safety is now quite concerned about the rise in motorbike accidents. The failure to wear a helmet is one of the main causes of these collisions. The suggested remedy for this problem is a smart helmet with an integrated control system. While wearing a helmet and obeying traffic laws are two ways to increase motorbike safety, there is still a need for modern gadgets to offer real-time support in emergency situations. Smart helmets have come to light as a potential improvement to motorcycle safety in recent years. These helmets have cutting-edge technologies like integrated sensors, cameras, and communication systems that let riders access real-time traffic, weather, and road condition information (Delbosc & Currie, 2012; Evans, 1994; Gehlot et al., 2021; Damani & Vedagiri, 2023).. For the purpose of preventing accidents, smart helmets have grown in popularity in recent years among riders of two-wheelers. These helmets come with technologies like heads-up displays, built-in cameras, Bluetooth connectivity, and GPS navigation. Additional safety features are also provided using tools like AI and deep learning algorithms. Smart helmets, for instance, can recognise when a rider is in an accident and instantly alert emergency authorities with the rider's GPS location. Additionally, they can keep an eye on the rider's vital signs and warn them if something seems off, such if the person is nodding off while they're moving. Some smart helmets come with sensors that can detect obstacles and alert the wearer to potential road hazards. Riders who are operating a vehicle in poor visibility or during inclement weather may find this to be especially helpful. In addition to these functions, some smart helmets have built-in speakers that enable the rider to receive calls or listen to music while they are on the road without the need for headphones that might make it difficult for them to hear their surroundings (Attal et al., 2018; Jalti et al., 2021). In general, smart helmets have the potential to dramatically increase two-wheeler users' road safety, especially in nations where the usage of helmets is not tightly enforced.

These helmets can give riders real-time information about their surroundings and warn them of potential hazards by integrating cutting-edge technology like AI and deep learning, thereby lowering the amount of collisions and fatalities on the road (Singh et al., 2018; Long et al., 2019; Li et al., 2017; Kim et al., 2021; P. N. R., 2016).

2. System Architecture

In this study we are going to implement a smart helmet with emergency SOS system. Our system will have primarily two sub systems, first one being the helmet side system which will be responsible for sensing whether the rider has worn the helmet, whether he/she has strapped the helmet strap properly, detect accident scenario in real-time and sending the same to the other sub-system using a nRF 433 transmitter as shown in the Figure 12.1. We are using an infrared emitter and receiver pair to identify whether the rider has worn the helmet, a pullup conductivity setting to identify if the rider has properly strapped the helmet strap and then a MPU6050 sensor is integrated which gives us the real-time accelerometer, gyroscope readings to detect accident, and at last a nRF 433 transmitter module to send the data to the other subsystem for further processing.

The second subsystem as depicted in Figure 12.2 is the vehicle mounted system,

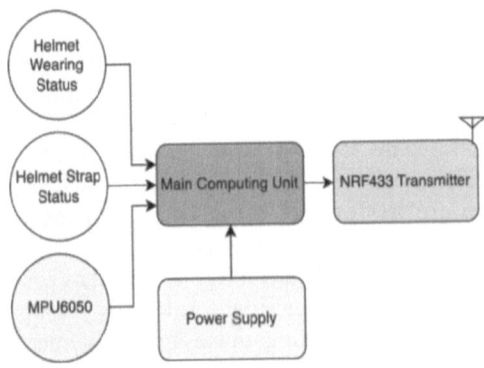

Figure 12.1: Helmet side system.

Source: Author.

which is responsible for receiving data from the helmet side system like when the user has worn the helmet and strapped the strap properly, it will consider the rider ready and will allow the ignition of the vehicle and also when any accident is detected by the helmet side system it will send emergency help message to the contacts along with rider's location information. It will also use its LCD to print the real time status like "System Not Ready", "System Ready", "Accident Occurred", GPS Coordinates, Message sent status, etc. for the rider's reference.

The overall system's workflow is depicted in the Figure 12.3, wherein once the system (Singh et al., 2022) is power up and ready, the helmet side system will first check if the rider has worn the helmet properly and strapped the helmet strap (Lakshmi et al., 2022), until and unless both these conditions are satisfied, the helmet side system will be in "System Not Ready" state and so will be the vehicle mounted system which will not let the rider start the ignition.

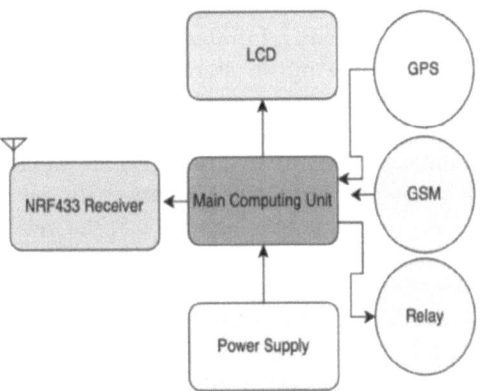

Figure 12.2: Vehicle mounted system.

Source: Author.

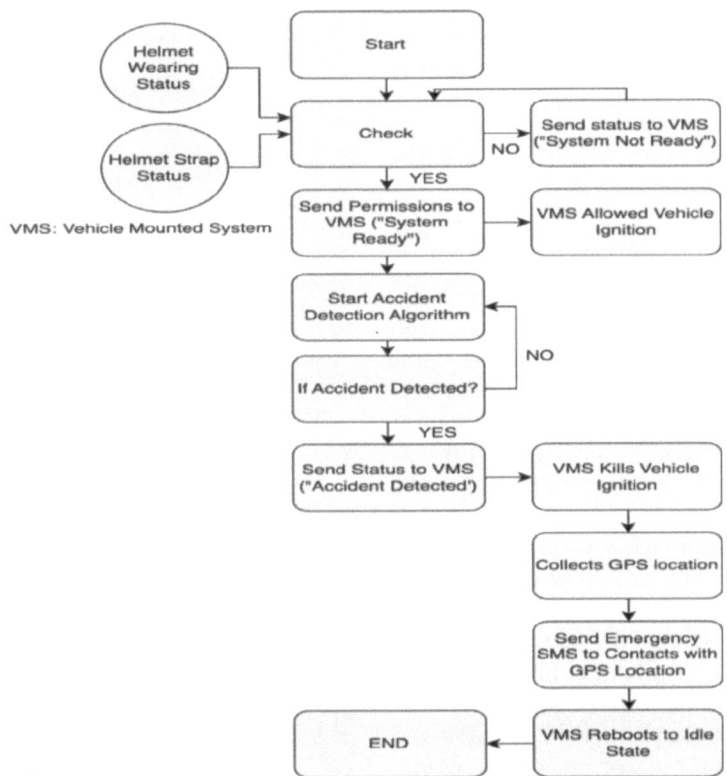

Figure 12.3: System workflow.

Source: Author.

Once both these conditions are satisfied, the vehicle mounted system will receive an updated state "System Ready" using its nRF 433 receiver, once this state is received the vehicle mounted system will start/allow the vehicle ignition. Simultaneously, the helmet side system will start the accident detection algorithm (Verma et al., 2022) which starts analysing the x, y, z axis gyroscope and accelerometer values from the MPU6050 and then passing these values through a complimentary filter to get a cumulative result and look for any sudden change in the values above a certain (Torres-Cruz et al., 2022) threshold. Once accident is detected by the system, it will send a status "Accident Occurred" to the vehicle mounted system. Following that, the vehicle mounted system immediately kills (Jain et al., 2022; Kannapiran et al., 2022; Mishra et al., 2022) the ignition and sends the emergency SOS message to the contacts along with the current GPS location of the vehicle. Once the message is sent successfully to all the listed contacts, the vehicle mounted system reboots to idle state.

3. Discussions

In this study we implemented a smart helmet and emergency SOS system, which had its two primary subsystems, one being mounted on the helmet of the rider with responsibilities including ensuring that the rider has worn their helmet properly and strapped the helmet strap as well, monitoring accidents scenario and sharing the current status with the other subsystem using an integrated nRF433 module. The second subsystem was a vehicle mounted system which is to be mounted in the rider's vehicle and will be responsible for receiving status updates from the helmet side system, control the vehicle (Petikam et al., 2023) ignition, like the ignition to be turned off until the helmet is worn properly by the rider and the strap is strapped, receiving accident scenario, extracting current GPS location and sending emergency message including the location to the listed emergency contacts.

As shown in the Figure 12.4, the helmet side system is being controlled using an Arduino nano board and is integrated with a nRF 433 transmitter (Joshi et al., 2022) module for data communication with the vehicle mounted system, a Infrared emitter-receiver pair for knowing the status of helmet status, a pullup conductivity node for knowing the strap status, and a MPU6050 sensor for detecting accidents by analysing the values of gyroscope and accelerometer.

As shown in the Figure 12.5, the vehicle mounted system's prototype was build using a Arduino uno board as the microcontroller, the system has a nRF433 receiver module for receiving data from the helmet side system, a GSM module (SIM800L) to send emergency messages in case of an accident to the listed emergency contacts, a GPS module (NEO-6M) to get the GPS coordinates of the vehicle to be sent in the emergency message in case of an accident,

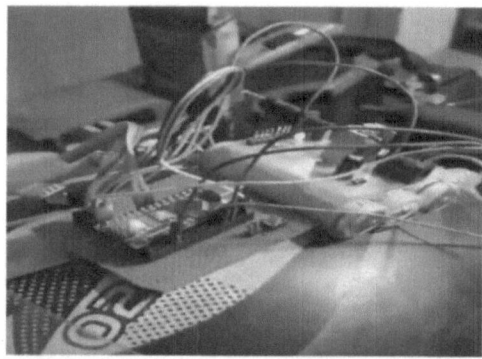

Figure 12.4: Helmet side system.

Source: Author.

Figure 12.5: Vehicle mounted system.

Source: Author.

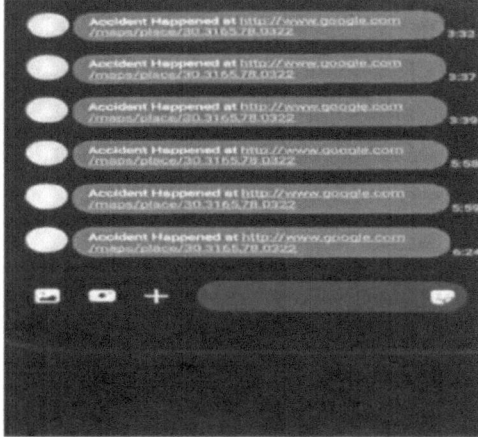

Figure 12.6: Vehicle mounted system.

Source: Author.

a single channel relay to control the vehicle ignition during various states, a 16*2 LCD to show the current state along with some other crucial information for the rider's reference and a voltage converter to supply exactly 4.0 V to the GSM module for proper functioning.

4. Conclusion

As shown in the Figure 12.6, in case of an accident, the vehicle mounted system sends the preset emergency message including the GSP location to the listed emergency contacts. The format being used in this system can be configured according to the needs, more information can also be added to help the emergency response authorities or emergency contacts get a hold of the whole situation beforehand. The GPS coordinates in the message is hidden for privacy reasons. To conclude, this implementation was found to be quite reliable and dependable, both the systems were found to be in proper sync and to work as expected. The accident detection algorithm was also found to be working really well wherein it was able to ignore the jerks the rider can experience in bumpy roads that would have raised false positive alarms. The thresholds adapted during testing were perfect for only detecting accident scenarios. Further improvements can also

be made to the system to make it more compact and rugged but, at this form or with improvements, the system can prove to be really beneficial for two wheeler vehicle riders and can save a lot of lives by ensuring helmet wearing amongst riders and by the emergency SOS system.

References

Agarwal, A., Shukla, V., Singh, R., Gehlot, A., & Garg, V. (2018). Design and development of air and water pollution quality monitoring using IoT and quadcopter. *Advances in Intelligent Systems and Computing*, 485–492. doi: 10.1007/978-981-10-5903-2_49.

Attal, F., Boubezoul, A., Samé, A., Oukhellou, L., & Espié, S. (2018). Powered two-wheelers critical events detection and recognition using data-driven approaches. *IEEE Transactions on Intelligent Transportation Systems*, 19(12), 4011–4022.

Buehler, R., & Pucher, J. (2012). Demand for Public Transport in Germany and the USA: An Analysis of Rider Characteristics. *Transport Reviews*, 32(5), 541–567. doi: 10.1080/01441647.2012.707695.

Damani, J., & Vedagiri, P. (2023). Predicting choice of filtering of motorized two wheelers in urban mixed traffic. *Transportation Research Record: Journal of the Transportation Research Board*, 036119812311583. doi: 10.1177/03611981231158320.

Delbosc, A., & Currie, G. (2012). Modelling the causes and impacts of personal safety perceptions on public transport ridership. *Transport Policy*, 24, 302–309.

Egset, K. S., & Nordfjærn, T. (2019). The role of transport priorities, transport attitudes and situational factors for sustainable transport mode use in wintertime. *Transportation Research Part F: Traffic Psychology and Behaviour*, 62, 473–482. doi: 10.1016/j.trf.2019.02.003.

Evans, A. W. (1994). Evaluating public transport and road safety measures. *Accident Analysis & Prevention*, 26(4), 411–428.

Gehlot, A., Alshamrani, S. S., Singh, R., Rashid, M., Akram, S. V., AlGhamdi, A. S., & Albogamy, F. R. (2021). Internet of things and long-range-based smart lampposts for illuminating smart cities. *Sustainability*, 13(11), 6398.

Gutierrez, M. I., & Mohan, D. (2020). Safety of motorized two-wheeler riders in the formal and informal transport sector. *International Journal of Injury Control and Safety Promotion*, 27(1), 51–60. doi: 10.1080/17457300.2019.1708408.

Jain, S., Patil, S., Dutt, S., Joshi, K., Bhuvaneswari, V., & Jayadeva, S. M. (2022, December). Contribution of Artificial intelligence to the Promotion of Mental Health. In *2022 5th International Conference on Contemporary Computing and Informatics (IC3I)* (pp. 1938–1944). IEEE.

Jalti, F., Hajji, B., & Mbarki, A. (2021, January). The potential outcomes of artificial intelligence applied to the powered two-wheel vehicle: analytical review. In *International Conference on Digital Technologies and Applications* (pp. 1595–1605). Cham: Springer International Publishing.

Joewono, T. B., & Kubota, H. (2006). Safety and security improvement in public transportation based on public perception in developing countries. *IATSS Research*, 30(1), 86–100. doi: 10.1016/s0386-1112(14)60159-x.

Joshi, K., Bhatt, S. S., Gehlot, A., Buddhi, D., Akram, S. V., & Bisht, Y. S. (2022, April). A education tracking approach using maps and geo-location method. In *2022 3rd International Conference on Intelligent Engineering and Management (ICIEM)* (pp. 727–730). IEEE.

Kannapiran, E., Joshi, K., Chougale, R. K., Rana, N., Neeraja, B., & Kaur, C. (2022, July). Smart electric vehicle charging station for residential complex. In *2022 International Conference on Innovative Computing, Intelligent Communication and Smart Electrical Systems (ICSES)* (pp. 1–4). IEEE.

Kim, Y., Baek, J., & Choi, Y. (2021). Smart helmet-based personnel proximity warning system for improving underground mine safety. *Applied sciences*, 11(10), 4342.

Kothakonda, H. (2021). Smart Helmet for Coal Miners *SSRN Electronic Journal*. doi: 10.2139/ssrn.3918263.

Lakshmi, P. S., Saxena, M., Koli, S., Joshi, K., Abdullah, K. H., & Gangodkar, D. (2022, April). Traffic response system based on data mining and internet of things (IoT) for preventing accidents. In *2022 2nd International Conference on Advance Computing and Innovative Technologies in Engineering (ICACITE)* (pp. 1092–1096). IEEE.

Li, J., Liu, H., Wang, T., Jiang, M., Wang, S., Li, K., & Zhao, X. (2017, February). Safety helmet wearing detection based on image processing and machine learning. In *2017 Ninth international conference on advanced computational intelligence (ICACI)* (pp. 201–205). IEEE.

Long, X., Cui, W., & Zheng, Z. (2019, March). Safety helmet wearing detection based on deep learning. In *2019 IEEE 3rd information technology, networking, electronic and automation control conference (ITNEC)* (pp. 2495–2499). IEEE.

Mishra, R., Joshi, K., & Gangodkar, D. (2022, December). Wireless Communications Network and Mobile Computing using Blockchain in Distributed Internet of Things. In *2022 11th International Conference on System Modeling & Advancement in Research Trends (SMART)* (pp. 832–836). IEEE.

NR, P. (2016). Smart pi cam based Internet of things for motion detection using Raspberry pi. *International Journal Of Engineering And Computer Science*.

Petikam, S., Sales, F. D. C. D., Suma, S., Gonzáles, J. L. A., Joshi, K., & Pant, B. (2023, January). Image Processing with Intelligence System Using Sensing in Cyber Security. In *2023 International Conference on Artificial Intelligence and Smart Communication (AISC)* (pp. 570–574). IEEE.

Singh, A., Joshi, K., Shuaib, M., Bharany, S., Alam, S., & Ahmad, S. (2022, December). Navigation and Speed Regulation Aimed at Travel through Immersive Virtual Environments: A Review. In *2022 IEEE International Conference on Current Development in Engineering and Technology (CCET)* (pp. 1–6). IEEE.

Singh, R., & Mishra, S. (2010, November). Temperature monitoring in wireless sensor network using Zigbee transceiver module. In *2010 International Conference on Power, Control and Embedded Systems* (pp. 1–4). IEEE.

Singh, R., Gehlot, A., Akram, S. V., Gupta, L. R., Jena, M. K., Prakash, C., ... & Kumar, R. (2021). Cloud manufacturing, internet of things-assisted manufacturing and 3D printing technology: reliable tools for sustainable construction. *Sustainability*, 13(13), 7327.

Singh, R., Gehlot, A., Mittal, M., Samkaria, R., & Choudhury, S. (2018). Application of icloud and wireless sensor network in environmental parameter analysis. *International Journal of Sensors Wireless Communications and Control, 7*(3), 170–177.

Talbot, R., Brown, L., & Morris, A. (2020). Why are powered two wheeler riders still fatally injured in road junction crashes?–A causation analysis. *Journal of safety research, 75*, 196–204.

Torres-Cruz, F., Tyagi, S., Sathe, M., Mary, S. S. C., Joshi, K., & Shukla, S. K. (2022, December). Evaluation of Performance of Artificial Intelligence System during Voice Recognition in Social Conversation. In *2022 5th International Conference on Contemporary Computing and Informatics (IC3I)* (pp. 117–122). IEEE.

Verma, S., Raj, T., Joshi, K., Raturi, P., Anandaram, H., & Gupta, A. (2022, June). Indoor real-time location system for efficient location tracking using IoT. In *2022 IEEE world conference on applied intelligence and computing (AIC)* (pp. 517–523). IEEE.

13 STM32 and LoRa-based wireless sensor network for precision agriculture

Asfar Siddiqui[1,a], Subramaniam Seshan Iyer[2,b], and Anup Ranade[1,c]

[1]Yeshwantrao Chavan College of Engineering, Nagpur, India
[2]Datta Meghe Institute of Management Studies, Nagpur, India

Abstract

Here, we are building and monitoring the soil level with current humidity and temperature using the features and technology of LoRa-based RF networks and IoT-based technology so that end users can monitor their precise farming and can efficiently irrigate their crops with the aid of an IoT mobile app. For communication between the agriculture node and the LoRa gateway node, this system uses LoRa RF technology. It has low power requirements, high data transfer rates, a long communication range to cover a bigger region, flexible and accurate measurement features, and real-time monitoring. It has the ability to send sensor data using encryption and decryption techniques for secure communication, and ESP8266 has the ability to send all data to the end user over a public network so that they can access the IoT mobile app Dashboard and control it manually or automatically.

Keywords: LoRa gateway, agriculture, ESP 8266, wireless sensor network

1. Introduction

Cultivation is the utilisation of fertile land for the raising animals, variety of crop, medicinal plants. Agricultural methods are now not just employed in rural areas. Urban agriculture helps cities maintain a sustainable way of life while offering food security as an outcome of urbanisation and population shifts towards urban areas. For these locations to utilise irrigation water effectively, proper irrigation techniques with monitoring are needed. The wireless sensor network and IoT are assisting to implement automatic and real-time irrigation system with the help of sensors data. WSNs offer beneficial facilities for agricultural applications through an affordable method that results in a boost in crop yield. It also enables farmers to reduce pesticide waste, control pests and diseases effectively, and provide a sufficient quantity of nutrients for precision agriculture.

The objective of IoT application is to offer interface for data visualisation, boost the installation of devices, Wi-Fi-based IoT, WSNs the primary technology. This study has implemented a solution that can precisely and accurately monitor the health of our soil and crops. The Agriculture Node

[a]asfar1811@gmail.com, [b]subramaniamcoeoffice@gmail.com, [c]drarranade@gmail.com

DOI: 10.1201/9781003567653-13

and the LoRa Gateway are two of the nodes in this research that work together to give farmers useful information. A strong STM32 board and a LoRa Module, which connects wirelessly to the LoRa Gateway, are included in the Agriculture Node. Additionally, it has sensors for temperature, humidity, and soil moisture that gather vital information about the crop and its surroundings. In order to ensure that the crop receives the ideal amount of water, an actuator is also placed to manage the water pump for agriculture.

The LoRa Gateway, on the other hand, is made to wirelessly ingest data from the Agriculture Node. Its components—a Power Supply Module, an STM32 board, a Lora module, and an ESP8266 Wi-Fi board—allow it to send data to the Blynk IoT app. With real-time data access provided by this app, farmers may assess the condition of their crops and make educated decisions about irrigation and other important issues. Overall, this idea offers a novel approach to modern agriculture, giving farmers a practical and affordable means to keep an eye on their crops and maximise their growth.

2. Review of Literature

LoRa and IoT based customised LPWAN Technologies are implemented in this study for update from remote centre. A mobile LoRaWAN gateway device is implemented with Raspberry Pi 3 B+ to measure humidity and temperature. Precision agriculture is implemented using WSN, the Message Queuing Telemetry Transport (MQTT) protocol, TSL2561, and sensors to monitor greenhouse humidity, temperature, and light intensity. Low power acquisition devices are implemented in vineyard to evaluate sensor with LoRa technology and also for overcoming the precision agriculture and precision viticulture issues related to power consumption

In this study, WSN is built using LoRa transceivers, and an innovative collision handling protocol is proposed to provide a reliable channel management mechanism that reduces energy consumption for data transfer. LoRa-based WSN has been developed and implemented with data communication (both way) to deliver autonomous sensing acts for the smart farming. A study has implemented two topologies for WSN based automatic monitoring of agriculture and the proposed is evaluated using metrics such as throughput, delay and load.

3. Block Diagram

In this, the block diagram of the two different hardware devices such as agriculture node, and gateway are discussed.

Power Supply, Microcontroller, Sensors, and Actuators are the four essential parts of the Agriculture Node in Figure 13.1. The Agriculture Node process information and operate the sensors, actuators by microcontroller component. A Moisture and DHT11 Sensor, which detect condition of soil, are included in Sensors component. A Water Pump Controller is part of the Actuators component and manages how the water pump functions.

Together, these elements gather and process information on environmental factors that affect crop development and health, such as soil moisture, humidity, and temperature. Based on this information, the water pump controller subsequently makes sure

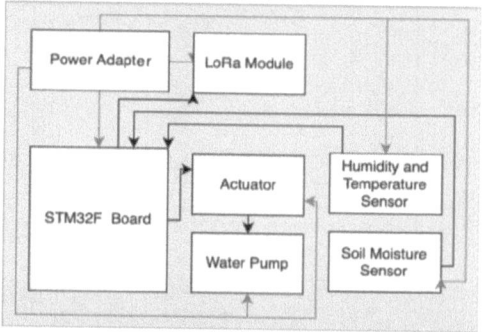

Figure 13.1: Block diagram of the agriculture node.

Source: Author.

that the right amount of water is provided to the crops. The LoRa Gateway receives the processed data after being wirelessly delivered there for additional analysis and decision-making.

Figure 13.2 shows how the LoRa Gateway with Blynk is made to wirelessly collect input given by Agriculture Node also send to BIOT app for ongoing monitoring and analysis. A Power Supply Module, Micro-controller, LoRa Module, and ESP8266 Wi-Fi Board are included in the Gateway Node block diagram. The Gateway Node is powered by Power Supply Module. The Micro-controller processing data that LoRa Module collect from the Agri-Node. The LO-RA Module receives data from the Agriculture Node using long-range, low-power wireless telecommunications-technology. The ESP-8266 WiFi Board then delivers data to Blynk IoT app over Wi-Fi after the processed data has been delivered there.

Farmers may get real-time information on the health of their crops and use the Blynk IoT app to make educated decisions regarding watering and other crucialaspects. The software can also send messages and alerts to farmers to let them know if there are any problems with their crops.

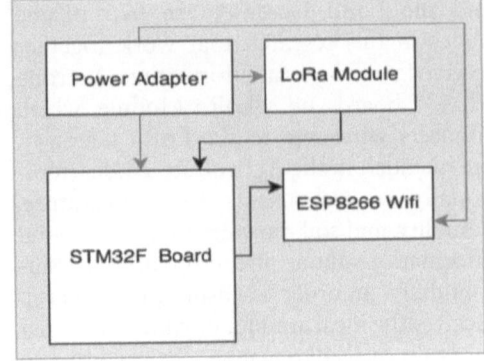

Figure 13.2: Block diagram of the LoRa gateway.

Source: Author.

4. Hardware Development

The Agriculture Node's link (Figure 13.3):

- Soil Moisture Sensor's pins(VCC, GND) join with STM32's +5V pin. The STM32 pin is connected to soil moisture sensor.
- DHT11 and GND connected with STM32, SGM32.
- STM32 Board are linked to the Single Channel Relay.
- The 5VDC wire linked to the Single Channel Relay's Common terminal and the 5V DC wire directly attached to the Relay's Normally Open terminal.

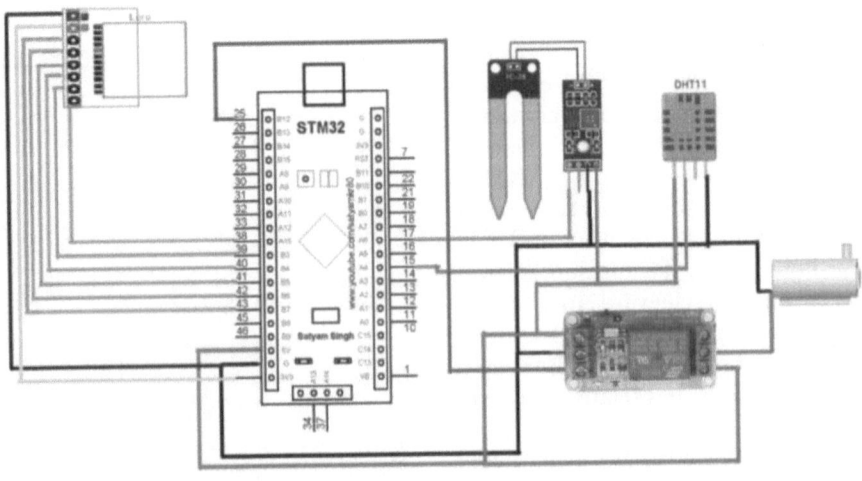

Figure 13.3: Connection diagram of Agriculture Node.

Source: Author.

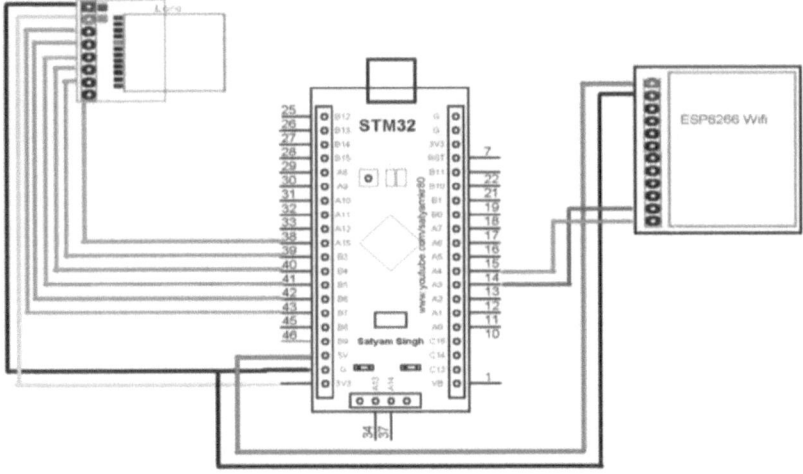

Figure 13.4: Connection diagram of LoRa gateway.

Source: Author.

The LoRa Gateway's (Figure 13.4):

- ESP8266 Wi-Fi and GND connected to STM32.
- TX pin of ESP8266 linked to 13 no pin STM32 board, Rx pin of the ESP8266 connected to t 14 no pin.
- STM32 Software SPI pins with Lora RF connected. STM32 3.3V & GND pin were linked to the Lora Vcc and GND pins, respectively. STM32 board connected to RST.

5. Implementations

5.1. Agri-Node

For real-time monitoring and analysis, the LoRa Gateway with Blynk is intended to wirelessly accept info from the Agriculture Node and transmit it to the BIoT app. A Power Supply Module, Micro-controllers, LoRa Module, and ESP8266 Wi-Fi Board are included in the Gateway Node block diagram. The Gateway Node is powered by the voltage and current supplied by the Power Supply Module. The Microcontroller is in charge of processing out-put from the Agri-culture Nodes. LoRa Module receives data from the Agriculture Node using long-range, low-power wireless communication technology.

The ESP8266 Wi-Fi Board then delivers the data to the BIoT apps over Wi-Fi after the processed data has been delivered there. The Blynk IoT app allows farmers to access updated data on their crops' condition, fix essential factors. The app can also generate alerts and notifications to warn farmers of any issues with their crops.

5.2. LoRa Gateway Node

Farmers may get real-time information on the health of their crops and use the Blynk IoT app to make educated decisions

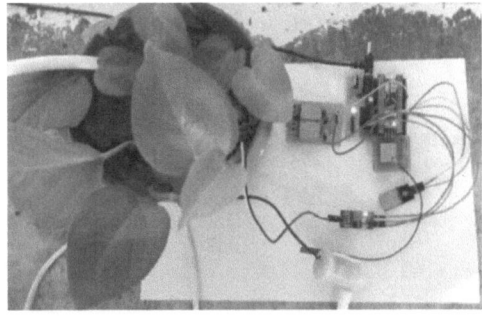

Figure 13.5: Agriculture node.

Source: Author.

regarding watering and other crucial aspects. The software can also send messages and alerts to farmers to let them know if there are any problems with their crops. Set the microcontroller up initially: The required libraries for the DHT11 sensor, soil moisture sensor, and LoRa module are imported after the microcontroller has been initialised. Set up the sensors: both sensor DHT11, soil moisture are then calibrated and initialised to get precise results. The information from the soil moisture sensor is then utilised to decide whether start or off water pump. The motor is activated if the quality of land level falls below a predetermined parameter. Otherwise water pump will be shut off. Sending sensor data through LoRa: After receiving the sensor data, the LoRa module is in charge of encoding it and sending it wirelessly to the LoRa gateway. For each

agriculture node, the LoRa module uses a different address to make sure the data is delivered to the right gateway.

The Agriculture Node's methodology entails initialising the microcontroller and sensors, reading the sensor data, regulating the borewell and wirelessly communicating the data to LoRa gateway for additional processing and analysis. The outcomes for the real-time data mapping and control can be observed for the Blynk IoT App's.

6. Conclusion

System for precise agriculture using STM32 and LoRa technology provides a potential answer for monitoring land fertility levels. The water pump may be remotely monitored and controlled by the Blynk IoT platform via the LoRa Gateway thanks to the system's sensors and actuator. In agriculture, the integration of wireless communication and IoT technology can boost production and efficiency while lowering water use and ultimately improving agricultural yields.

By enabling accurate and effective irrigation control based on real-time sensor data, this technology has the potential to enhance agricultural operations. It exemplifies how IoT, and wireless communication may be used in agriculture to increase productivity, decrease water waste, and eventually improve crop yields.

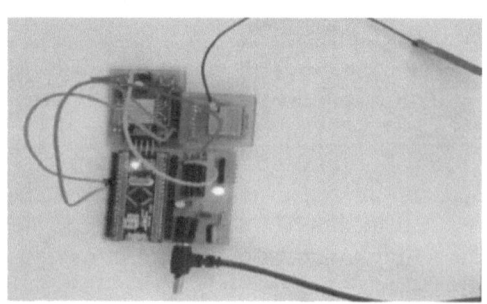

Figure 13.6: LoRa gateway.

Source: Author.

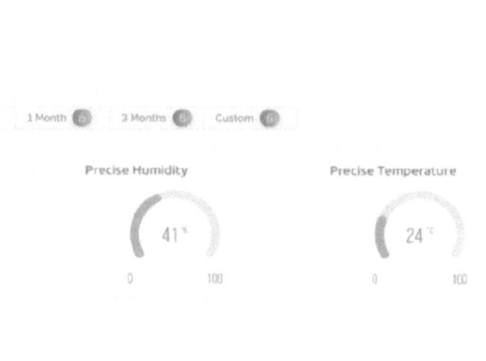

Figure 13.7: Blynk 2.0 web dashboard.

Source: Author.

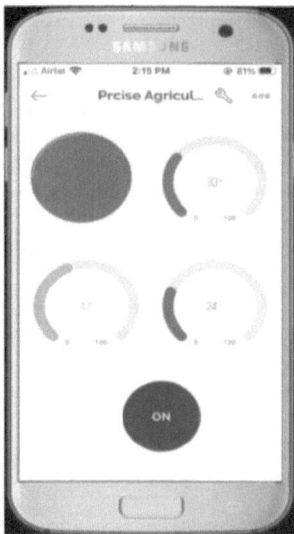

Figure 13.8: Blynk2.0 IoT mobile app dashboard.

Source: Author.

References

Abdollahi, A., Rejeb, K., Rejeb, A., Mostafa, M. M., and Zailani, S. (2021). Wireless sensor networks in agriculture: Insights from bibliometric analysis. Sustainability, 13(21), 12011.

Abu, N. S. et al. (2022). Internet of Things applications in precision agriculture: A review. Journal of Robotics and Control, 3(3), 338–347.

Anita, R. Singh, S. Choudhury, and Singh, B. (2015). Wireless disaster monitoring and management system for dams. Procedia Computer Science, 48, 381–386. doi: 10.1016/j.procs.2015.04.197

Caruso, A., Chessa, S., Escolar, S., Barba, J., and López, J. C. (2021). Collection of data with drones in precision agriculture: Analytical model and LoRa case study. IEEE Internet of Things Journal, 8(22), 16692–16704.

Chawla, N. and Anitha, G., (2019). Wireless sensor network using LoRa for smart agriculture. International Journal for Research in Applied Science and Engineering Technology, 7(6), 1503–1510.

Gehlot, A. et al. (2021). Internet of Things and long-range-based smart lampposts for illuminating smart cities. Sustainability, 13(11), 6398. doi: 10.3390/su13116398

Gresl, J., Fazackerley, S., and R. Lawrence (2021). Practical Precision Agriculture with LoRa based Wireless Sensor Networks. In SENSORNETS, 131–140.

Gutiérrez, S., Martínez, I., Varona, J., Cardona, M., and R. Espinosa ((2019)). Smart mobile LoRa agriculture system based on internet of things. 2019 IEEE 39th Central America and Panama Convention (CONCAPAN XXXIX), pp. 1–6.

Keshtgary, M. and Deljoo, A. (2012). An efficient wireless sensor network for precision agriculture. Can. J. Multimed. Wirel. Networks, 3(1), 1–5.

Kim, W. S., Lee, W. S., and Kim, Y. J. (2020). A review of the applications of the Internet of Things (IoT) for agricultural automation. Journal of Biosystems Engineering, 1–16. doi: 10.1007/s42853-020-00078-3.

Ma, Y.-W. and Chen, J.-L. (2018). Toward intelligent agriculture service platform with lora-based wireless sensor network. 2018 IEEE International Conference on Applied System Invention (ICASI), pp. 204–207.

Mishra, R., Joshi, K. and Gangodkar, D. (2022). Wireless communications network and mobile computing using blockchain in distributed Internet of Things. 2022 11th International Conference on System Modeling & Advancement in Research Trends (SMART), Moradabad, India, pp. 832–836, doi: 10.1109/SMART55829.2022.10047182.

Silva, N. et al. (2019). Low-cost IoT LoRa® solutions for precision agriculture monitoring practices. Progress in Artificial Intelligence: 19th EPIA Conference on Artificial Intelligence, EPIA 2019, Vila Real, Portugal, September 3–6, 2019, Proceedings, Part I 19, pp. 224–235.

Singh, D. K., Sobti, R., Jain, A., Malik, P. K., and Le, D. (2022). LoRa based intelligent soil and weather condition monitoring with internet of things for precision agriculture in smart cities. IET Communications, 16(5), 604–618.

Singh, R. and Mishra, S. (2010). Temperature monitoring in wireless sensor network using Zigbee transceiver module. ICPCES 2010 - International Conference on Power, Control and Embedded Systems. doi: 10.1109/ICPCES.2010.5698701.

Singh, R. K., Aernouts, M., De, M. Meyer, Weyn, M., and Berkvens, R. (2020). Leveraging

LoRaWAN technology for precision agriculture in greenhouses. Sensors, 20(7), 1827.

Swain, M., Zimon, D., Singh, R., Hashmi, M. F., Rashid, M., and Hakak, S. (2021). LoRa-LBO: An experimental analysis of LoRa link budget optimization in custom build IoT test bed for agriculture 4.0. Agronomy, 11(5). doi: 10.3390/agronomy11050820

Syafarinda, Y., Akhadin, F., Fitri, Z. E., Widiawan, B., and Rosdiana, E. (2018). The precision agriculture based on wireless sensor network with MQTT protocol. in IOP Conference Series: Earth and Environmental Science, 207(1), p. 12059.

Teja, G. N. L. R., Harish, V. K. R., Khan, D. N. M., Krishna, R. B., Singh, R., and Chaudhary, S. (2014). Land slide detection and monitoring system using wireless sensor networks (WSN). 2014 IEEE International Advance Computing Conference (IACC), 149–154

Thakur, A. K. et al. (2022). Advancements in solar technologies for sustainable development of agricultural sector in India: A comprehensive review on challenges and opportunities. Environmental Science and Pollution Research, 29(29), 43607–43634. doi: 10.1007/s11356-022-20133-0.

Wang, C.-Y., Tsai, C.-H., Wang, S.-C., Wen, C.-Y., Chang, R. C.-H., and Fan, C.-P. (2021). Design and implementation of LoRa-based wireless sensor network with embedded system for smart agricultural recycling rapid processing factory. IEICE Trans. Inf. Syst., 104(5), 563–574.

Yu, C. (2016). Low cost locating method of wireless sensor network in precision agriculture. Cybernetics and Information Technologies, 16(6), 123–132.

14 Comparison of αβ0 and dq0 transform for power transformer protection

Vijay Kumar Sahu[1,a], Yogesh Pahariya[2,b], Gaurav Gadge[3,c], and Neha Sahu[4,d]

[1]Electrical Engineering, Dr. B.B.A. Government Polytechnic, Karad, India
[2]Vice Chancellor, Rai Technology University, Bangalore, India
[3]Electrical Engineering, St. Vincent Pallotti College of Engineering and Technology, Nagpur, India
[4]Electrical Engineering, Government Engineering College, Daman, India

Abstract

Power transformer is one of the most important as well as essential element of power system. Hence the protection of such an element is utmost necessary. There are various methods for the protection of the transformer. The most widely used method is the differential protection. Since the development of this protection various methods are being developed to improve its reliability, sensitivity, selectivity and various other functional characteristics. In this paper, such two methods are being compared for various faults in the transformer. The first approach is called the αβ0 transform; the trip signal is obtained with the aid of a fuzzy system after the three phase current components of the primary and secondary, received from the current transformer, are transformed into αβ0 components. Second method to be compared is the dq0 transform, in this transform also the three phase current components from both sides of the transformer are obtained for the current transformer and converted into dq0 component and finally with the help of fuzzy system the trip signal is obtained. By the end of the paper, considering most of the conditions the better method is concluded.

Keywords: Differential protection, power transformer, αβ0 transform, dq0 transform, fuzzy logic

1. Introduction

One of the most crucial pieces of equipment in the electrical power system is the power transformer. The failure of such equipment can lead to bad impact and can cost high maintenance and even black outs in the power system. So to deliver reliable and stable energy various protection schemes for power transformer are available (Sahu and Pahariya, 2021).

The most widely used protection technique is the percentage differential logic for the transformer protection. Many methods are developed to improve this protection used for power transformer. Ashrafian,

[a]vjsahu3003@gmail.com, [b]ypahariya@yahoo.com, [c]gsgadge106@gmail.com, [d]sahuneha1294@gmail.com

DOI: 10.1201/9781003567653-14

et al. (2014) have implemented by transform method for the improvement; Etumi and Anayi, (2021) have enhanced the reliability by differentiating between inrush current and internal faults. Some methods by Sanati et al. (2020) developed to avoid, Naseri (2019) to detect and Ajaei et al. (2011) for correction/compensation of CT saturation.

In this paper comparison is being done between αβ0 and dq0 transform for power transformer protection.

2. Fuzzy Logic

A superset of traditional (Boolean) logic that has been expanded to incorporate the idea of partial truth—truth values that fall between "completely true" and "completely false"—is known as fuzzy logic defined by Kumar et al. (2015). Fuzzy logic operates in a similar way as humans as the base of decisions are the conditions present. It can be said fuzzy system operates on a bunch of IF-THEN statements.

3. αβ0 Transform

In the study by Sahu and Pahariya, (2022) this transform was implemented in MATLAB Simulink and the flowchart of the relay is shown in Figure 14.1. First, the primary and secondary current transformers in the transformer are used to obtain the current signals that need to be safeguarded. These transform components added from primary and secondary, respectively, are feed to the fuzzy system.

This transform is effective for phasor values along with instantaneous values. The transform distinguishes between internal defects, over-excitation, energization, and normal operation. The data acquired is pre-processed and converted into αβ0 components.

There are 12 different rules used in fuzzy system that is being used. Table 14.1 shows the rules that are being used for the protection purpose where Δα, Δβ and Δ0 are αβ0 components as mentioned above. Table 14.2 gives the specification for the proposed relay.

Table 14.1: Fuzzy rules using αβ0 transform

Rule	Inputs			Output
	Δα	Δβ	Δ0	
1	Low	Low	Low	Fault
2	Low	Low	Medium	Fault
3	Low	Low	High	Fault
4	Low	Medium	Low	Fault
5	Low	Medium	Medium	Fault
6	Medium	Low	Low	Fault
7	Medium	Medium	Medium	Steady State
8	Medium	High	High	Fault
9	High	Low	Low	Fault
10	High	Low	High	Fault
11	High	Medium	Medium	Fault
12	High	Medium	High	Fault

Source: Sahu and Pahariya (2022).

Table 14.2: Specification of the proposed relay using αβ0 transform

Measurements	Transformer primary and secondary currents
Relay Output	Trip signal
Threshold of operation	Fuzzy output 0.5

Source: Sahu and Pahariya (2022).

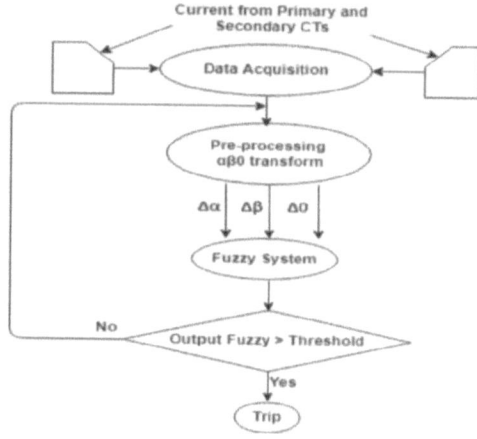

Figure 14.1: αβ0 transform basic relay algorithm.

Source: Sahu and Pahariya (2022).

4. dq0 Transform

This transform also was implemented in MATLAB Simulink as the previous transform, and the flowchart of the relay is shown in Figure 14.2. This transform is also effective for phasor values along with

Table 14.3: Fuzzy Rules using dq0 transform

Rule	Inputs			Output
	Δα	Δβ	Δ0	
1	Medium	High	High	Fault
2	High	High	Low	Fault
3	Low	Low	medium	Fault
4	High	High	Medium	Fault
5	High	Medium	Medium	Fault
6	Low	High	Low	Fault
7	Low	High	Medium	Fault
8	Medium	Medium	Medium	Steady State
9	High	High	High	Fault
10	Low	Medium	Medium	Fault
11	Low	Low	Low	Fault
12	Medium	High	Medium	Fault
13	Very Low	Very Low	Very High	Fault and CT saturation
14	Very Low	Very Low	Very Low	Fault and CT saturation
15	Very High	High	Very High	Fault and CT saturation
16	Very High	Very Low	Very High	Fault and CT saturation
17	Very High	Very High	Very High	Fault and CT saturation
18	Very Low	Very High	Very High	Fault and CT saturation
19	Very Low	Very High	Very Low	Fault and CT saturation

Source: Sahu and Pahariya (2021).

Table 14.4: Specification of the proposed relay using dq0 transform

Measurements	Transformer primary' and secondary' currents
Relay Output	Trip signal with CT saturation detection
Threshold of operation	Fuzzy output 0.5 (Both)

Source: Sahu and Pahariya (2021).

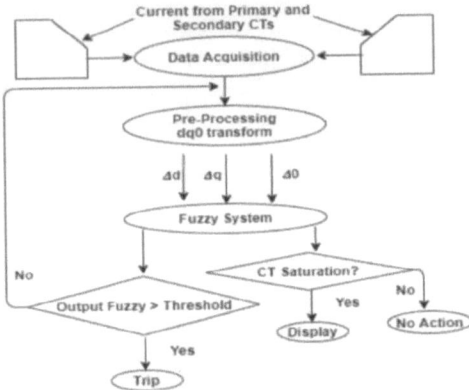

Figure 14.2: dq0 transform Basic relay algorithm.

Source: Sahu and Pahariya (2021).

instantaneous values. Normal operation, internal defects, over-excitation, CT saturation, and energization are all distinguished by the transform.

There are 19 different rules used in fuzzy system that is being used in dq0 transform. Table 14.3 presents the rules that are being used for the protection purpose. Table 14.4 gives the specification for the proposed relay.

5. The Simulated Power System

The MATLAB software was used to simulate the electrical system used for comparing the result of the two different transform. Figure 14.3 show the simulated power system developed to obtain the results for comparison. It is an

overall differential protection scheme used for power transformer. Table 14.5 lists the electrical system's components. The secondary winding of the power transformer incorporates a star connection, whereas the primary winding of the transformer has a delta connection. CTs connected for protection are star on the primary side and delta on the secondary side, in line with the winding.

6. Comparison Between αβ0 Transform and dq0 Transform for The Protection of Transformer

All the faults in primary as well as secondary were developed at 0.2 second and the fault detection time were obtained for comparison. Table 14.6 gives the comparison between the two transform.

Table 14.5: Electrical system composition.

S. No.	Components	Rating
1.	Equivalent Source	138kV and 90MVA
2.	Three Phase Power Transformer	138:13.8 kV and 25 MV A
3.	The Current Transformer On Pri-mary Side	Nominal power- 25VA, Frequency- 50Hz and 200A:5A current rating
4.	The Current Transformer On Secondary Side	Nominal Power - 25VA, Frequency- 50Hz and 3400A:5A current rating
5.	Load	10 MVA with 0.92 inductive power factor

Source: Sahu and Pahariya (2022).

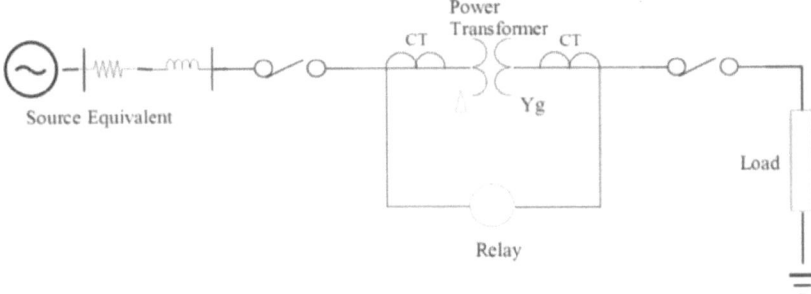

Figure 14.3: Line diagram of power system considered.

Source: Sahu and Pahariya (2022).

Table 14.6: Comparison between αβ0 transform and dq0 transform

S. No.	Condition for comparison	αβ0 transform	dq0 transform
1	Total number of roles for the fuzzy system	12	19
2	Incorrect tripping during energization	No	No
3	Incorrect tripping during over-excitation below 135%	No	No
4	Incorrect tripping during over-excitation above 135%	Yes	No
5	% of faults detected early compared to the next transform	40	60
6	% of faults detected along with CT saturation	0	100

Source: Author.

7. Conclusion

Performance comparison of αβ0 transform and dq0 transform for the protection of the power transformer is discussed. The αβ0 transform and dq0 transform were both able to detect all the fault in absence of CT saturation, but dq0 transform had an upper hand in the results as it was able to detect 60% of fault faster than compared to αβ0 transform. During over-excitation αβ0 transform was effective for 135% over-excitation, and over this value it shows incorrect tripping, whereas the dq0 trans-form is effective of over-excitation above 135%. When CT saturation was present, faults could still be detected by the dq0 transform; however, the αβ0 transform was not able to identify any at all.

The dq0 transform has advantage over other methods of protection of power transformer. Other methods require an additional system to detect or correct the CT saturation but by adding certain rule in the fuzzy system CT saturation can be detected.

References

Ajaei, F. B., Sanaye-Pasand, M., Davarpanah, M., Rezaei-Zare, A., and Iravani, R. (2011). Compensation of the current-transformer saturation effects for digital relays. IEEE Transactions on Power Delivery, 26(4), 2531–2540.

Ashrafian, A., Vahidi, B., and Mirsalim, M. (2014). Time–time-transform application to fault diagnosis of power transformers. IET generation, transmission & distribution, 8(6), 1156–1167.

Etumi, A. A. A., and Anayi, F. J. (2021). Current signal processing-based methods to discriminate internal faults from magnetizing inrush current. Electrical Engineering, 103(1), 743–751.

Kari, T., Gao, W., Zhao, D., Abiderexiti, K., Mo, W., Wang, Y., and Luan, L. (2018). Hybrid feature selection approach for power transformer fault diagnosis based on support vector machine and genetic algorithm. IET Generation, Transmission & Distribution, 12(21), 5672–5680.

Kim, Y., Park, T., Kim, S., Kwak, N., and Kweon, D. (2019). Artificial intelligent fault diagnostic method for power transformers using a new classification system of faults. Journal of Electrical Engineering & Technology, 14, 825–831.

Kumar, M., Misra, L., and Shekhar, G. (2015). A survey in fuzzy logic: an introduction. International Journal for Scientific Research & Development, 3(6), 822–824.

Naseri, F., Kazemi, Z., Farjah, E., and Ghanbari, T. (2019). Fast detection and compensation of current transformer saturation using extended Kalman filter. IEEE Transactions on Power Delivery, 34(3), 1087–1097.

Noori, M., Effatnejad, R., and Hajihosseini, P. (2017). Using dissolved gas analysis results to detect and isolate the internal faults of power transformers by applying a fuzzy logic method. IET Generation, Transmission & Distribution, 11(10), 2721–2729.

Sahu, V. K. (2021). A review of various protection schemes of power transformers. Turkish Journal of Computer and Mathematics Education (TURCOMAT), 12(9), 3220–3228.

Sahu, V. K., and Pahariya, Y. (2021). Transformer Protection Improvement Using Fuzzy Logic. Design Engineering, 4914–4926.

Sahu, V. K., and Pahariya, Y. (2022). Power transformer protection based on fuzzy logic system. Emerging Electronics and Automation: Select Proceedings of E2A 2021 (pp. 195–205). Singapore: Springer Nature Singapore.

Samet, H., Ghanbari, T., and Ahmadi, M. (2015). An auto-correlation function based technique for discrimination of internal fault and magnetizing inrush current in power transformers. Electric Power Components and Systems, 43(4), 399–411.

Sanati, S., and Alinejad-Beromi, Y. (2020). Avoid current transformer saturation using adjustable switched resistor demagnetization method. IEEE Transactions on Power Delivery, 36(1), 92–101.

Zhou, L., and Hu, T. (2020). Multifactorial condition assessment for power transformers. IET Generation, Transmission & Distribution, 14(9), 1607–1615.

15 Hybrid solar photovoltaic-thermal technologies for sustainable future

Madhuri S. Bhagat[1], Bhojraj N. Kale[2,a], V. G. Arajpure[2], Rahul D. Gorle[2], Sriram Ughade[2], and Robin Babu[2]

[1]Civil Engineering Department, Yashwantrao Chavan College of Engineering, Nagpur, India
[2]Mechanical Engineering Department, Suryodaya College of Engineering and Technology, Nagpur, India

Abstract

The two solar thermal as well as photovoltaic technologies represent cutting-edge and promising methods of obtaining solar energy. Integrating the two technologies into a single system is a desirable approach to maximize available space and maybe raise total solar energy consumption. Tragically, the efficiency of photovoltaics decreases at high temperatures, which makes integrating them into hybrid systems difficult. A review of hybrid photovoltaic/thermal technology is provided herein. The main definitions of efficiency for both systems are the first topics covered in the article, along with some possible methods for computing the combination's efficiency. The remaining material concentrates on methods for designing complex systems, highlighting some of the drawbacks and advantages of each strategy.

Keywords: Solar, PV, hybrid, sustainability

1. Introduction

Generally speaking, there are two separate categories of solar energy depending on the method used to convert receiving photons. The immediate transformation of absorbed solar energy toward electrical energy is the main goal of photovoltaics (PV). The process of turning photons toward heat or other beneficial thermal energy is the primary goal of solar thermal power. Photons under and well over the bandgap result in substantial thermal energy going to waste in the PV element. PV devices function by collecting and transforming incoming photons having energy levels over the bandgap of the cell constituents into electron–hole pairs. The way solar thermal device's function is by immediately converting incoming photons, which usually come from the whole solar spectrum, into thermal energy, which is then collected by a heat transfer medium.

A strong case can be presented for combining a PV device using a solar thermal collector to create a hybrid system, known as a photovoltaic/thermal (PV/T) collector, because of the quantity of thermal energy produced in PV devices and the need to

[a]bnkale.phd2018.me@nitrr.ac.in

DOI: 10.1201/9781003567653-15

maintain low temperatures during operation (Chow, 2010). Figure 15.1 depicts a straightforward schematic depiction of a flat plate PV/T collector. The PV cells in the configuration depicted in Figure 15.1 receive incident sunlight flux, which they collect, convert part of it directly into electrical power, and waste heat energy in the PV cell components. Then be transferred to a thermal power load.

2. Efficiency of Solar PV System

According to Markvart (2009), the PV collector's efficiency is commonly described as follows:

$$\eta_{pv} = \frac{V_{oc}I_{sc}\,FF}{GA_{pv}} \tag{1}$$

where, Voc- Voltage (Open-Circuit); Isc-Current (short-circuit); FF-Fill-factor; G-Solar Irradiation; Apv- Area (PV Pannel). According to the useful thermal energy that the working fluid has collected, the thermal collector's efficiency is commonly defined as follows (Duffie and Beckman, 2013):

$$\eta_{thermal} = \frac{\dot{m}c_p\,(T_{out}-T_{in})}{GA_{thermal}} \tag{2}$$

where $\dot{m}$- mass flow rate; Cp-specific heat (working fluid's); Tin, Tout- inlet and exit temperatures, and $A_{thermal}$- thermal collector area. In many instances, especially in PV/T systems with no sun quantity, the values and correspond to the same region. These regions may be the same in PV/T systems with concentration, but they are probably going to differ. Under those circumstances, disclosing the efficiency must be done with the utmost care.

The overall efficiency of a hybrid PV/T collector is a crucial factor to take into account. As a result, PV/T systems that use concentration, or that have distinct areas for thermal and PV systems, should report efficiency on the basis of the total aperture area. The most straightforward way of calculating the overall effectiveness is to simply add the PV plus thermal efficiencies:

$$\eta_{overall} = \eta_{thermal} + \eta_{PV} \tag{3}$$

Although this method of determining total efficiency is straightforward, it probably oversimplifies the value of a hybrid PV/T collector because it makes the assumption that the values of electrical as well as thermal energy are equal. Furthermore, the thermal efficiency is frequently far higher than the PV efficiency, to the point where it can frequently outweigh the overall efficiency, making it challenging to contrast system performance across different designs. Using a factor of weighting for the relative electrical as well as thermal energy quantities is one very helpful method. This can be found in the following equation (Huang et al., 2021):

$$\eta_{overall,weighted} = \omega\eta_{thermal} + \eta_{PV} \tag{4}$$

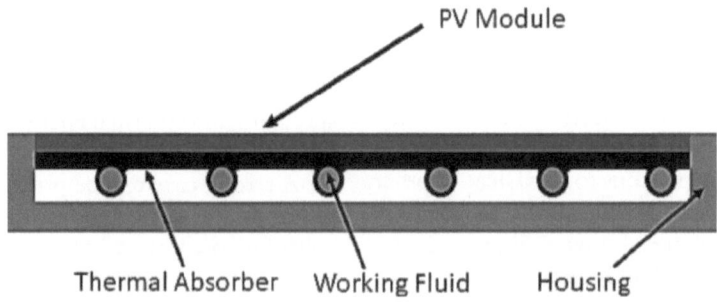

Figure 15.1: Conceptual schematic representation of a hybrid PV/T collector connected to a PV module.

Source: Bierman, et al., 2016.

Where, ω -weighting function that reflects the thermal energy value to the electrical energy value.

This method is very helpful since it may define the value using thermodynamics if factors like cost, environmental strategies which enable for an elevated level of flexibility, or the conversion of thermal energy to working temperatures are taken into account. If this method is employed, it is crucial to understand how the value is defined and ascertained.

Using the total exergetic efficiency is additional helpful strategy, especially if the thermal energy will be used to produce electrical energy. This can be done by using the equation as follows (Widyolar et al., 2018):

$$\eta_{exergy} = \eta_{thermal} X \eta_{Carnot} + \eta_{PV} \quad (5)$$

where η_{Carnot} Carnot is the Carnot efficiency of a heat engine. This method incorporates the effects of high-temperature thermal energy collecting and harnesses the capacity of thermal energy to generate beneficial mechanical work. No matter whatever overall efficiency strategy is applied, reporting the total energy generated by the thermal system and PV system used in a particular design is also helpful. The detrimental impact of temperature is a crucial component of any photovoltaic system.

3. Advanced Solar PV Designs

It has been demonstrated that rising temperatures have a negative influence on PV cell efficiency, as illustrated by the equation that follows (Markvart, 2009):

$$\eta_{PV} = \eta_{ref}\left[1 + \beta_{ref}\left(T_{PV} - T_{ref}\right)\right] \quad (6)$$

where η_{ref}- Efficiency of PV at referred temperature; β_{ref}-coefficient of cell efficiency temperature; Tpv-cell temperature, and Tref-referred temperature of cell. Temperature has a significant impact on PV cell efficiency, as shown by Eq. (6). Since of this, the implementation seen in Figure 15.1 unmistakably

highlights a problem with PV/T systems since the thermal collector has a significant impact on the PV cell temperature. Poor PV efficiency results from higher PV cell temperatures needed to capture greater quantities of heat at higher operational temperatures. This results in two methods for configuring the design. The initial method involves cooling the PV cell using the thermal recipient's working fluid. This is possible when the PV cell temperature is limited while optimizing the beneficial temperature prosper as Figure 15.1 illustrates. Using a high-velocity operating fluid to proactively cool the PV cell by lowering its temperature is an alternate strategy. The drawback of this strategy is that it also reduces the fluid's usable temperature rise (Mittelman et al., 2007). This method is usually used in PV/T systems that receive focused solar radiation.

Separating the thermal receiver from the PV receiver is the second design principle that reduces the temperature gain of PV cells. This permits the thermal receiver to operate at a greater temperature and may allow the PV receiver to operate at a lower temperature. Figure 15.2 provides an overview of thermal decoupling techniques; however, it is not all-inclusive.

The main objective of thermal dissociation is to either selectively absorb/transmit light (Goel et al., 2020) or divide the

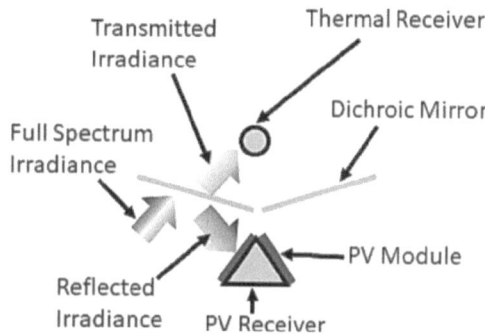

Figure 15.2: Conceptual schematic representations of hybrid PV/T receivers that are thermally coupled.

Source: Bierman, et al., 2016.

incoming sunlight via spectral splitter (Ju et al., 2017; Mojiri et al., 2013). This arrangement, as illustrated on the left side of Figure 15.2, involves inserting a dichroic mirror through the full spectrum irradiance beam path. It allows for partial light reflection to the PV receiver, that offers the option of active cooling, and transmission of the remaining light to the thermal recipient (Stanley et al., 2016; Wingert et al., 2020). A thermal receiver which is positioned in the complete spectrum beam path in Figure 15.2 has the capacity to either transmit as well absorb part of the spectrum in a spectrum-wide manner while sending the remaining component to the PV receiver (Hassani et al., 2016; Otanicar et al., 2018).

The optical characteristics of the spectrum splitting technology play a crucial role in deciding the overall amount of energy allocated to each technology, hence both thermal decoupling ideas rely heavily on it. The wavelength (bandgap energy) impacting on the module is crucial to PV technology's ability to efficiently absorb and convert solar radiation into electricity. Insofar as the absorbers are extremely dark, thermal imaging techniques are generally wavelength agnostic. More recent research has demonstrated that the optimal spectral attributes are strongly reliant on the weighting factor (), which maximizes the total efficiency, along with to the wavelength of the spectrally selective element being extremely reliant on the overall efficacy (Brekke et al., 2016; Bierman et al., 2016; Huang et al., 2021).

4. Results and Discussion

As hybrid nanofluids have ideal thermal and optical characteristics, solar systems based on them have demonstrated outstanding performance (over 200%). It has been discovered that the use of hybrid nanofluids changes performance determination parameters for solar energy systems, such as solar weighted absorption, extinction coefficient, and photothermal conversion efficiency. Up

to a certain point, hybrid nanofluids' optical performance is seen to improve as the concentration of nanoparticles increases.

In high radiation areas, this solar heating technology t is used in industrial operations and is cost-competitive with fossil fuel alternatives. This is the low-temperature method of use that is most economically feasible. The ability of these systems to produce thermal and electrical energy concurrently leads to a larger overall energy production per unit area, which is one of its key advantages.

Utilizing nanofluids widely can have a significant impact on reducing emissions that harm the environment and the high cost of solar energy conversion systems. As the blended nanofluid gives an elevated level of thermal conductivity at a cheap price, hybrid nanofluids are more cost-effective than mono-nanofluids.

The high heat conductivity of CNTs is responsible for the effectiveness of nanofluids. However, the problem lies in the CNTs' hydrophobic nature, which subsequently causes early sedimentation. Because of their unique tube structure, CNTs have a larger surface area, which increases their thermal efficiency.

5. Conclusions

As global decarbonization efforts expand, hybrid photovoltaic/thermal technologies are poised for further market penetration. Furthermore, a lot of nations are working toward high levels of electrification, which may continue to expand as long as carbon dioxide emissions are decreased and the waste heat from a conventional photovoltaic system becomes more profitable. Selecting which of the several PV/T and output thermal temperature approaches to use is one of the major business issues. As a result, there will likely be differences in the working fluid temperature as well as the weighting method used to compare the amount of heat to electricity among the various client segments in this market. Furthermore, it will be essential for future developments and implementation to critically consider how to assess overall efficiency and to explicitly disclose the methodology used to arrive at that conclusion.

References

Bierman, D. M., Lenert, A., and Wang, E. N. (2016). Spectral splitting optimization for high-efficiency solar photovoltaic and thermal power generation. Applied Physics Letters, 109(24), 243904.

Brekke, N., Otanicar, T., DeJarnette, D., and Hari, P. (2016). A parametric investigation of a concentrating photovoltaic/thermal system with spectral filtering utilizing a two-dimensional heat transfer model. Journal of Solar Energy Engineering, 138(2), 021007.

Chow, T. T. (2010). A review on photovoltaic/thermal hybrid solar technology. Applied Energy, 87(2), 365–379.

Duffie, J. A., and Beckman, W. A. (2013). Solar engineering of thermal processes. John Wiley and Sons.

Goel, N., Taylor, R. A., and Otanicar, T. (2020). A review of nanofluid-based direct absorption solar collectors: Design considerations and experiments with hybrid PV/thermal and direct steam generation collectors. Renewable Energy, 145: 903–913.

Hassani, S., Taylor, R. A., Mekhilef, S., and Saidur, R. (2016). A cascade nanofluid-based PV/T system with optimized optical and thermal properties. Energy, 112, 963–975.

Huang, G., Wang, K., and Markides, C. N. (2021). Efficiency limits of concentrating spectral-splitting hybrid photovoltaic-thermal (PV-T) solar collectors and systems. Light: Science & Applications, 10(1), 28.

Ju, X., Xu, C., Han, X., Du, X., Wei, G., and Yang, Y. (2017). A review of the concentrated photovoltaic/thermal (CPVT) hybrid solar systems based on the spectral beam splitting technology. Applied Energy, 187, 534–563.

Markvart, T., Ed. (2009). Solar electricity. John Wiley and Sons.

Mittelman, G., Kribus, A., and Dayan, A. (2007). Solar cooling with concentrating photovoltaic/thermal (CPVT) systems. Energy Conversion and Management, 48(9), 2481–2490.

Mojiri, A., Taylor, R., Thomsen, E., and Rosengarten, G. (2013). Spectral beam splitting for efficient conversion of solar energy—A review. Renewable and Sustainable Energy Reviews, 28, 654–663.

Otanicar, T., Dale, J., Orosz, M., Brekke, N., DeJarnette, D., Tunkara, E., Roberts, K., and Harikumar, P. (2018). Experimental evaluation of a prototype hybrid CPV/T system utilizing a nanoparticle fluid absorber at elevated temperatures. Applied Energy, 228: 1531–1539.

Stanley, C., Mojiri, A., Rahat, M., Blakers, A., and Rosengarten, G. (2016). Performance testing of a spectral beam splitting hybrid PVT solar receiver for linear concentrators. Applied Energy, 168, 303–313.

Widyolar, B., Jiang, L., and Winston, R. (2018). Spectral beam splitting in hybrid PV/T parabolic trough systems for power generation. Applied Energy, 209, 236–250.

Wingert, R., O'Hern, H., Orosz, M., Harikumar, P., Roberts, K., and Otanicar, T. (2020). Spectral beam splitting retrofit for hybrid PV/T using existing parabolic trough power plants for enhanced power output. Solar Energy, 202(15), 1–9.

16 An overview of induction motors with soft starts that have speed controls and monitoring systems

Trivelli Shekhar Naidu[a], Saksham Chimurkar[b], Gaurav Gawai[c], Avishkar Kamble[d], and Sanskar Raut[e]

Department of Electrical Engineering, St. Vincent Pallotti College of Engineering and Technology, Nagpur, India

Abstract

Even now, Asynchronous motors are the most often utilized kind of motor in industrial settings. In many applications, operation depends on monitoring and adjusting an induction motor's properties. There are several ways to do this. This paper focuses on monitoring and controlling one-phase induction motor. A module of sensors and transducers is used to monitor factors such as the induction motor's temperature, current, and voltage. Following acquisition, the data processing unit receives the acquired data and displays the server settings. By utilizing a server gateway, the system prevents possible system errors by using both automated and manual control methods to start or stop the induction motor. The installation of this plan is expected to improve the machine's operational efficiency through constant monitoring to avoid malfunctions and promote preventative maintenance.

Keywords: Arduino controller, LCD, parameter monitoring, induction motor, parameter controlling

1. Introduction

Prior to the introduction of induction drives, DC motors were dominant in servicing industrial demands. However, due to their better performance characteristics, induction motors have replaced DC motors in manufacturing automation. The performance with an induction motor is inextricably connected to Electrical, mechanical and environmental factors. As a result, control methods for induction motors with AC are heavily dependent on these motor characteristics. Thus, monitoring induction motor characteristics is critical for continuous operation and estimating the pre-fault state will help in minimizing breakdowns.

This model combines sensors to monitor the properties of an induction motor, including voltage, current, acceleration, and

[a]trivellinaidu@gmail.com; tnaidu@stvincentngp.edu.in, [b]himurkarsaksham@gmail.com, [c]gauravgawai108@gmail.com, [d]kambleavishkar0@gmail.com, [e]sanskarraut.fy20@stvincentngp.edu.in

DOI: 10.1201/9781003567653-16

temperature. The speed for the induction motor may be readily changed by adjusting the power input frequency. Regular tracking of these parameters promotes continued production in industries, improving motor dependability and thereby increasing industrial output.

2. Literature Review

In recent years, there has been a surge in research focusing on the development and implementation of IoT-based environmental monitoring systems. These systems leverage advancements in sensor technologies, wireless communication, and data analytics to enable real-time monitoring and analysis of various environmental parameters. Several studies have explored the utilization of platforms such as Arduino UNO and Thingspeak for this purpose. Evan (2011) throws light on IoT and its application in modern world.

Deekshath et al. (2016) introduced an IoT-based environmental monitoring system utilizing Arduino UNO and Thingspeak platform. While this approach offers the advantage of easy implementation and integration, it is limited in its scalability and may not support extensive sensor networks or complex data processing requirements.

Similarly, Sharmad Pasha (2016) proposed a sensing and monitoring system for IoT using Thingspeak with MATLAB analysis. While this system provides robust data analysis capabilities, it may face challenges in terms of real-time responsiveness and scalability, particularly in scenarios with high data volume.

Darbastwar et al. (2016) presented an IoT-based environmental factor sensing and monitoring system over wireless sensor networks. While wireless sensor networks offer flexibility and scalability, they may encounter issues related to power consumption, data transmission reliability, and security vulnerabilities.

In addition to environmental monitoring, research has also focused on efficiency estimation and energy management of induction motors. Lu et al. (2008) proposed a nonintrusive motor-efficiency estimation method based on air-gap torque, which provides valuable insights into motor performance without interrupting operations. However, this method may have limitations in accurately estimating efficiency under varying operating conditions.

Amaro and Ferreira (2010) developed a low-cost wireless sensor for in-field monitoring of induction motors, offering a cost-effective solution for monitoring motor performance. Nonetheless, the reliability and accuracy of such sensors may be compromised in harsh industrial environments or under dynamic operating conditions.

Moreover, Li and Yu (2007) explored energy management of induction motors based on non-intrusive efficiency estimation techniques. While non-intrusive methods offer advantages in terms of ease of implementation, they may lack precision compared to intrusive methods, especially in applications requiring high accuracy.

Furthermore, Nagendrappa and Bure (2010) proposed an energy audit and management approach for induction motors using genetic algorithms. Vermesan and Friess (2014) discussed the transition of IoT from research and innovation to market deployment, highlighting challenges and opportunities in this evolving landscape.

Overall, while existing research has made significant strides in advancing IoT-based environmental monitoring and motor efficiency estimation, there remain various challenges and limitations that need to be addressed to realize the full potential of these technologies

3. Methodology and Model Specifications

3.1. Hardware Specifications

5V 30A SMPS, Single Phase IM (1HP, 6500 rpm, 220V, 0.75Ah), Opto-isolated 4 channel TRIAC Board, The Voltage

Detection Sensor Module 25V , 5A range of single-phase AC current sensor module, Temperature Sensor–LM 35, Speed sensor, Capacitive load, Arduino Uno (ATMega328), JHD162A LCD Screen, Rectifier, Voltage Regulator(7805 IC Module), 12 V Power supply unit.

3.2. Software Specifications

Arduino Compiler, Micro-Controller Programming Language: Embedded C

3.3. Methodology

An Arduino controller controls the circuit's functionality. Its goal is to replace the frequently utilized TRIAC firing angle. The circuit is intended to power a single-phase induction motor. The circuit diagram Figure 16.1, depicts the integration of the four sensors for tracking critical metrics including voltage, current, speed, and temperature. These sensors play an important role for tracking the motor's condition.

The Arduino Uno receives the induction motor's current state and presents pertinent information via the LCD display module. The temperature, current, voltage, and speed of the induction motor are among the parameters that are monitored and shown on the LCD display.

This paper takes a unique approach for acceleration control, emphasizing efficiency and employs sensors for thorough motor parameter monitoring. The addition of an Arduino microprocessor and an LCD display improves the general functionality and user experience for the system.

The voltage transformer is used to measure the motor supply voltage. Its output is regulated and fed to the microcontroller. A current transformer is used to drive the motor, a filtered stepped current is sent to the microcontroller. Temperature of the drive's body is measured using the LM35 sensor. The output of the sensor is fed to the Arduino via a double low-pass filter. Arduino is used to measure speed and vibration.

The Arduino starts with a 1 second delay and then counts all the pulses through the proximity sensor using a timer/counter of 0 multiplied by 30 and then sends to Arduino

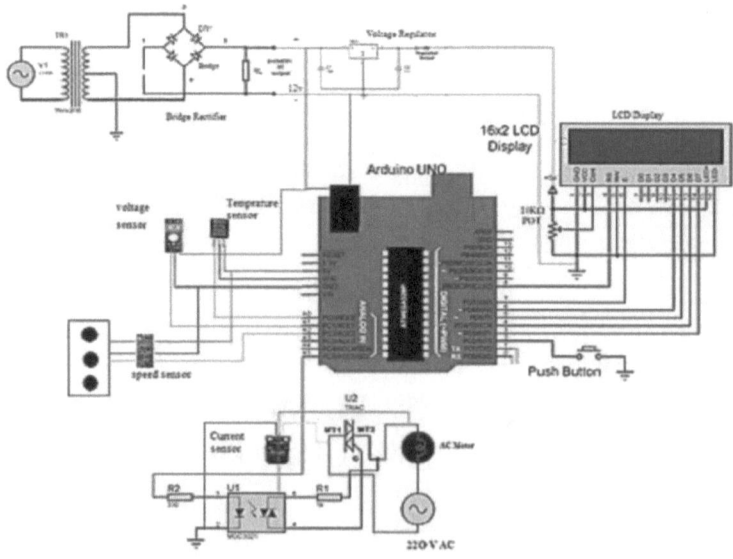

Figure 16.1: Circuit diagram of the Prototype.

Source: Author.

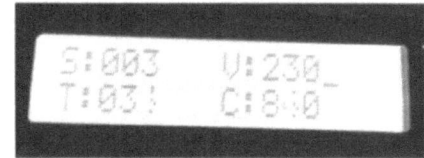

Figure 16.2: Prototype of induction motors with soft starts that have speed controls and monitoring systems and actual readings of model.

S = Speed (Level). V = Voltage, T = Temperature (°C), C = Current (mA).

Source: Author.

ports b and d. Optrons isolate the proximity sensor, speaker and Microcontroller circuit. Ethernet Shield is connected to the Tx/Rx pin of the Arduino for communication between the IoT platform and the hardware. In addition, the IoT platform is ready to monitor motor parameters.

4. Empirical Results

In this approach for monitoring and early detection of motor systems is presented with the help of sensors and Arduino controller. The system's ability to integrate different parameter readings in real time enhances the ability to identify a variety of issues.

Speed, temperature, voltage, and consumption are only a few of the data measured when motor systems are monitored. So, this model has more data sources that allow for perturbations than previous systems that rely simply on temperature.

The image of the Prototype of Induction Motors with soft starts that have Speed controls and Monitoring systems and Actual Readings of Model are given below in Figure 16.2.

5. Conclusion

This paper aims to create induction motor parameter tracking system. The created system will be capable of performing tasks such as starting and halting the motor, as well as keeping track of characteristics such as temperature, speed, voltage, or current. The values that have been recorded for these parameters are efficiently transmitted to the interface's LCD display.

References

Darbastwar, S. S., Sagare, S. C., and Khetade, V. G. (2016). IoT based environmental factor sensing and monitoring system over wireless sensor networks. International Journal of Advanced Research in Computer Science and Software Engineering Research Paper, 6(12), December, ISSN: 2277 128X.

Deekshath, R., Dharanya, P., Dimpil Kabadia, K. R., and Deepak Dinakaran, G. (2018). IoT based environmental monitoring system using arduino UNO and Thingspeak. IJSTE – International Journal of Science Technology & Engineering, 4(9). ISSN (online): 2349–784X,

Evans, D. The Internet of Things-how the next evolution of the internet is changing everything. Cisco Internet Business Solutions Group (IBSG). http://www.cisco.com/web/about/ac79/docs/innov/IoT_IBSG_0411FINAL.pdf

Lu, B., Habetler, T. G., and Harley, R. G. (2008). A nonintrusive and in-service motor-efficiency estimation method using air-gap torque with considerations of condition monitoring" IEEE Transactions on Industry Applications, 44, 1666–1674.

Nagendrappa, H., and Bure, P. (2010). Energy audit and management of induction motor using Field test and genetic algorithm. ACEEE International Journal on Electrical and Power Engineering, 1, 1. doi: 01.ijepe.01.01.01.

Pasha, S. (2016). Thingspeak based sensing and monitoring system for IoT with matlab analysis. International Journal of New Technology and Research (IJNTR), 2(6), 19–23, ISSN: 2454-4116.

Pedro Amaro, J., and Ferreira, F. J. T. E. (2010). Low-cost wireless sensor for in field monitoring of induction motor. IEEE Transactions on Industry Applications, 44(6), 1666–1674.

Vermesan, O., and Friess, P. (2014). Internet of Things – From research and innovation to market deployment. River's Publication. ISBN: 978-87-93102-94-1 (Hard copy) 978-87-93102-95-8 (Ebook)

Yanfeng Li, and Haibin Yu (2007). Energy management of induction motors based on non-intrusive efficiency estimation. Proceeding of International Conference on Electrical Machines and Systems.

17 Experimental and statistical investigation to study effect of machining conditions on behaviour of medium carbon steel under cyclic loading

Ghanshyam Boob[a], Shripad Deo[b], and Pramod Bhagat[c]

Assistant Professor,SVPCET Nagpur, St. Vincent Pallotti College of Engineering and Technology, Nagpur, India

Abstract

Experimental and statistical investigations are carried out to study influence of varying machining parameters on fatigue performance of widely used medium carbon steel grade, SAE-1026. Design of experiment study is carried out by selecting three machining parameters namely cutting speed (Vc), feed rate (f), and depth of cut (dc) varied at three levels and experimental design matrix comprising of fifteen test runs is obtained. As per ASTM standard for material fatigue hour glass shape test specimens with threaded grip were prepared. Influence of varying turning process parameters is induced in specimen gauge length by setting cutting condition as per DOE. Tension compression strain controlled fatigue is carried out on test samples by setting constant strain rate of $\pm 0.3\%$ at frequency of 0.3 Hz. Servo-hydraulic computer controlled axial fatigue machine is employed for testing. As a result of fatigue test it was observed that stress-life (S-Nf) behaviour over a set of specimens is not consistent. Further test results were analysed using statistical tools and mathematical model for estimation of fatigue life of selected steel grade inclusive of influence of machining parameters was deduced.

Keywords: Cyclic loading, turning process parameters, strain controlled fatigue, DOE, steel

1. Introduction

Medium carbon steel is used as base material for variety of fabrication works/ power transmission elements etc. Behaviour of material under cyclic or fatigue loading is one the prime factor to be considered in design of components, machine frame, roll-cage etc. Machining or metal removal and surface finishing is commonly used manufacturing process on steel. Fatigue crack initiation is surface dependent phenomenon. Quality of surface produced after machining contributes a lot in performance of material under fatigue loading.

Considerable quantum of studies were carried out to investigate effect of varying cutting conditions during machining of steel on produced surface roughness parameters (Lopes et al., A. Javidi et al., 2008). Also,

[a]gboob@stvincentngp.edu.in, [b]sdeo@stvincentngp.edu.in, [c]pbhagat@stvincentngp.edu.in

DOI: 10.1201/9781003567653-17

some studies were further extended and concluded that surface roughness predominantly influences performance of material under cyclic loading (Davim, 2001). General trend observed in research literature was that, better surface finish give better fatigue performance but, later some disagreements were reported in some research outcomes (Novovic et al., 2004, Koster, 1991). The concept of surface integrity characterised by surface microstructure, hardness, roughness and residual stress was introduced.

During machining thermo-mechanical stresses are generated causing plastic deformation of work material and affect surface quality (BorjaCoto et al., 2011, Dahlman, Gunnberg and Jacobson, 2004). Machining produced surface quality is characterised as surface roughness, induced residual stress, surface hardness and microstructure (Koster, 1991). Influence of varying cutting conditions employed during turning operation on individual properties defining surface quality was examined in different research studies (Dogra, 2012, Griffiths, 2001, Sasahara, 2005).

Surface roughness and residual stress were identified to have considerable effect on material fatigue performance (Javidi et al., 2008, Sasahara, 2005). According to Novovic et al., 2004, residual tensile stress induced on machined surface layer has detrimental effect on material fatigue life whereas compressive nature of induced residual stress improves fatigue performance. Rise in temperature during machining cause's work hardening at surface layer and increase in surface hardness increases fatigue life (Javidi et al., 2008). Residual stress and work hardening are interdependent and it is very complex to evaluate the resultant cascaded effect of varying surface properties and residual stress on material fatigue behaviour. Micro-level examination of surface integrity factors and there further effects on material fatigue performance is very tedious and difficult to analyse. Genel, 2004, Roessle, Fatemi, 2000, proposed mechanical properties based fatigue life prediction model for various steel grades.

Resultant influence of various surface characteristics produced after machining on material fatigue life is very difficult to evaluate. Hence, it is proposed to develop correlation model between employed cutting condition and material fatigue life on the basis of experimental results.

2. Experimental Programme

2.1. Preliminary Investigations of Selected Test Material

Medium carbon steel SAE1026 is selected as test material for this study. Spectroscopic analysis as per ASTM E-415:2008 is performed on supplied steel material as a outcome of which chemical composition as given in Table 17.1 is recorded and steel grade is confirmed. Further, tension and hardness tests were carried out on test material to get various mechanical properties as recorded in Table 17.2.

Table 17.1: Recorded chemical composition and mechanical properties of test material

C	Si	Mn	P	S	Cr	Ni	Mo	Cu	Al
0.234	0.190	0.524	0.075	0.055	0.019	0.023	0.001	0.034	0.006

Source: Author.

Table 17.2: Recorded mechanical properties of test material

Sut(MPa)	Syt(MPa)	%age Elongation	%Age Reductionin Area	E(GPa)	Poisson's Ratio (μ)	Rockwell Hardness No.(HRC)
490	415	14	40	180	0.27	20

Source: Author.

2.2. Fatigue Test Specimen Preparation

Box-Behken experimental design was implemented and three level three factorial experimental design matrix comprising of three centre points and one block is prepared (Box, Hunter and Hunter, 1978, Cebeci, 2007). Experimental matrix comprising of total fifteen test specimens to be prepared by varying cutting conditions as given in Table 17.3 is obtained.

Hour glass shape strain controlled fatigue test specimens as per ASTM standard with threaded grip are prepared considering experimental factors as per DoE matrix given in Table 17.3. Complete machining sequence was programmed on CNC lathe machine for required setting of cutting conditions. Finishing cuts on specimen gauge length are given by setting cutting conditions as per DoE. Actual specimen and its geometry with dimensional details are shown in Figure 17.1.

2.3. Fatigue Testing

Axial fatigue testing is performed as per ASTM E-23 standard, and specimens are tested under constrained strain amplitude

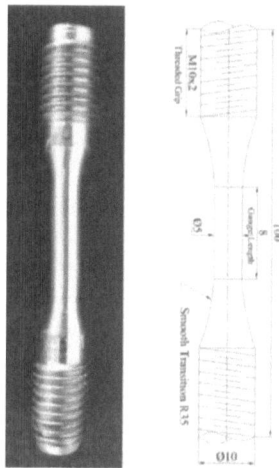

Figure 17.1: Dimensional details.

Source: Author.

Table 17.3: Design of experiment matrix

Run Order No.	Vc (rpm)Actual (coded)	f (mm/rev)Actual (coded)	dc(mm)Actual (coded)
1	1250 (0)	0.2 (−1)	0.1 (−1)
2	1500(+1)	0.2 (−1)	0.3 (0)
3	1250(0)	0.3 (0)	0.3 (0)
4	1000 (−1)	0.3 (0)	0.5(+1)
5	1250 (0)	0.3 (0)	0.3 (0)
6	1250 (0)	0.2(−1)	0.5(+1)
7	1000 (−1)	0.4(+1)	0.3 (0)
8	1000 (−1)	0.2 (−1)	0.3 (0)
9	1000 (−1)	0.3 (0)	0.1 (−1)
10	1500(+1)	0.3 (0)	0.1 (−1)
11	1250 (0)	0.4 (+1)	0.1 (−1)
12	1500 (+1)	0.4 (+1)	0.3(0)
13	1250 (0)	0.3 (0)	0.3(0)
14	1500 (+1)	0.3 (0)	0.5(+1)
15	1250 (0)	0.4 (+1)	0.5(+1)

Source: Author.

of 0.3% strain rate at cyclic frequency of 0.3 Hz. Servo-hydraulic computer controlled axial fatigue testing machine of 15 KN capacity as shown in Figure 17.2 is employed. Perfect alignment of specimen axis with reference plunger and actuator stroke is ensured using self-aligning LCF grips. Applied strain rate during test running is measured by extensometer of 12.5 mm gauge length which is precisely positioned centrally on specimen gauge surface using elastic rubber bands. Stress produced corresponding to applied strain is measured using load cell.

Real-time stress-strain hysteresis loop is displayed on PC connected with machine. Smooth transition of loop without spikes is ensured to confirm perfect alignment of specimen and mounting of extensometer. VB-LCF software provided with machine process strain-strain data and determine various fatigue properties of test material. For every specimen tested, stress-life behaviour (S-Nf) showing variation in stress level to produce set stain rate of 0.3% against number of reversals is obtained. Number of reversals recorded just before start of steep drop in stress amplitude as visualised in 'S-Nf' behaviour is considered as fatigue life (Nf) of test specimen.

3. Experimental Results and Discussions

Experimental data of fatigue life recorded over a set of fifteen specimens is given in Table 17.4 and combined stress-life behaviour (S-Nf) plot is shown in Figure 17.3. Stress-life plots show broader inconsistency in behavior under fatigue over a set of fifteen specimens. Also, recorded experimental value of fatigue life, as given in Table 17.4 shows much wider range of variation over a set of 15 test runs. Maximum fatigue life of 6709 cycles is recorded for specimen ID A-1 and minimum fatigue life of 2300 cycles is recorded for specimen ID A-12 which is approximately 65% variation over a set of selected test runs.

Hence, it can be stated that experimental factors i.e. turning process parameters which are only varying parameters over a set of test runs are responsible for variation in fatigue performance of test specimens. Statistical analysis of test result is performed to study individual and interaction effects of varying experimental factors on material fatigue life as given in next section.

3.1. *Statistical Analysis of Test Results*

Response surface regression is used to analyse individual and interaction effect of experimental factors of material fatigue performance. Turning process parameters (Vc,f,dc) are input to regression model while number of reversals recorded (Nf) i.e. fatigue life is output. The evolution of the relation between test input–output parameters is carried out by computer simulation programming using 'Mini Tab-17'statistical analysis software. Second order regression comprising of individual, squared and interaction effect of experimental factors is

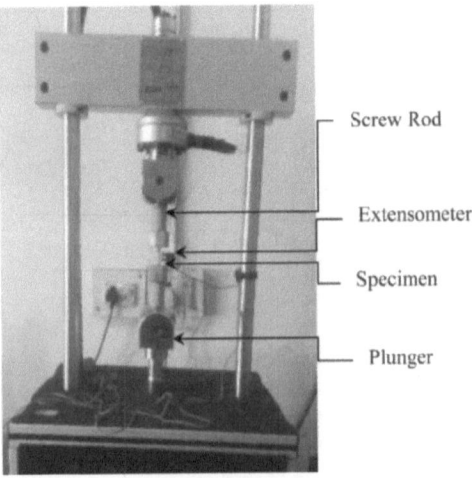

Screw Rod

Extensometer

Specimen

Plunger

Figure 17.2: Experimental setup for axial fatigue test.

Source: Author.

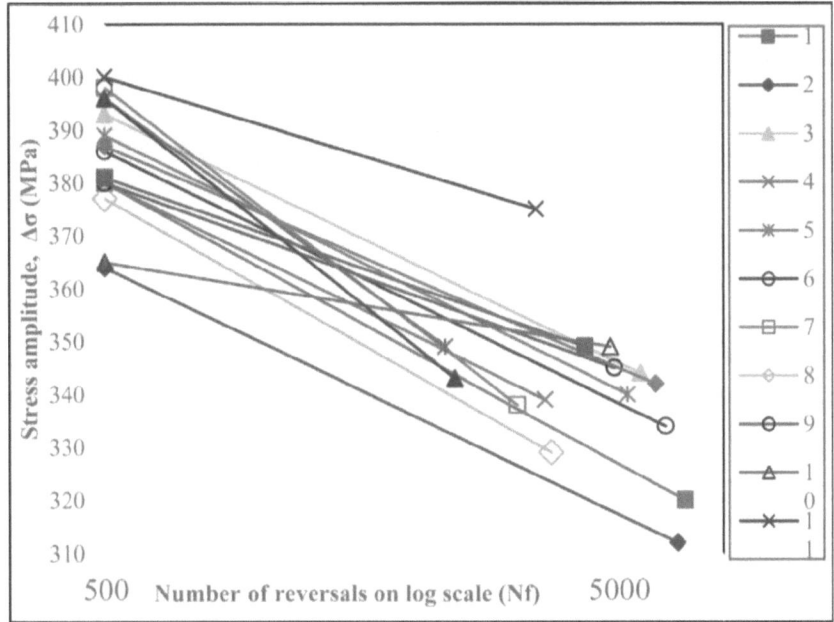

Figure17.3: S-Nf plots of set of fifteen specimen of test material.

Source: Steel SAE-1026.

Table 17.4: Summary of fatigue test results (experimental results)

Vc	f	dc	ID	Nf
0	−1	−1	A-1	6709
+1	−1	0	A-2	6486
0	0	0	A-3	5500
−1	0	+1	A-4	3600
0	0	0	A-5	5200
0	−1	+1	A-6	6149
−1	+1	0	A-7	3180
−1	−1	0	A-8	3700
−1	0	−1	A-9	4904
+1	0	−1	A-10	4816
0	+1	−1	A-11	3460
+1	+1	0	A-12	2300
0	0	0	A-13	5887
+1	0	+1	A-14	4300
0	+1	+1	A-15	2400

Source: Author.

obtained, generalised response prediction regression model is given by Eq.(1).

$$Nf'=5529+315.Vc-1463.f-430.dc-943.$$
$$Vc*Vc-669.f*f-181.dc*dc-917.Vc*f$$
$$+197.Vc*dc -125.f *dc \qquad (1)$$

3.2. Individual and Interaction Effects of Experimental Factors

From response surface regression model individual and interaction effect plots as obtained as shown in Figures17.4 and 17.5.

From individual and interaction effect plots following important observations are drawn:

1. Increase in cutting speed up to certain limiting mid value has positive influence on material fatigue life (Nf).
2. Fatigue life decreases drastically with increase in feed rate. It is found to be one of the most dominating factors to affect test material fatigue performance.

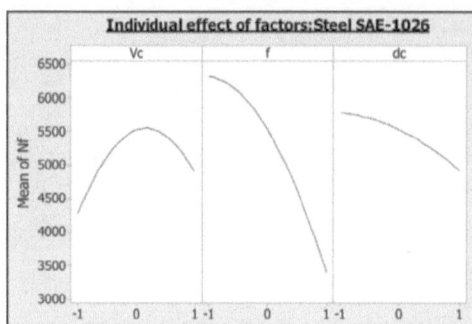

Figure 17.4: Individual effect plot.

Source: Author.

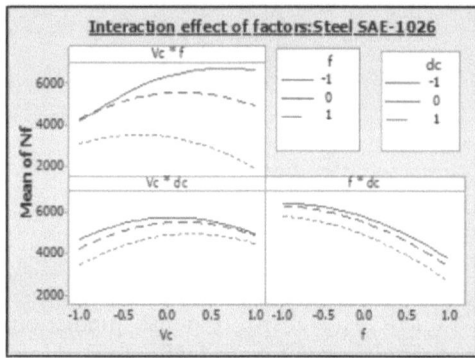

Figure 17.5: Interaction effect plot.

Source: Author.

3. Effect of varying depth of cut is not as significant as that of cutting speed and feed rate.
4. Effect of simultaneous variation of cutting speed-feed rate is more significant to influence fatigue life.
5. Hence, it can be stated that by setting the cutting condition as: low feed rate and depth of cut and average cutting speed maximum fatigue life of turned finished steel material can be achieved.

4.. Conclusion

Major contributions and outcomes of the present research investigation are summarised as follows:

1. Outcome of current research study is competent enough to predict influence of employed turning process parameters; cutting speed, feed rate and depth of cut on fatigue response behaviour of steel material.
2. Fatigue life prediction model as a function of turning process parameters is developed. It provides major contribution in this area of research which was till date limited to development of prediction model as a function of mechanical properties of unprocessed material.
3. Using proposed optimised setting of turning process parameters (Vc,f,dc) negative effect of metal cutting process will be reduced and considerable improvement in fatigue performance of turned finished steel material can be achieved.
4. Finally, it can be stated that overall outcomes of research investigation presented in the current study will be of great help to industries and academia while designing and manufacturing of the components of various engineering applications to be employed under fatigue loading conditions.

References

Aslan, N., and Cebeci, Y. (2007). Application of Box–Behnken design and response surface methodology for modelling of some Turkish coals. International Journal of Fuels, 86, 90–97.

Box, G. E. P., Hunter, W. G., and Hunter, J. S. (1978). Statistics for experimenters, Wiley.

Coto, B. et al. (2011). Influences of turning parameters in surface residual stresses in AISI 4340 steel. The International Journal of Advanced Manufacturing Technology, 53(9–12), 911–919.

Dahlman, P., Gunnberg, F., and Jacobson, M. (2004). The influence of rake angle, cutting feed and cutting depth on residual stresses in hard turning. Journal of Materials Processing Technology, 147(2), 181–218.

Davim, J. P. (2001). A Note on the determination of optimal cutting conditions for surface finish obtained in turning using design of experiments. The Journal of Materials Processing Technology, 116, 305–308.

Dogra, M. (2012). Surface integrity a key issue in hard turning – A review. International

Journal of Machining and Machinability of Materials, 12(1–2), 88–116.

Genel, K. (2004). Application of artificial neural network for predicting strain-life fatigue properties of steels on the basis of tensile tests. International Journal of Fatigue, 26(10), 1027–1035.

Griffiths, B. (2001). Manufacturing Surface Technology-Surface Integrity & Functional Performance. Manufacturing engineering modular series, Taylor & Francis, ISBN 1-5603-2970-X.

Javidi, A. et al. (2008). The effect of machining on the surface integrity and fatigue life. International Journal of Fatigue, 30(10–11), 2050–2055.

Kim, K. S. (2002). Estimation methods for fatigue properties of steels under axial and torsional loading. International Journal of Fatigue, 24(7), 783–793.

Koster, W. P. (1991). Effect of residual stress on fatigue structural alloys. Practical Applications of Residual Stress Technology, Conference Proceedings, Indianapolis, Indiana, 15–17, pp. 1–9.

Lopes, K. S. S. et al. (2008). Influence of machining parameters on fatigue endurance limit of AISI 4140 steel. Journal of the Brazilian Society of Mechanical Sciences and Engineering, 30(1), 79–83.

Novovic, D. et al. (2004). The effect of machined topography and integrity on fatigue life. The International Journal of Machine Tools and Manufacture, 44(2–3), 125–134.

Roessle, M. L., and Fatemi, A. (2000). Strain-controlled fatigue properties of steels and some simple approximations. International Journal of Fatigue, 22(6), 495–511.

18 Machine learning based model for rainfall pattern prediction: a comprehensive overview

Nilesh Dhannaseth[a] and Sanjay Yedey[b]

P.G. Department of Computer Science and Technology, DCPE, HVPM, Amravati, India

Abstract

One of the most challenging things to do is to predict the pattern of rainfall in the weather forecasting process. It is now more challenging than ever to precisely predict rainfall patterns due to extreme climate fluctuations. By uncovering hidden trends in historical weather data, predictive machine learning approaches can forecast patterns of precipitation. Picking a suitable categorisation method for forecasting purposes is a challenging undertaking. Using machine learning approaches, this study suggests a new actual time downpour trend forecasting method for smart cities. In this research an empirical approach is used over the historical data to predict the rainfall pattern and establish its relationship to various atmospheric variables. The Multivariate Linear Regression methodology can be applied on the dataset for pattern analysis. Statistics are used to improve the accuracy of the result, resulting in a result with higher level of accuracy. With this, the consequences of unplanned and excessive rains affecting farmers can be significantly reduced. The suggested system makes use of well-liked supervised machine learning techniques.

Keywords: Rainfall pattern prediction, machine learning, decision tree, regression, historical data

1. Introduction

More precise estimation models are required for advance warning, as they can improve agricultural management and reduce the hazards to people and property, because unexpected and unanticipated precipitation can have a variety of effects, including losses in crops and harm to property. Farmers stand to gain the most from this projection, and water resources may be utilised more effectively. Forecasting rainfall is a challenging task, and the results must be accurate. Based on meteorological information including temperature, humidity, and pressure, for the purpose of forecasting rainfall, various physical tools are available. Because traditional methods are less effective, machine learning algorithms like Decision Trees, Multivariate Linear Regression, and Random Forest are used to produce accurate results. By looking at historical rainfall data and predicting rainfall for upcoming seasons, it may be done

[a]nileshdhannaseth@gmail.com, [b]sanjayeyedey@gmail.com

DOI: 10.1201/9781003567653-18

with ease. Various methodologies, including classification and regression, may be used depending on the needs. These methods calculate the precision and error of the actual and predicted values. It is crucial to select the appropriate method and model it in accordance with the requirements because different approaches yield varying degrees of accuracy.

2. Literature Review

Various methods for rainfall analysis are examined. These include statistical techniques such as frequency analysis, time series analysis, and spatial interpolation. Moreover, the review explores the utilisation of advanced technologies like remote sensing, machine learning, and data mining for robust and accurate analysis.

Climate change's impact on rainfall patterns is a significant focus. The review investigates how shifting climate conditions alter rainfall distribution, intensity, and seasonality. This section underscores the importance of adapting analysis techniques to account for these changes.

Applications of rainfall analysis are explored, spanning flood prediction, water resource management, urban planning, and agricultural strategies. Real-world case studies are used to illustrate the practical implications of rainfall analysis in diverse scenarios.

Challenges are acknowledged, such as data quality issues, uncertainties, and the need for improved spatial and temporal resolution. The review also points towards emerging technologies and methodologies that hold promise for enhancing the accuracy and scope of rainfall analysis.

In conclusion, the review underscores the multifaceted nature of rainfall analysis, demonstrating its value in understanding natural variability, predicting extreme events, and informing decision-making across various sectors. In order to meet the shifting problems brought about by a changing environment, it emphasises the necessity

of ongoing research and innovation in this area of study.

Many researchers have concentrated on improving the accuracy of machine learning methods used in weather forecasting during the last 20 years. Here, some related studies are covered. Sawale and Gupta (2013) described an ANN-based method to forecast environmental circumstances. A variety of meteorological characteristics, such as humidity, temperature, and wind speed, were included in the dataset utilized for the prediction.

Joseph (2013) suggested a hybrid strategy for predicting rainfall by combining extraction of features and forecasting algorithms. The National Oceanic and Atmospheric Administration (NOAA) provided the dataset for the experiment, which included data on humidity, pressure, temperature, and wind speed over a period of more than 50 years. Based on a predetermined training set, Neural networks were employed to categories the cases into low, middle, and upper classes.

Nikam and Meshram (2013), using a Bayesian mathematical technique, researchers proposed a data-driven model for precipitation forecasting. The India Meteorological Department provided data for the experiment, and seven of the thirty-six attributes were deemed to be the most crucial. Pre-processing and transformation stages were carried out in order to guarantee efficient processing prior to prediction. Using moderate computing resources, the suggested method demonstrated good precipitation forecast accuracy, in contrast to meteorological facilities that use powerful computing ability for weather forecasting.

Liyew and Melese (2021) explore the application of machine learning techniques to predict daily rainfall amounts. The research is driven by the increasing need for accurate weather forecasting methods that can help in planning and decision-making processes related to agriculture, water resource management, and disaster preparedness.

Gnanasankaran and Ramaraj (2020) the authors explain the selection of ML models for comparison, which might include Linear Regression, Random Forest, Support Vector Machines, and Neural Networks, among others. The criteria for evaluating and comparing the models' performance, such as accuracy, precision, recall, and Mean Squared Error (MSE), are also defined.

Pirone et al. (2023) proposes a machine learning model for probabilistic rainfall now casting at 10 min intervals for short lead times—from 30 min up to 6 h. The model employs cumulative rainfall fields from station data as inputs for a feed forward neural network to predict rainfall interval and the corresponding probability of occurrence. Cumulative rainfall depths from station data were used to overcome the lack of temporal memory of the feed forward neural networks. In this way, using only the current rain field as input, the model exploited pattern recognition techniques combining both temporal cumulative rainfall depth and spatial cumulative rainfall field information

3. Proposed Methodology and Model Specifications

Multivariate Linear Regression: Regression is a statistical empirical data mining technique with a broad range of applications, including business, biology, and climate prediction. It is evolving into a useful tool for society.

Multidimensional linear regression is a guided machine learning approach that examines many data variables. An extension of multiple regressions, a multidimensional regression has several variables that are independent with a single dependent variable. We make an effort to predict the outcome by counting the number of independent variables.

The multidimensional regression model's generalised equation is shown below: $y = \alpha 0 + \alpha 1.x1 + \alpha 2.x2 +. + \alpha n.xn$

The formula denotes the quantity of variables that are distinct (n), the coefficients ($\alpha 0$-αn), and the variable that is independent (x1=xn).

The multidimensional model helps us understand and compare factors throughout the output. Because of its tiny cost function, multidimensional linear regression is the better model in this instance.

As per finding from research gaps in previous papers, following parameters can be used precipitation, sunshine, temperature, humidity, wind speed with Multivariate Linear Regression model to predict rainfall pattern (Figure 18.1).

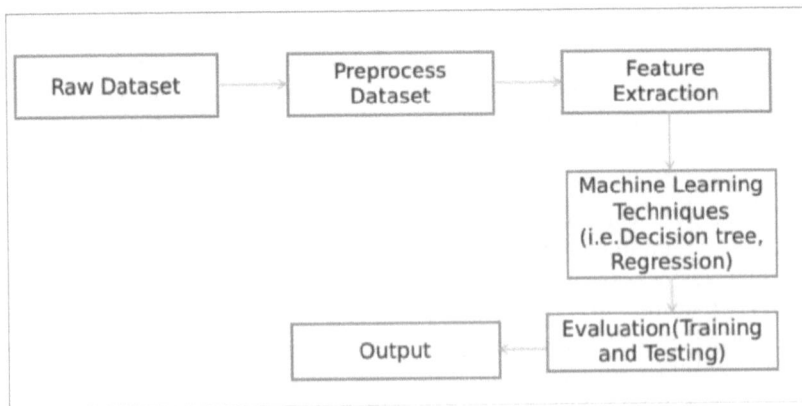

Figure 18.1: Architectural flow.

Source: Author.

4. Conclusion

Rainfall forecasting is crucial since excessive and erratic precipitation can have a number of negative effects, including crop loss and property damage. Therefore, a more accurate forecasting model is required for early warnings that can lower risks to people and property while simultaneously enhancing agricultural field management. Farmers stand to gain the most from this outlook, and effective water utilization can be accomplished.

References

Chowdari, K., Girisha, R., and Gouda, K. (2015). A study of rainfall over India using data mining. International Conference on Emerging Research in Electronics, Computer Science and Technology.

Diez-Sierra, J., and del Jesus, M. (2020). Long-term rainfall prediction using atmospheric synoptic patterns in semi-arid climates with statistical and machine learning methods. Journal of Hydrology, 586, 124789. doi: 10.1016/j.jhydrol.2020.124789

Gnanasankaran, N, and Ramaraj, E. (2020). A multiple linear regression model to predict rainfall using Indian meteorological data. International Journal of Advanced Science and Technology, 29(8), 746–758.

Joseph, J. (2013). Rainfall prediction using data mining techniques. International Journal of Computers and Applications, 83, 11–15.

Kumar, V., Yadav, V. K., and Dubey, E. S. (2022). Rainfall prediction using machine learning. International Journal for Research in Applied Science and Engineering Technology, 10(5), 2494–2497.

Liyew, C. M., and Melese, H. A. (2021). Machine learning techniques to predict daily rainfall amount. Journal of Big Data, 8(1). doi: 10.1186/s40537-021-00545-4.

Nikam, V. B., and Meshram, B. B. (2013). Modelling. rainfall prediction using data mining method: A Bayesian approach. Proceedings of the International Conference on Computational Intelligence, Modelling and Simulation, Bangkok, Thailand, 24–25 September 2013, pp. 132–136.

Paras and Mathur, S.(2012). A simple weather forecasting model using mathematical regression. Department of Electronics & Communication Engineering, College of Technology, G.B. Pant University of Agriculture & Technology, Pantnagar, (India) 263 145.

Pirone, D., Cimorelli, L., Del Giudice, G., and Pianese, D. (2023). Short- term rainfall forecasting using cumulative precipitation fields from station data: A probabilistic machine learning approach. Journal of Hydrology, 617, 128949. doi: 10.1016/j.jhydrol.2022.128949

Prabakaran, S., Kumar, P. N., and Tarun, P. S. M. (2017). Rainfall prediction using modified linear regression. ARPN: Journal of Engineering and Applied Sciences, 12(12), 3715–3718.

Rahman, A. U., Abbas, S., Gollapalli, M., Ahmed, R., Aftab, S., Ahmad, M., Khan, M. A., and Mosavi, A. (2022). Rainfall prediction system using machine learning fusion for smart cities. Sensors, 22(9), 3504.

Ridwan, W. M., Sapitang, M., Aziz, A., Kushiar, K. F., Ahmed, A. N., & El-Shafie, A. (2021). Rainfall forecasting model using machine learning methods: Case study Terengganu, Malaysia. Ain Shams Engineering Journal, 12(2), 1651–1663.

Sawale, G. J., and Gupta, S. R. (2013). Use of artificial neural network in data mining for weather forecasting. International Journal of Computer Science and Applications, 6, 383–387.

Selvaraj, R. S., and Raajalakshmi Statistical method of predicting the Northeast rainfall of Tamil Nadu. Universal Journal of Environmental Research and Technology, 1(4), 557–559.

Siddiqua L. A., and Senthilkumar, N. C. (2019). Heavy rainfall prediction using Gini index in decision tree. International Journal of Recent Technology and Engineering (IJRTE), 8(4), 4558–4562.

Vijayan, R., Mareeswari, V., Mohankumar, P., Gunasekaran, G., and Srikar, K. (2020). Estimating rainfall prediction using machine learning techniques on a dataset. International Journal of Scientific and Technology Research ;9(06), 440–405.

Zeelan, C., and Nagulla, B. (2020). Rainfall prediction using machine learning & deep learning techniques. IEEE Xplore.

19 Digital twins in metaverse

Shabana Pathan[a], Sumedh Dubey[b], Anurag Katre[c], and Yash Sharma[d]

Department of Information Technology, St. Vincent Pallotti College of Engineering and Technology, Nagpur, India

Abstract

A digital twin in the metaverse is a dynamic, data-driven virtual replica of physical objects, environments, or systems. Digital twins facilitate simulation, analysis, and modelling of real-world behaviors and interactions, empowering users to experiment with different scenarios and predict outcomes. Through an extensive review of existing literature and case studies, this paper elucidates the advancements, methodologies, and tools employed in combining digital twins with AR and VR technologies. It also investigates the potential applications, benefits, challenges, and future directions of utilizing digital twins in AR and VR. Furthermore, it discusses the implications of this integration for industries such as manufacturing, healthcare, urban planning, and beyond. The paper concludes with insights into the opportunities and limitations of digital twin technology in AR and VR environments, along with recommendations for further research and development.

Keywords: 3D modelling, augmented & virtual reality, digital twining

1. Introduction

A digital twin is a virtual model that accurately reflects a physical object. The object being studied—for example, a wind turbine—is outfitted with various sensors related to vital areas of functionality. These sensors produce data about different aspects of the physical object's performance, such as energy output, temperature, weather conditions, and more. This data is then relayed to a processing system and applied to the digital copy. Once informed with such data, the virtual model can be used to run simulations, study performance issues, and generate possible improvements; all to generate valuable insights—which can then be applied back to the original physical object. The integration of digital twins with AR and VR environments holds immense potential to revolutionize numerous sectors, including manufacturing, healthcare, urban planning, education, and entertainment. By harnessing the synergies between these technologies, organizations can unlock new opportunities for innovation, efficiency, and decision-making. For instance, in manufacturing, digital twins coupled with AR and VR enable engineers to virtually design, simulate, and optimize production processes, leading to enhanced productivity and reduced costs. Moreover, in healthcare, the combination of digital

[a]spathan@stvincentngp.edu.in, [b]sumedhdubey872@gmail.com, [c]anuragkatre36@gmail.com, [d]yashsharma19788@gmail.com

DOI: 10.1201/9781003567653-19

twins with AR and VR facilitates medical training simulations, patient monitoring, and surgical planning, thereby improving healthcare outcomes and patient safety. In urban planning, city planners can use digital twins integrated with AR and VR to visualize and simulate urban environments, enabling better urban design, infrastructure management, and disaster preparedness. Despite the promising potential, the integration of digital twins with AR and VR presents several challenges, including data integration, synchronization, visualization, and user interface design. Furthermore, ethical considerations such as data privacy, security, and accessibility must be carefully addressed to ensure responsible deployment and adoption of these technologies. This paper aims to provide a comprehensive review of the integration of digital twin technology with AR and VR environments. It explores the applications, benefits, methodologies, tools, challenges, and future directions of this integration, drawing insights from existing literature, case studies, and real-world implementations. By shedding light on the opportunities and limitations of combining digital twins with AR and VR, this research seeks to inspire further exploration and innovation in this exciting interdisciplinary field. Colleges and universities are increasingly embracing digital transformation to enhance learning experiences, foster collaboration, and streamline administrative processes. Model the equipment, workstations, and infrastructure of each sublab to facilitate realistic simulations and experiments. Integration of AR Technology: Establish an Augmented Reality (AR) lab within the digital twin environment, equipped with AR-enabled devices and interactive interfaces. Develop AR applications to overlay digital content, such as instructional materials, virtual experiments, and interactive simulations, onto physical objects within the lab. The digital twin to include tutorial rooms, providing additional spaces for collaborative learning and hands-on activities.

2. Literature Review

The use of VR in architectural visualization and design review processes is discussed in the paper. Academic articles discussing the impact of VR walkthroughs on architectural decision-making, user experience, and design comprehension (Howard & Davis, 2022).

Books on 3D modelling techniques, rendering methods, and VR applications in architecture, such as "Architectural Rendering with 3Ds Max and V-Ray" by Markus Kuhlo and Enrico Eggert. Case studies highlighting successful VR architectural walkthrough projects and their methodologies (https://sketchfab.com/3d-models/basic-structure 10d4bdcd56f-f4220aa 5522c623aa0).

The Rise of AR, VR, MR Technology articles or conference papers on techniques for converting AR content into VR environments. Studies focusing on the integration of AR and VR technologies to create mixed reality experiences, discussing challenges, and potential solutions (Malik & Bilberg, 2018).

Journals like the "International Journal of Human-Computer Studies" or "IEEE Transactions on Visualization and Computer Graphics" often publish research on mixed reality and AR/VR convergence. Reports or whitepapers from tech companies working on AR/VR convergence projects (https://youtube/OkB-mtkVrHg?si=jPT1q6nG-oslJdrb).

Shao, et al. (2022) explores the historical evolution, current trends, and future prospects of AR, VR, and MR technologies Review articles in journals like "Virtual Reality" or "IEEE Computer Graphics and Applications" covering advancements in sensing technologies utilised in AR/VR/MR systems. Papers discussing sensor fusion, tracking systems, haptic feedback, and other sensory aspects influencing the user experience in AR/VR environments (https://www.livehome3d.com/useful-articles/ar-in-home-design).

The C# programming in Unity for VR/AR, like "Learning C# Programming with Unity 3D" by Alex Okita. Academic papers discussing scripting and development practises specifically tailored for VR/AR applications using Unity. Online tutorials, forums, and community resources related to C# scripting and Unity development for VR/AR. Reviews, and case studies showcasing the usage of assets and prefabs from platforms like Articles or blog posts discussing best practices for selecting, implementing, and customising prefabs/assets (Sudhansu Rajan, 2023).

3. Methodology

3.1. *Define Objectives*

The purpose and objectives of creating a digital twin within the metaverse. Whether it's simulating real-world scenarios, enhancing design processes, or optimising operations, having a clear objective is crucial. Regarding the data collection, several methods can be employed to gather the necessary information for creating and maintaining the digital twin:

3D Scanning: Utilise 3D scanning technologies to capture the geometry and visual appearance of the physical object or environment. This can involve techniques such as laser scanning, photogrammetry, or LiDAR.

Sensor Data: Install sensors in the physical system to collect real-time data on parameters such as temperature, pressure, vibration, or fluid flow. This data is essential for updating the digital twin and monitoring the system's behaviuor.

IoT Devices: Integrate IoT devices to gather data from various components of the physical system. These devices can communicate with the digital twin to provide information on performance, status, and operational conditions.

Manual Input: In some cases, manual input may be necessary to supplement automated data collection methods. This can involve inputting metadata, annotations, or qualitative observations into the digital twin interface.

Identify Assets: Identify the physical assets, systems, or processes that you want to replicate in the metaverse. This could range from buildings and infrastructure to manufacturing processes or even entire cities.

3.2. *Data Acquisition Methods*

The collection of relevant data about the physical assets or systems. This could include architectural plans, sensor data, CAD models, historical performance data, and any other relevant information needed to accurately recreate the object or system in the digital realm. modelling and simulation tools to create a virtual representation of the physical object or system.

Designing and Development: Ensure interoperability between the digital twin and other systems or platforms within the metaverse. This may involve integrating with IoT devices, data analytics platforms, or other virtual representations to enable seamless interaction and data exchange. Validate the accuracy and performance of the digital twin through testing and simulation.

Deployment and Maintenance: Once validated, deploy the digital twin within the metaverse environment. Regularly maintain and update the digital twin to reflect any changes or updates to the physical object or system it represents.

Continuous Improvement: Continuously monitor the performance of the digital twin and identify opportunities for improvement. This could involve refining simulation models, optimising algorithms, or incorporating new data sources to enhance the accuracy and utility of the digital twin over time. The methodology section of a research paper outlines the procedures and techniques used to conduct the research.

Research Design: Begin by explaining the overall design of the research. Are you conducting experimental research, observational research, or a combination of both? In this case, one may engage in both observational research (collecting data on the college campus) and experimental research (creating and testing the digital twin).

3.3. Data Collection Methods

Architectural Plans: Obtain architectural plans and blueprints of the college campus. These documents provide essential information about the layout, dimensions, and features of buildings.

Photographs and Videos: Take photographs and videos of the campus from various angles and viewpoints. These visual assets can serve as references for creating detailed 3D models.

Environmental Data: Gather data on environmental factors such as climate, terrain, and vegetation. This information can influence the design and simulation aspects of the digital twin.

Foot Traffic Patterns: Study foot traffic patterns on the campus to understand how students and faculty move around. This data can inform the design of pathways and optimise the layout of buildings.

3D modelling Software: Choose suitable 3D modelling software for creating the digital twin. Consider software packages like Blender, Autodesk Maya, or SketchUp, which offer robust tools for modelling and rendering.

Familiarise with the selected software and its capabilities. This may involve training or experimentation to ensure proficiency in creating detailed 3D models of buildings, landscapes, and infrastructure.

AR/VR Functionality: The AR/VR features and interactions implemented in the application, such as object recognition, gesture control, spatial mapping, and virtual object manipulation. Provide technical details on how these features are implemented using Unity3D, AR/VR SDKs, and custom scripting. Identify performance bottlenecks and optimization techniques employed to improve frame rate, reduce latency, and enhance overall performance. The techniques such as occlusion culling, LOD (Level of Detail) management, and asset optimization.

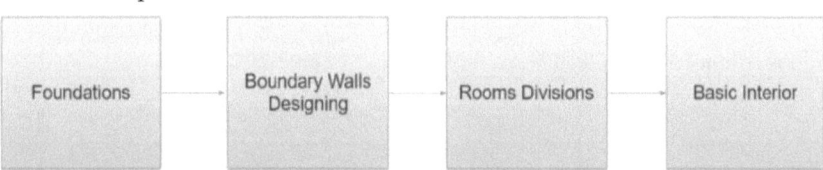

Figure 19.1: Flowchart of proposed Digital Twin system.

Source: Author.

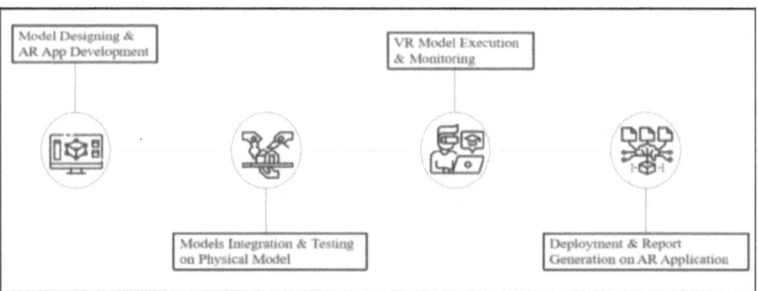

Figure 19.2: System flow diagram.

Source: Sudhansu Rajan, 2023.

Figure 19.1 gives the flowchart of proposed digital Twin System

Sudhansu Rajan (2023) describes the System Flow diagram for AR application development in Figure 19.2.

4. Conclusion

In conclusion, our AR/VR project on digital twins has provided valuable insights into the potential applications and benefits of this technology. Through our exploration, we have demonstrated how digital twins can revolutionise various industries, including manufacturing, healthcare, and urban planning. One of the key takeaways from the paper is the power of immersive technologies like AR and VR in enhancing the visualization and interaction with digital twins. As technology continues to evolve, one can anticipate further advancements in areas such as augmented analytics, predictive maintenance, and autonomous systems. In conclusion, our AR/VR project has not only expanded our understanding of digital twins but also underscored their transformative potential in driving innovation, improving decision-making, and enhancing operational efficiency across various sectors. As we continue to explore this exciting field, we are optimistic about the endless possibilities it holds for shaping a smarter, more connected world.

References

Howard, M. C., and Davis, M. M. (2022). A meta-analysis and systematic literature review of mixed reality rehabilitation programs: Investigating design characteristics of augmented reality and augmented virtuality. Computer in Human Behaviour, 130(c). doi: 10.1016/j.chb.2022.107197

Malik, A. A., and Bilberg, A. (2018). Digital twins of human robot collaboration in a production setting. Procedia Manufacturing, 17, 278–285. doi: 10.1016/j.promfg.2018.10.047

Shao, L. et al. (2022). Augmented reality calibration using feature triangulation iteration-based registration for surgical navigation. Computers in Biology and Medicine, 148(1). doi: 10.1016/j.compbiomed.2022.105826

Sudhansu Rajan. (2023). 10 amazing examples of digital twin technologies for industries. https://www.toobler.com/blog/digital-twin-examples (accessed February 5, 2024)

https://sketchfab.com/3d-models/basic-structure 10d4bdcd56ff4220aa 5522c623aa0

https://youtube/OkB-mtkVrHg?si=jPT1q6n G-oslJdrb

https://www.livehome3d.com/useful-articles/ar-in-home-design

20 Design and analysis of a brake disc for formula SAE race car

Sushant S. Satputaley[1,a], Himanshu Sawarkar[1,b], Manav Raut[1,c], Mandar Dhopte[1,d], Aditya Warhade[1,e], and Vikalp Shendekar[2,f]

[1]Department of Mechanical Engineering, St. Vincent Pallotti College of Engineering and Technology, Nagpur, India
[2]College of Engineering Computing and Cybernetics, The Australian National University, Canberra, Australia

Abstract

This research paper presents the comprehensive design and analysis of an optimised brake disc for use in a Formula SAE race car. The focus is on achieving a lightweight, high-performance disc that effectively dissipates heat under demanding braking conditions. Key considerations include calculating essential braking parameters, material selection focusing on strength and thermal properties, and thorough CAD modelling. Finite Element Analysis (FEA) is used for static structural and steady-state thermal simulations to evaluate the design's ability to withstand braking forces and manage heat buildup. The design process is iterative, with optimization based on FEA results to ensure a robust and efficient brake disc. This research provides a valuable framework for designing and analyzing high-performance brake discs for Formula SAE applications.

Keywords: Brake disc, formula SAE, Ansys workbench, computer-aided design (CAD), finite element analysis (FEA)

1. Introduction

Brake discs are critical components of any vehicle's braking system. Their primary function is to convert the vehicle's kinetic energy into heat through friction, effectively decelerating the car. In the high-performance world of Formula SAE racing, brake discs must withstand extreme conditions – repeated hard braking, high operating temperatures, and the demand for minimal weight. Careful design and analysis are crucial to achieving a brake disc that delivers robust performance within the unique constraints of a Formula SAE race car (Reynolds et al., 2015).

The design of a Formula SAE brake disc begins by gathering essential requirements such as vehicle weight, target speeds, and braking performance goals. Material selection is a crucial step, balancing the need for high strength, excellent heat dissipation, and low weight. Typically, high-performance brake discs utilize either lightweight aluminium alloys or steel alloys in cases where budget allows.

[a]ssatputaley@stvincentngp.edu.in, [b]himanshusawarkar003@gmail.com, [c]rautmanav0509@gmail.com, [d]mandardhopte21@gmail.com, [e]adityawarhade4923@gmail.com, [f]u7517877@anu.edu.au

DOI: 10.1201/9781003567653-20

The geometry of the brake disc significantly influences its performance. The disc's diameter and thickness directly affect the available contact area for braking force. The design of the ventilation channels, if present, plays a key role in heat dissipation. Drilled holes on the disc surface aid with gas evacuation under extreme braking conditions and can improve wet-weather braking performance (Mora, 2018).

2. Design of Brake Disc

Firstly, the vehicle's braking performance needs to be quantified. This involves calculating the maximum deceleration achievable given the car's weight and target race speeds. The required pedal ratio, brake pressure, and braking torque are determined using the vehicle's specifications and known relationships. This information, along with the desired disc diameter and thickness, allows the calculation of the friction force needed between the disc and brake pads (Reynolds et al., 2015).

For the calculations, vehicle parameters were taken. The car weighs 270 kg, with a total weight of 2648.7 N considering gravity (9.81 m/s²). The wheelbase is 1.55 m, and the centre of gravity (C.G.) is 0.2286 m high, positioned 0.62 m ahead of the rear axle. The front and rear brake discs share the same dimensions, with an outer diameter of 156.21 mm and an inner diameter of 96.21 mm. The braking system utilises a 6:1 pedal ratio, with a 15.875 mm diameter master cylinder bore and 30 mm diameter calliper pistons. The coefficient of friction between the tyres and the road is assumed to be 0.6, with the vehicle travelling on a flat surface (θ = 0°). The wheels have a radius of 0.2286 m.

The provided information details the calculations involved in determining braking dynamics for a vehicle (Kate et al., 2022; Ishwar Gupta et al., 2013), such as

- Weight transfer: During braking, weight transfers forward, placing a load of 48.9% on the front wheels compared to 51.1% on the rear.

- Deceleration: The calculated deceleration is 5.88 m/s² (0.6g).
- Braking forces: The total braking force is 1562.73 N, distributed between the front (776.32 N) and rear wheels (812.9 N) based on the friction coefficient (0.6).
- Required Torque: The required torque on the front and rear wheels is 177.46 Nm and 185.83 Nm, respectively.
- Master cylinder force: The force required on the master cylinder piston is 2100 N, divided between the front (1029 N) and rear (1071 N) based on the braking force distribution.
- Brake Pressure: The pressure generated in the front and rear master cylinders is around 5.2 MPa and 5.4 MPa, respectively.
- Calliper friction force: The friction force generated by the callipers on the front and rear discs is approximately 2940 N and 3060 N, respectively.
- Wheel torque: The final torque applied to the wheels is around 204 Nm for the front and 212 Nm for the rear.

Based on the previous calculations of braking forces and considering the following:

- Vehicle speed: 45 km/hr (12.5 m/s)
- Relative velocity of vehicle: 4.27 m/s
- Friction coefficient between the pads and discs: 0.4
- The heat flux (rate of heat transfer per unit area) on each of the front and rear discs was calculated (Vidiya & Singh, 2017; Kowal et al., 2012):
- Front disc: 2510.9 W/m²
- Rear disc: 2613.4 W/m²

On calculating the braking dynamics, it's crucial to select an appropriate brake disc material that can withstand the mechanical stresses and thermal loads imposed during braking.

2.1. *Material Selection*

Selecting the optimal material for a Formula SAE brake disc requires careful consideration of several key properties. It demands high strength to resist braking forces, excellent thermal conductivity to manage heat, and low weight for better handling (Maleque et al., 2010). While cast iron offers strength and wear, it's heavy. Aluminium alloys are lighter but may struggle with peak temperatures. Though Steel has a higher density than aluminium but has optimum thermal conductivity to manage the disc temperature. EN19 steel is a popular choice for Formula SAE brake discs due to its well-rounded properties. It offers a good balance between structural strength and wear resistance, crucial for withstanding the demanding braking forces and repeated use in racing conditions. While not boasting the highest thermal conductivity compared to some advanced materials, its cost-effectiveness makes it a viable option for student racing teams. Additionally, EN19's good machinability allows for easier fabrication of complex disc geometries, further contributing to its suitability for Formula SAE applications (Gurumurthy et al., 2016).

A few materials were taken into consideration for selecting brake disc material. The properties of those materials are Table 20.1.

2.2. Design Objectives

Once the material is chosen, the design phase begins with a 3D Computer-Aided Design (CAD) model of the disc. The model incorporates the chosen dimensions, ventilation channels, and potential features like drilled holes. The initial step was to choose the type of disc to be designed. The major objectives while designing brake disc are:

- High strength
- Excellent heat dissipation
- Weight reduction

Design constraints taken into consideration while designing brake disc are:

- Since the disc will be assembled on the wheel hub inside the wheel rim, sufficient clearances are to be given in the design.
- Provision of drill and slot patterns are provided for better thermal management as it improves airflow through the disc.
- Giving sufficient thickness to endure the clamping and friction force developed while braking.

Based on the design objectives and design constraints, disc geometry is finalised, given below Table 20.2.

Table 20.1: Comparison of material properties

Materials	UTS (MPa)	YTS (MPa)	Density (gm/cm³)	Thermal Conductivity (W/mK)	Specific Heat (J/kg.K)
Al 7075	572	503	2.81	130	796
SS 321	515	205	8.02	16.1	500
EN 24	745	470	7.8	43	470
EN 19	655	415	7.7	42.6	473

Source: Maleque, et al., 2017.

Table 20.2: Disc geometry

Front Brake Disc	Values	Rear Brake Disc	Values
Outer Diameter	156.21 mm	Outer Diameter	156.21 mm
Inner Diameter	96.21 mm	Inner Diameter	96.21 mm
Thickness	4 mm	Thickness	4 mm
Pitch Circle Diameter	75 mm	Pitch Circle Diameter	75 mm

Source: Author.

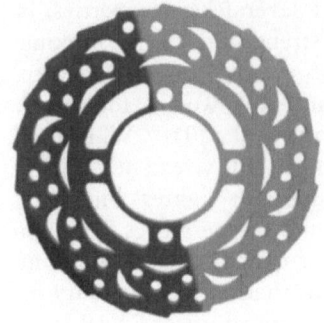

Figure 20.1: CAD model of brake disc.

Source: Author.

3. FEA Simulation

The next phase involves evaluating the designed brake disc's performance through Finite Element Analysis (FEA) simulations. Static structural analysis evaluates the disc's ability to withstand braking forces without excessive deformation or failure. Steady-state thermal analysis investigates the temperature distribution within the disc under braking conditions, helping identify potential thermal issues like brake fade. Based on the FEA results, the design can be iteratively optimised by adjusting disc geometry, material distribution, or ventilation features until the desired performance is achieved (Belhocine et al., 2016).

The braking calculations indicate that the front and rear discs experience almost similar loads. Therefore, to optimise the disc design, we can analyse them under the same conditions. To ensure robustness, we will consider higher-than-expected loads in our analysis. This will lead to a more conservative design that can handle potential variations in braking performance (Pasqual et al., 2020; Eshaan Gupta et al., 2022).

3.1. Steady State Thermal Analysis

Under this analysis, a constant heat flux of 5000 W/m² simulates the braking heat, while a convective heat transfer coefficient of 23 W/m² and an ambient temperature of 25°C represent heat dissipation to the surrounding air.

3.1.1. Results

The steady-state thermal analysis predicted a maximum temperature of 144.44°C

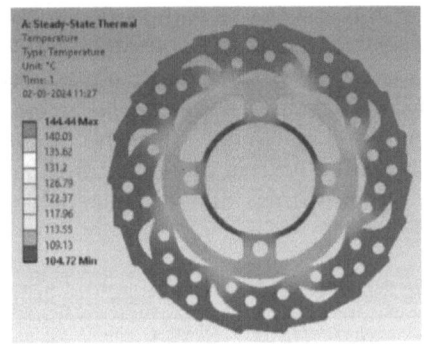

Figure 20.2: Max temperature.

Source: Author.

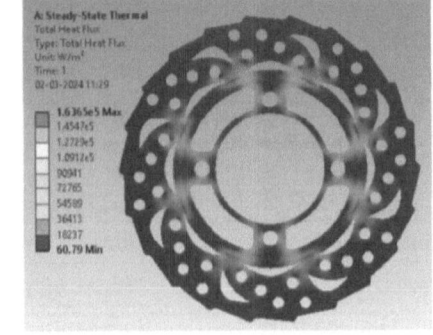

Figure 20.3: Total heat flux.

Source: Author.

within the brake disc. The total heat flux across the disc surface reached 1.636e5 W/m². These results will be crucial in evaluating the disc's thermal performance and potential for brake fade.

3.2. Static Structural Analysis

This analysis will simulate two opposing forces of 5000 N each applied where the brake pads clamp onto the disc's outer diameter. Additionally, a fixed support will be applied at the disc's mounting holes, showing its connection to the vehicle. To account for potential braking torque during braking, a 400 Nm moment will be applied at the disc's outer edges.

3.2.1. Results

The static structural analysis revealed a maximum deformation of 0.02 mm and an

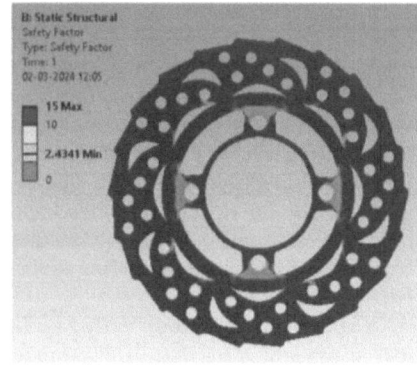

Figure 20.6: F.O.S.

Source: Author.

equivalent stress of 170.5 MPa within the disc. This translates to a safety factor (FOS) of 2.4. These results will be assessed to ensure the disc's structural integrity under braking loads.

4. Conclusion

This research paper presented a comprehensive approach to designing and analysing a brake disc for a Formula SAE race car. Following the determination of braking performance requirements and material selection, a brake disc design was developed, focusing on achieving a lightweight solution with optimal heat dissipation. Finite Element Analysis (FEA) simulations played a crucial role in evaluating the disc's performance under static structural and steady-state thermal loading conditions. The analysis results provided valuable insights into stress distribution, deformation, and temperature distribution within the disc. Through an iterative process informed by FEA results, the brake disc design was optimised to ensure it meets the demanding performance requirements of a Formula SAE car, balancing factors like structural strength, thermal management, and weight reduction. This research provides a valuable framework for designing and analysing high-performance brake discs for Formula SAE applications.

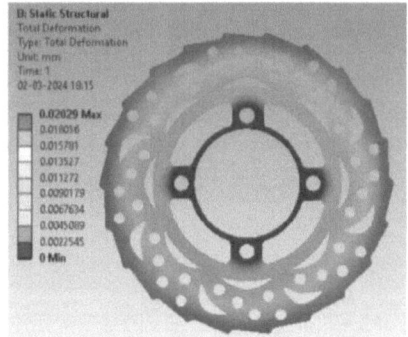

Figure 20.4: Total deformation.

Source: Author.

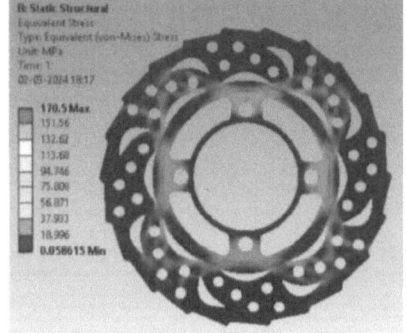

Figure 20.5: Equivalent stress.

Source: Author.

References

Belhocine, A., Bakar, A. R. A., and Bouchetara, M. (2016). Thermal and structural analysis of disc brake assembly during single stop braking event. Australian Journal of Mechanical Engineering, 14(1), 26–38.

Gupta, E., Singh Bora, D. K., Rammohan, A. (2022). Design and analysis of brake system for FSAE race car. Engineering Research Express.

Gupta, I., Saxena, G., and Modi, V. (2013). Thermal analysis of rotor disc of disc brake of baja Sae 2013 car through finite element analysis. International Journal of Engineering Research and Applications (IJERA), 1, 324–329.

Gurumurthy, S. S., Sharma, G., Sowmith, A., and Devar, A. (2016). Study of Mechanical Properties of Dual Phase EN19 Steel (AISI4140). Indian Journal of Science and Technology, 9(37), 1–6.

Kate, U. S., and Patil, R. M. (2022). Design and analysis of FSAE brake disc. International Journal of Engineering Research & Technology (IJERT), 11(01), 7–10.

Kowal, L., Turewicz, K., and Kruczek, T. (2012). Measurements of temperature of brake disks in hoisting machines of mine shaft hoists. 11th International Conference on Quantitative InfraRed Thermography.

Maleque, M.A., Dyuti, S., and Rahman, M.M. (2010). Material selection method in design of automotive brake disc. Proceedings of the World Congress on Engineering 2010.

Mora, L. A. (2018). Design of a FSAE braking system.

Pasqual, G. V. F., and Malcher, L. (2020). Thermal analysis of discs for BAJA SAE vehicle. 18th Brazilian Congress of Thermal Sciences and Engineering.

Reynolds, A., Angeliu, K., Moser, T., and Martins, B. (2015). Design and optimization of FSAE vehicle.

Vidiya, M., and Singh, B. (2017). Experimental and Numerical Thermal Analysis of Formula Student Racing Car Disc Brake Design. Journal of Engineering Science and Technology Review, 10(1), 138–147.

21 Smart solar-powered agricultural system

N. Dhote[1,a], A. Gawande[1], B. Badage[1], S. Panekar[1],
J. Mallaravapu[1], H. Chincholkar[1], K. Vaidya[1], and
J. Meshram[2,b]

[1]St. Vincent Pallotti College of Engineering and Technology, Nagpur, India
[2]Principal Scientist, ICAR-Central Institute for Cotton Research, Nagpur, India

Abstract

Agriculture serves as the foundation of our nation, with a wide array of activities now undertaken in this sector. However, mastering manual methods for these tasks can be challenging. Thus, there is a focus on developing a machine capable of efficiently applying fertilisers and pesticides, mowing grass, and planting seeds with minimal human intervention. This machine, powered by batteries and utilising solar energy, sprays pesticides, effectively manages weeds and facilitates seed planting. It incorporates features such as rotating knives for grass cutting and precise seed distribution mechanisms. By harnessing solar power, the machine reduces operational costs and enables versatile functionality, thereby enhancing efficiency. Controlled by a microcontroller and responsive to WIFI commands, the machine can be operated remotely via an Android application. Adopting this technology can significantly lower labour expenses and increase time savings for farmers.

Keywords: Smart agriculture, solar energy, robotic system, microcontroller

1. Introduction

Agriculture stands as a cornerstone of human survival. This article delves into addressing challenges within artisanal farming, such as mowing, seeding, and pesticide application. The machine operates on solar batteries, with DC motors propelling its wheels. Control over all processes is managed by the AT mega microcontroller. The aim of the robot is to streamline operations and enhance precision, replacing manual tasks. It excels in mowing, precise seed placement, and optimal seed compaction, ensuring proper soil depth and timed seed dispersal.

The machine employs a curved bar at its rear to cover clusters with soil. Pesticides are dispensed from tanks to safeguard farmers and crops from harmful chemicals. Various tasks including shovelling, planting, grazing, mowing, storage, and watering are routine in crop fields. Among these, mowing, seeding, and pesticide spraying are pivotal daily activities for growers.

To bolster the economy, agricultural productivity and quality must be enhanced. Automation simplifies work processes and minimises errors. The machine's small wheels and lightweight prevent soil compaction, contrasting with cumbersome conventional methods. Advancements in

[a]nitindhote@gmail.com, [b]j.h.meshram@gmail.com

DOI: 10.1201/9781003567653-21

agricultural technology are imperative for increased productivity and reduced farmer stress. However, agricultural automation is still nascent due to a lack of sophisticated equipment, technology, and technical expertise, impacting the prosperity of developing nations where agriculture employs a significant portion of the workforce.

2. Literature Review

This article proposes to design a machine that runs on solar energy, which is considered a clean energy source. Operations are performed using an Android application. The goal of this process is to implant two seeds of different sizes. The Bluetooth device can then be used to act as a link between two or more microcontrollers, such as an Arduino, or with any Smartphone device, such as a laptop or mobile. This increases the efficiency of the system and eliminates problems encountered during manual planting. A study (Ranjitha et al., 2019) showed that through the Bluetooth/Android application, we can send signals about the robot's required movements and necessary mechanisms. This type of paper automatically sprays pesticides with minimal human assistance. Solar energy can be stored as electrical or thermal energy. It can also store mechanical energy in the form of a flywheel. The sprayer can spray liquid quickly and easily, reducing the farmer's effort. Zigbee is used for wireless monitoring and control applications. A powerful, flexible, and cost-effective microcontroller is used to control the machine (Dhote et al., 2021). The study on IoT-based multipoint pesticide spraying machines (Dhote & Edgar, 2022) proposes a modern solution to agricultural challenges. Despite its innovation, limitations include the need for robust connectivity infrastructure and potential cybersecurity risks associated with IoT devices. The study (Sahu & Sendhil Kumar, 2018) investigating solar-assisted automated pesticide sprayers addresses sustainability. However, limitations may include the weight and bulkiness

of the sprayer due to the incorporation of solar panels, potentially impacting usability and manoeuvrability. The study on autonomous seed-sowing agricultural robots (Jayakrishna et al., 2020) has several limitations, including high initial investment costs, technical complexity, limited versatility in application, and significant energy and resource requirements.

The study (Murthy et al., 2017; Rajesh et al., 2016) on the design and fabrication of a solar pesticide sprayer addresses advancements in agricultural technology. Limitations may include dependency on adequate sunlight for operation, potentially limiting functionality in regions with inconsistent sunlight or during cloudy weather. The study (Sarath et al., 2019) investigated a solar-based pesticide sprayer with Bluetooth control, the system may rely on Bluetooth connectivity, which could pose challenges in areas with weak or unreliable network coverage. Additionally, battery life and power consumption associated with Bluetooth communication may affect operational efficiency. The conventional methods of mowing, spraying pesticides, and sowing seeds face many problems. Conventional techniques continue to rely on human power and outdated methods; they take more time and effort. Agriculture requires the use of highly skilled labour. This process can be automated to reduce labour needs.

3. Smart Agricultural Robot

In the proposed system, the automatic robot has a simple structure and is used to reduce human labour and save time in agricultural activities. The block diagram of the proposed model is shown in Figure 21.1. The Android application is used to send the commands to the microcontroller for the required operation and movement of the machine. The commands are transferred using Wi-Fi technology. The AT Mega 328 microcontroller is used. The proposed system uses solar panels. Electrical radiation is converted into an electrical signal, which

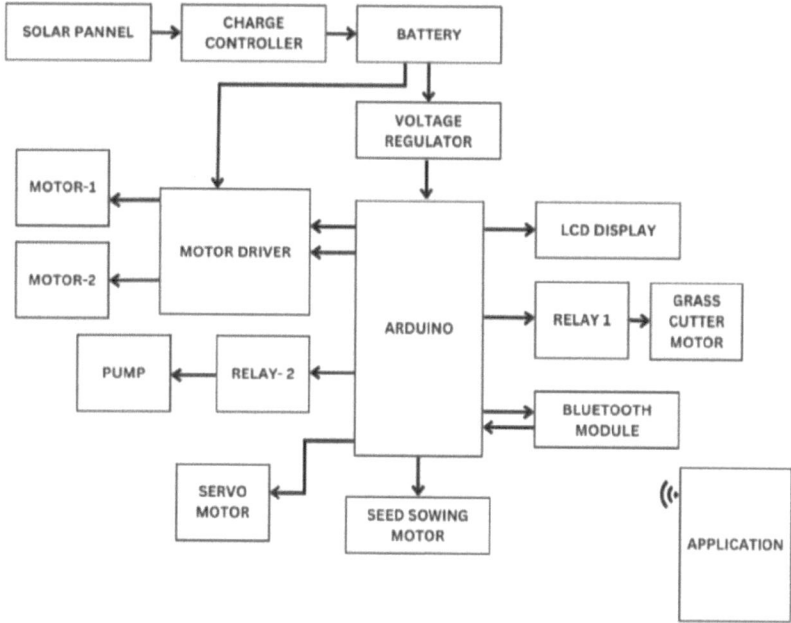

Figure 21.1: Block diagram.

Source: Author.

would then be preserved in the system. The machine is driven by the battery. The motor driver is a device that regulates the operation of two motors at the same time which is used to move the machine. When the command is transmitted to the micro-controller using Wi-Fi, the relay is made to perform such operations as pesticide spraying, weed cutting, and seed sowing. A prototype of a smart agriculture robot is shown in Figure 21.2.

The ESP8266 Wi-Fi Connector may be a self SOC that would give you access to your Wi-Fi hotspot towards any micro-computer. The ESP8266 may either get to organise or offload all Wi-Fi organising functionalities to a dedicated application processor. AT instruction set framework could be a software which is pre--modifies the interior of each ESP8266 unit, which suggests you merely have to plug it into your Arduino move to procure by way of numerous capacities as the Wi-Fi extension board. The ESP8266 module could be an exceptionally beneficial advancement

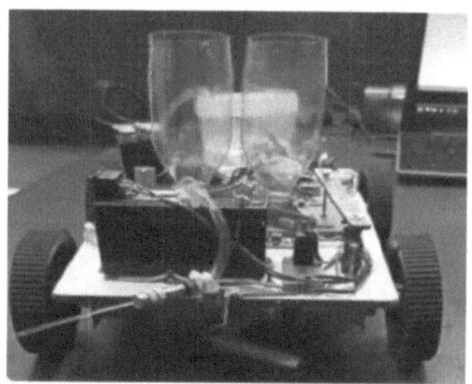

Figure 21.2: Smart agriculture robot.

Source: Author.

board employing an expansive mounting office. PV cells are superconducting materials that can also be observed on spacecraft and gadgets. As it suggests (photos mean "light" plus "voltaic" means energy), solar power is unswervingly converted into electricity. A component is a set of batteries (usually called solar panels) that are electrically connected and encapsulated in a frame. Photovoltaic arrays are an excellent approach to eliminate or at least minimize our home's carbon impact by cutting off the power that everyone wants to live without. This goal is made achievable by solar panels. Photovoltaic (PV) cells are used in solar panels to generate electrical energy. A coordinate current (electrical) engine may be a machine that changes electrical input into mechanical control. The wheels associated with the ESP8266 Wi-Fi Connector may be a self SOC to the gadget and are administered by two engine drives. The 200 rpm Economy Arrangement DC Engine may be a high-quality, low-cost DC-adapted engine with a middle shaft. Steel gears and pinions empower a longer benefit life and higher wear and weakness quality. From the interior, the gears are joined to the gearbox. Despite the reality that the motor runs at rpm at 12V, it runs superbly from 4V to 12V and contains a wide cluster of Speed and execution. The table underneath appears a legitimately solidified steel shaft. A voltage regulator is an electronic appliance that provides a stable DC voltage regardless of load current, climate, or AC line multiple requirements. A simple feed-forward architecture or negative feedback can be used in the regulator. Electromechanical mechanisms or circuit boards can be used. It can be used to modulate one or more AC or DC voltages, depending on the manufacturer. To control the DC engines, a 12V DC battery is required. The unit's controller, LCD, and transfer utilises a 5V control.

4. Conclusion

This machine is implied to function naturally within the field of horticulture. The vital thing is to automate operations like pesticide splashing, grass cutting, and seed sowing, in arrange to get a more noteworthy surrender and help the farmers to move towards the utilising of progressed innovation. It makes a difference for them to accomplish globalisation. The seeds were put within the right area, coming full circle in appropriate vegetative development. This headway in the agrarian division is very conceivable to accomplish a more noteworthy efficiency rate and reduce the control utilisation and work prerequisite. This Keen machine is assigned to extend efficiency and diminish human endeavours. It is the culmination apparatus that does not contribute to climate alteration since it begins running on sun-based vitality.

References

Dhote, N., Choudhary, S., and Kalbande, K (2021). IoT based multipoint Pesticide spraying machine. IEEE 6th International Conference on Inventive Computation Technologies. CFP21F70-ART; ISBN: 978-1-7281-8501-9: 432–36.

Dhote, N., and Edgar, A. (2022). IoT based multipoint pesticide spraying machine. International Research Journal of Modernization in Engineering Technology and Science, 4(4), 2450–2454.

Jayakrishna, P. V. S., Suryavamsi Reddy, M., Jaswanth Sai, N., Susheel, N., and Peeyush K. P. (2020). Autonomous seed sowing agricultural robot. IEEE International Conference on Advances in Computing, Communications, and Informatics. 978-1-5386-5314-2/18:2332:36.

Murthy, K., Kanwar, R., Yadav, I., and Das, V. (2017). Solar pesticide sprayer. International Journal of Latest Engineering Research and Applications, 2(5), 82–88.

Rajesh, R., Vimal Kingsley, V., Selva Pandi, M., Niranjan, G. and Varun Harshath, G. (2016). Design and fabrication of solar pesticide sprayer. International Journal of Innovative Research in Science, Engineering and Technology, 5(8), 131–135.

Ranjitha, B., Nikhitha, M. N., Aruna K., Afreen, and Venkatesh Murthy, B. T. (2019). Solar powered autonomous multipurpose agricultural robot using bluetooth app. IEEE Conference on International Electronics Communication and Aerospace Technology. 45616; IEEE Xplore ISBN: 978-1-7281-0167-5; 872–77.

Sahu, Susant Kumar, and Sendhil Kumar, N. (2018). Design and development of solar assisted automated pesticides sprayer. International Journal of Engineering Research & Technology, 6(16), 1–4.

Sarath, R., Arun Kumar, S., Kumar, M., and Balaram Sai, R. (2019). Solar based pesticide sprayer using bluetooth control. International Journal of Electronics Engineering, 11(1), 352–355.

22 Single axis solar power tracker

Pankaj Ramtekkar[a] and Harikumar Naidu[b]

Electrical Engineering, G H Raisoni College of Engineering Nagpur, India

Abstract

The present era has noticed a surge in the adoption of solar panels to efficiently convert solar energy into electrical power. These panels can be deployed in various capacities, either as standalone systems or integral components within larger solar configurations connected to electricity grids. It is estimated that the Earth receives an estimated 80 Terawatts of power, while our global daily consumption hovers around 10 Terawatts, there is a concerted effort to harness more energy from the sun. To optimize the conversion of solar energy into electricity, it is essential to position solar panels perpendicularly to the sun. This necessitates accurate tracking of the sun's position. The primary goal of project is that to develop an system with automatic tracking, capable of precisely determining sun location. A proposed single-axis solar tracking system will dynamically adjust orientation of panel of solar to consistently align it perpendicular to the sun, thus maximizing energy conversion throughout the day. Sunlight hits the solar panel, the LDR sensor promptly measures the intensity of light. Subsequently, the Arduino issues precise commands to the motor based on the detected light intensity, resulting in the calculated rotation of the solar panels. The tests conducted gave satisfactory results, which will pay the way to the R&D Professionals and Industrial experts to commercialize this endeavour.

Keywords: Solar panels, Arduino UNO, light sensing system

1. Introduction

Solar cells offer a cost-efficient and highly effective solution for electricity generation, especially when contrasted with hydraulic generators. The working of panels is intricately tied to intensity for sunlight. The traditional grid system, unfortunately, lacks adaptability to fluctuating light levels. Addressing this limitation, a new solution is presented with an Automatic Solar Power Tracker. (Barsoum, 2009) This system integrates LDR sensors and Arduino technology, ensuring constant monitoring of light levels and dynamic adjustment of the panel orientation to optimize exposure to maximum light intensity. This dynamic tracking capability significantly boosts overall efficiency (Masih and Odinaev, 2019). The Light Dependency Register (LDR) is a pivotal element, specifically designed for measuring sunlight intensity. In conjunction with an Arduino-controlled GSM module, the system is equipped to send notifications to the user. The automatic solar tracking system not only maximizes energy capture but also enhances overall system efficiency (Ramtekkar and Haigune, 2021). Solar panels, acting as collectors of sunlight, consist of multiple solar units that serve as the system's power hub. A controller within this system actively monitors and adjusts the panel orientation based on varying light intensity (Ramtekkar et al.,

---

[a]pankaj.ramtekkar@raisoni.net, [b]harikumar.naidu@raisoni.net

DOI: 10.1201/9781003567653-22

2021). This approach proves to be both cost-effective and efficient, with a straightforward installation process (Dudhe et al., 2022). It's important to note, however, that these tracking systems are more intricate compared to their fixed solar counterparts (Ramtekkar et al., 2022).

2. Literature Review

Solar cells outperform hydroelectric generators in terms of cost- effectiveness and efficiency (Jagtap and Chandrakar, 2021). The effectiveness of a solar module is contingent on harnessing the maximum intensity of solar radiation. In today's grid system, tracking light intensity is easily achieved with a solar tracker, but this challenge is efficiently addressed through the implementation of an Light Dependency Registers (LDRs) play a pivotal role in measuring maximum sunlight intensity, with Arduino overseeing a GSM module for user notifications (Dudhe, Ramtekkar and Jagtap, 2018). Single-axis solar power tracking monitoring systems not only facilitate the optimization of solar energy but also enhance overall efficiency. Solar panels actively capture sunlight and efficiently convert it into electricity, relying on a utility composed of distinct solar-oriented units. (Khandelwal, Ramtekkar, Chauhan, Bhute and Kouthekar, 2022). The central control unit for this solar system is a comprehensive controller that diligently monitors and adjusts light intensity as needed (Mahalaxme et al., 2020). The installation of a solar power tracker proves to be a judicious and efficient choice (Walde and Ramtekkar, 2022) While setting up a single-axis PV system is straightforward, it's important to note that trackers, due to their increased complexity compared to fixed solar systems, offer a more sophisticated solution (Adware and Chandrakar, 2022).

3. Components

The key components of this system include a light-sensing mechanism, motor driver, gear motor system, Arduino UNO, and the solar panel.

3.1. Gear Motor

The gear motor encompasses a spectrum of designs, ranging from a fundamental setup where a motor is paired with a straightforward gear to more sophisticated constructions such as the Nord Gear unit that incorporates bevel gears, resulting in a 90-degree hollow-shaft output. In essence, a geared motor is a mechanism crafted to fine-tune the speed of the connected motor, ensuring precise operation at a designated speed.

3.2. Motor Driver

This IC is an type of chip which is integrated circuit used as a motor controlling device in autonomous robots and embedded circuits.

3.3. LDR Sensor

Its resistance increase or decreases depending on the amount of intensity of light.

3.4. Solar Panel

The panel functions as a device that captures solar radiation and transforms it into electrical energy, constituting a vital component of the photovoltaic system, also referred to as the solar module. Comprising multiple solar cells, these panels are combined to produce an electricity and have various rectangular types of shapes. PV panels have an important role in photovoltaic type systems by concentrating solar energy, converting light energy in to electrical energy that can further transformed into electrical energy, stored in the form of heat or battery is another option. Whether termed a solar panel, photovoltaic panel, photovoltaic (PV) module and solar panel, it typically consists of an array of pv cells arranged within a structure, usually rectangular types of shape. An organized grouping of PV modules is known as a PV array (Tembhare et al., 2020).

3.5. *ARDUINO Uno*

Arduino Uno stands as a microcontroller board cantered around ATmega328P. Boasting 14 inputs of digital type and also in output pins (with 6 designated for PWM outputs), along with 6 analogy inputs, a 16 megahertz quartz type crystal, USB type connection, jack for power purpose, ICSP header, and a reset type button. It comprehensively assemble all essential elements for microcontroller support. This board offers the flexibility to connect to a laptop or computer via a USB cable or, alternatively, to given power by an adapter AC-DC converter.

4. Methodology

The flow chart shows the process flow adopted in this paper as in Figure 22.1 Solar panels, designed for capturing sunlight and transforming it into either electricity or heat, consist of solar (or photovoltaic) cells arranged in a grid pattern on the panel's surface. A single-axis solar power tracker employs LDR sensors strategically placed at each end of the solar panel as shown in Figure 22.2. These sensors gauge incoming light intensity, prompting the panel to reposition itself with the assistance of a servomotor. The rotational movement of the solar panels is executed through servo motors, with control managed by a motor driver. The project architecture involves an LDR sensor detecting maximum solar power, relaying the information to the Arduino, which digitizes the LDR output. The active sensors, known as Light Dependent Resistors (LDR), continuously monitor sunlight and adjust the solar panel toward the direction with the highest sunlit intensity. The Hardware circuit diagram shown in Figure 22.3, which was tested and the results was satisfactory.

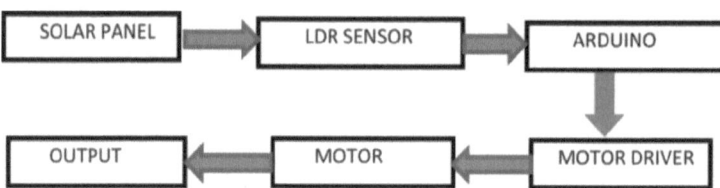

Figure 22.1: Flow chart.

Source: Author.

5. Circuit Diagram

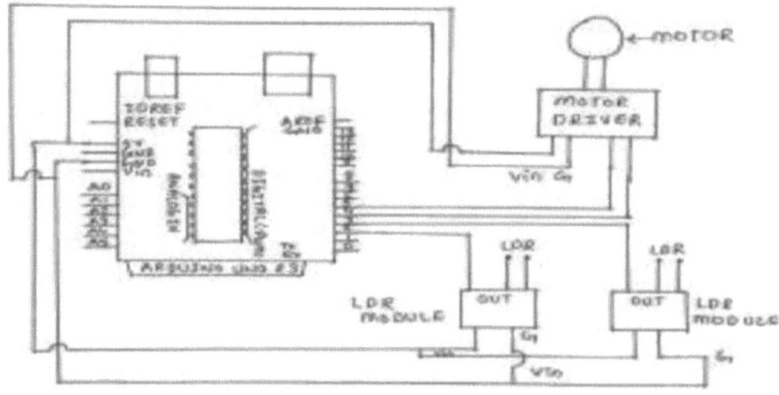

Figure 22.2: Circuit diagram.

Source: Author.

Figure 22.3: Hardware prototype.

Source: Author.

6. Result and Discussions

The tests conducted showed that the Single Axis Solar Power trackers dynamically adjust the positioning of a panel is in the direction of the Sun movement across the sky. This strategic alignment ensures the solar panel remains perpendicular to the Sun, optimizing sunlight absorption, reducing reflection, and enhancing energy absorption for efficient power conversion.

References

Adware, R. and Chandrakar, V. (2022). "Power Quality Enhancement through Reactive Power Compensation Using Hybrid STATCOM," 2022 Second International Conference on Power, Control and Computing Technologies (ICPC2T), Raipur, India, pp. 1–5, doi: 10.1109/ICPC2T53885.2022.9777006.

Barsoum, N. (2009). "Implementation of a Prototype for a Traditional Solar Tracking System," 2009 Third UKSim European Symposium on Computer Modeling and Simulation, pp. 23–30, doi: 10.1109/EMS.2009.59.

Dudhe, S., Ramtekkar, P., and Jagtap, P. (2018). "PLC ladder trouble shooting using bitwise scanning method,"2018 International Conference on Smart Electric Drives and Power System (ICSEDPS), pp. 362- 364, doi:10.1109/ICSE.

Dudhe, S. S., Khawse, A., Ramtekkar, P., and Gupta, A. (2022). Implementation of Scalable Low Cost Industrial IoT GATEWAY for Uninterrupted Monitoring and Data Acquisition of Industrial Machines. In Smart Technologies for Energy, Environment and Sustainable Development, (Vol 1, pp. 487–493). Springer, Singapore.

Jagtap, P. and Chandrakar, V. (2021). "Comparative Study of UPFC Controllers to Improve Transient and Dynamic Stabilty of Power System," 2021 IEEE 4th International Conference on Computing, Power and Communication Technologies (GUCON), pp. 1–7, doi: 10.1109/GUCON50781.2021.9573804.

Khandelwal, V., Ramtekkar, P., Chauhan, M., Bhute, Y., and Kouthekar, R. (2022). "Sensor Based Vibration Analysis of Motor Using MATLAB Software," 2022 10th International Conference on Emerging Trends in Engineering and Technology - Signal and Information Processing (ICETET-SIP-22), Nagpur, India,, pp. 1–4, doi: 10.1109/ICETET-SIP-2254415.2022.9791823.

Mahalaxme, S., Khubalkar, S., and Bharadwaj, S. (2020) "Low voltage Distribution Box Monitoring- New Way to Monitor Power in Industry," 2020 4th International Conference on Trends in Electronics and Informatics (ICOEI)(48184), Tirunelveli, India, pp. 214–216, doi: 10.1109/ICOEI48184.2020.9142937.

Masih, A. and Odinaev, I. (2019). "Performance Comparison of Dual Axis Solar Tracker with Static Solar System in Ural Region of Russia," 2019 Ural Symposium on Biomedical Engineering, Radioelectronics and Information Technology (USBEREIT), pp. 375–378, doi: 10.1109/USBEREIT.2019.8736642.

Ramtekkar, P., Naidu, H., and Dudhe, S. (2021). "A Novel Wireless Fire Containment and Extinguishing System to Save Life and Destruction of Property," 2021

International Conference on Computational Intelligence and Computing Applications (ICCICA), pp. 1–3, doi: 10.1109/ICCICA52458.2021.9697245.

Ramtekkar, P., Dudhe, A., Gupta, A., & Meshram, M. (2022). Distance Relaying with Power Swing Detection in the Presence of Distributed Resources. In Smart Technologies for Energy, Environment and Sustainable Development, Vol 2 (pp. 543–548). Springer, Singapore

Ramtekkar, P. and Haigune, K. (2021). "A Novel Approach for Overload and Short Circuit Protection System," 2021 International Conference on Advancements in Electrical, Electronics, Communication, Computing and Automation (ICAECA), pp. 1–3, doi: 10.1109/ICAECA52838.2021.9675696

Tembhare, M., Naidu, H., and Kokate, P. (2020). "A Review Study on the Multiple and Useful Application of Fiber Optic Illumination System," 2020 Fourth International Conference on Computing Methodologies and Communication (ICCMC), Erode, India, pp. 919–924, doi: 10.1109/ICCMC48092.2020.ICCMC-000170.

Walde, K. and Ramtekkar, P. (2022). "Designing Parameters and MATLAB Model of 11 KV / 440-volt Mobile Distribution Station," 2022 2nd Asian Conference on Innovation in Technology (ASIAN-CON), Ravet, India, pp. 1–5, doi: 10.1109/ASIANCON55314.2022.9909054.

23 Revolutionising healthcare monitoring: an intelligent machine learning approach to real-time body posture identification in bedridden patients

Kaushik Roy[a] and Ankit Mahule[b]

Department of CSE (Cyber Security), Shri Ramdeobaba College of Engineering and Management (RCOEM), Nagpur, India

Abstract

Introducing a groundbreaking approach in intelligent healthcare monitoring, our methodology employs cutting-edge machine learning techniques to achieve real-time identification of body postures in bedridden patients. The system seamlessly integrates a network of strategically positioned sensors and cameras, meticulously capturing and analysing patients' body postures. A bespoke machine learning model is intricately crafted to categorise a spectrum of postures, including those indicative of discomfort, distress, or potential health issues. Through the synergistic application of sophisticated computer vision and machine learning algorithms, the system attains an unprecedented level of precision in posture recognition. This paper delves into the forefront of research in remote patient monitoring, shedding light on the escalating impact of machine learning in optimising healthcare strategies. The amalgamation of advanced technologies showcased in our methodology holds promise for revolutionising healthcare practices, ensuring timely and accurate detection of patient conditions through innovative posture analysis.

Keywords: Healthcare monitoring, machine learning, body posture detection, alerting system, bedridden patients, remote patient monitoring, computer vision

1. Introduction

The healthcare landscape is witnessing a significant transformation, with a growing emphasis on leveraging technology to enhance patient care, particularly in the realm of monitoring bedridden patients for their well-being and safety (Pardeshi et al., 2017). Despite advancements in Remote Patient Monitoring (RPM) systems, there remains a notable gap in existing approaches concerning the continuous monitoring of

---

[a]roykr@rknec.edu, [b]mahuleaa1@rknec.edu; ankitmahule2@gmail.com

DOI: 10.1201/9781003567653-23

patients at risk due to prolonged immobility, such as those in long-term care or post-surgery recovery (Dhamodaran et al., 2020; Dhinakaran et al., 2022). Traditional methods often lack real-time monitoring capabilities and may not offer comprehensive insights into patient conditions beyond periodic check-ups (Alaziz et al., 2017; Amendola et al., 2015; Davoudi et al., 2019). In response to this gap, the proposed methodology aims to introduce an innovative approach to intelligent healthcare monitoring driven by machine learning for real-time detection of body postures in bedridden patients. By addressing the limitations of current approaches, our methodology seeks to enhance patient care by providing continuous monitoring and timely interventions, thereby improving outcomes and reducing healthcare costs. This paper explores the evolution of remote healthcare, highlighting the pivotal role of machine learning and sensor technologies in enabling effective patient monitoring beyond traditional healthcare settings. Moreover, it delves into related work in remote patient monitoring, underscoring the significance of integrating machine learning and cloud computing for continuous data storage and analysis (Malche et al., 2022). Furthermore, recent studies have showcased promising advancements in RPM systems, such as the introduction of AIoT-based patient activity tracking systems (Malche et al., 2022) and monitoring systems for body movements in bedridden patients using wearable sensors (Dhinakaran et al., 2022; Tuli et al., 2022). These innovations underscore the growing importance of leveraging technology to address the evolving needs of patient care, particularly in the context of remote healthcare delivery. However, there remains a critical need to bridge the gap in existing approaches by developing robust methodologies that can seamlessly integrate machine learning techniques with sensor technologies to enable continuous monitoring of bedridden patients. In essence, this paper aims to contribute to the discourse on intelligent healthcare monitoring by proposing a novel methodology that leverages machine learning for real-time detection of body postures in bedridden patients. By filling the existing gap in continuous monitoring approaches, our methodology seeks to enhance patient care, promote early disease identification, and improve overall healthcare efficiency in remote settings.

2. Related Work

In the realm of remote patient monitoring (RPM), the literature emphasises the pivotal role of machine learning techniques and technological integration in healthcare enhancement. For instance, (Dhinakaran et al., 2022) presents a comprehensive RPM system that employs diverse machine learning algorithms. However, a detailed discussion on the specific algorithms used and their performance metrics would facilitate a more insightful comparative analysis. Similarly, (Malche et al., 2022) introduces an AIoT-based patient activity tracking system that utilises machine learning for activity discernment and vital sign monitoring. While this integration of machine learning with the Internet of Things (IoT) holds promise for comprehensive patient monitoring, a more thorough evaluation of algorithm performance would strengthen its credibility. Regarding sensor technologies in RPM, (Wu et al., 2022) focuses on monitoring body movements in bedridden patients using compact wearable sensors. Although the use of wearable sensors allows for real-time monitoring, further discussion on sensor accuracy and reliability would enhance the understanding of its practical applicability. Meanwhile, (Alaziz et al., 2017) introduce an In-Bed Body Motion Detection and Classification System, emphasising precision in detecting movements. However, there is a need for exploration into the system's scalability and compatibility with diverse patient demographics for broader adoption. In terms of posture detection systems, (Davoudi et al., 2019; Panini & Cucchiara,

2003) propose a machine learning-based approach for human posture detection primarily targeting intelligent home automation. While efficient in detecting postures, its applicability to healthcare monitoring requires validation in clinical settings. Additionally, (Muneeswaran et al., 2020) contribute with a body movement detection system for coma patients using Wireless Sensor Networks (WSN), enabling real-time monitoring. However, further exploration is needed to assess the system's adaptability to different patient conditions and clinical scenarios. The synthesised literature underscores the convergence of machine learning, sensor technologies, and advanced monitoring systems in driving healthcare transformation, particularly in RPM.

3. Methodology

The proposed methodology integrates advanced technology, utilising the Media-Pipe library for real-time pose detection and Support Vector Machines (SVM) for high-dimensional data analysis. Support Vector Machines (SVM) were chosen for their ability to handle high-dimensional data effectively while maintaining robust performance

in classification tasks. SVMs provide a clear margin of separation between classes, making them suitable for discerning complex patterns in healthcare monitoring data. The methodology employs Convolutional Neural Networks (CNNs) for pose detection and classification tasks. CNNs are well-suited for extracting spatial features from images, making them suitable for analysing human postures in real-time. This approach aims to discern and classify body postures, crucial for bedridden patient monitoring, through sophisticated computer vision techniques. A 'Stillness Check' follows, evaluating the significance of detected movement. The methodology dynamically progresses based on user interaction, leading to the activation of the display mechanism for relevant information.

The high-level architecture of the Healthcare Monitoring System incorporates sensors, cameras, a Data Fusion Module, Machine Learning Module, Classification Component, and an Alerting System (Figure 23.1). This intelligent system provides real-time patient data, supporting personalised health analytics and posture analysis. Importing essential libraries involves OpenCV for computer vision tasks

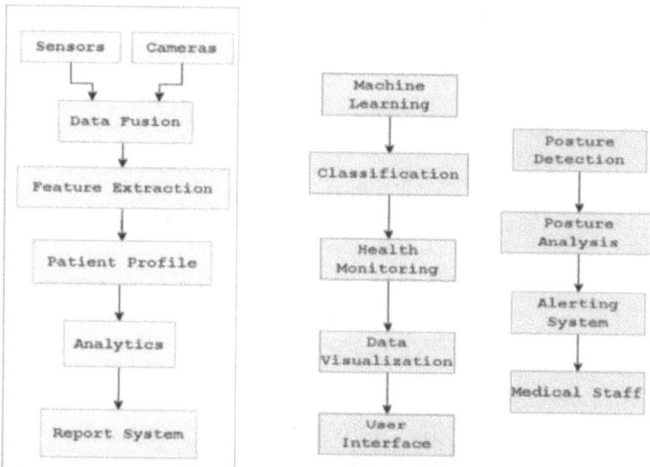

Figure 23.1: Healthcare monitoring system architecture.

Source: Author.

and MediaPipe for pose detection. Movement detection includes landmark tracking, movement calculation, thresholding, and stillness confirmation. Functions such as Calculate Movement and *is_person_still* contribute to the real-time analysis of body postures.

The visualisation aspect ensures effective representation of analysis results on the video frame, enhancing user engagement. Specific functions, *Detect_upright_punching* and *Detect_sleeping*, exemplify the system's ability to recognise nuanced postures. User interaction incorporates a 'q' key exit trigger, providing a user-friendly termination mechanism. A full-screen window optimises visual real estate, presenting live video frames and analysis results in an immersive manner. The proposed methodology's technical foundations rely on SVM for high-dimensional data analysis, showcasing its versatility in capturing complex patterns. Notable libraries include OpenCV for computer vision, MediaPipe for pose detection, and SVM for data handling. In essence, the methodology offers a sophisticated and versatile approach to real-time pose detection and movement analysis, leveraging advanced technologies to enhance healthcare monitoring for bedridden patients.

4. Results and Discussions

The implementation of the intelligent healthcare monitoring system, has demonstrated impressive outcomes in real-time body posture detection, specifically focusing on movements of bedridden patients. Accuracy metrics for movement detection, as presented in Table 23.1, showcase the robust design of the algorithm, achieving a commendable overall accuracy of 97.55%. The system exhibits proficiency in discerning true positives, true negatives, false positives, and false negatives, highlighting its capability in evaluating dynamic changes in posture.

Table 23.1: Accuracy metrics for movement detection

Metrics	Accuracy
True Positives	95.00%
True Negatives	98.00%
False Positives	02.00%
False Negatives	01.00%
Overall Accuracy	97.55%

Source: Author.

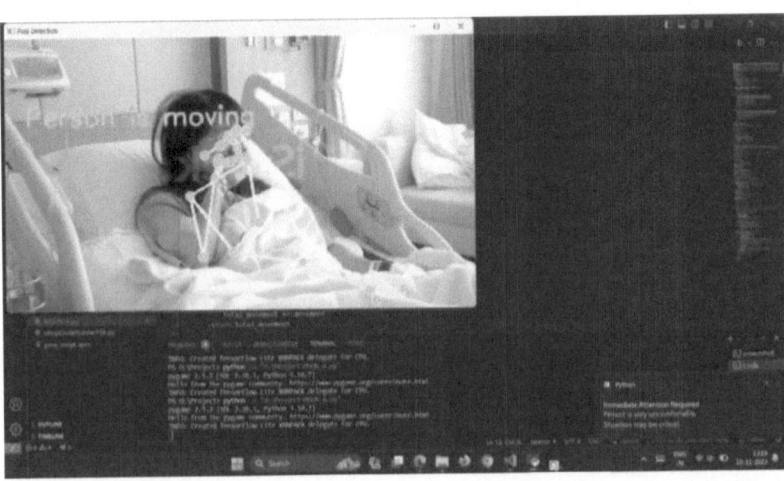

Figure 23.2: Sleeping moment detection.

Source: Author.

The algorithm employs endpoint verification to validate movement detection accuracy, minimising false positives and negatives. Additionally, a stillness confirmation mechanism ensures precise identification of sustained immobility, distinguishing transient movements from prolonged stillness. Real-time feedback is visually presented in a full-screen window, showcasing detected postures, movement status indicators, and relevant annotations. Figures 23.2 and 23.3 exemplify the system's effectiveness in detecting various movements, including sleeping moments and real-time detection of person movement in bed.

In practical healthcare scenarios, the system has proven applicable for continuous monitoring of bedridden patients, allowing timely identification of potential complications. The algorithm's balanced sensitivity ensures responsiveness to subtle shifts and pronounced activities, catering to diverse patient conditions. The integration of a user-friendly interface, with features like the 'q' key for program exit, enhances system responsiveness, ensuring seamless interaction and user convenience. The results highlight the system's efficacy in real-time body posture detection, yet it's crucial to address potential limitations encountered during implementation, such as sensor placement issues or computational resource requirements. Additionally, incorporating feedback from healthcare professionals or end-users on usability and practicality would provide valuable insights. Addressing these aspects would enrich the discussion, enhancing understanding of the system's real-world applicability and usability in diverse healthcare settings.

5. Conclusion and Future Scopes

The implementation of the proposed intelligent healthcare monitoring system, spearheaded by the innovative contributions of the authors, signifies a notable advancement in remote patient monitoring. Focused on detecting patient movements and alerting healthcare providers to anomalous postures, the system holds significant implications, especially for bedridden patients requiring continuous surveillance.

5.1. Key Strengths of the Methodology

1. **Movement Detection:** Achieves high accuracy (97.5%) through landmark tracking.
2. **Stillness Confirmation:** Effectively minimises false positives/negatives, enhancing reliability.
3. **User Interaction:** Simplified with the inclusion of user-friendly features like 'q' for exit.
4. **Technical Foundations:** Utilises robust libraries (OpenCV, MediaPipe) and Support Vector Machines (SVM) for high-dimensional data analysis.

Figure 23.3: Real-time detection of person movement in bed.

Source: Author.

The strategic integration of OpenCV and MediaPipe provides a robust technological backbone. Movement detection intricacies, encompassing landmark tracking, movement calculation, and thresholding, showcase the precision embedded in the methodology. Key functions, from movement calculation to posture detection, form a cohesive suite for real-time posture analysis. The visualisation aspect, including a full-screen window and dynamic overlay annotations, ensures an intuitive user experience. User-friendly features, exemplified by the 'q' key for program termination, enhance the system's usability. The technical foundation, including Support Vector Machines (SVM), fortifies the machine learning aspect, highlighting adaptability in capturing complex patterns. The project's adaptability extends to broader healthcare applications, particularly in monitoring patient movements and detecting unusual postures. This resonates with the critical need for continuous surveillance, offering timely interventions for bedridden patients and averting potential health risks. However, future research directions could include ethical considerations related to patient privacy, data security, and potential biases in the machine learning algorithms while also exploring potential areas for further improvement or refinement in the system's design and functionality.

References

Alaziz, M., Jia, Z., Howard, R., Lin, X., and Zhang, Y. (2017). MotionTree: A tree-based in-bed body motion classification system using load-cells. 2017 IEEE/ACM International Conference on Connected Health: Applications, Systems and Engineering Technologies (CHASE), 127–136. doi: 10.1109/CHASE.2017.71

Amendola, S., Bianchi, L., and Marrocco, G. (2015). Movement detection of human body segments: Passive radio-frequency identification and machine-learning technologies. IEEE Antennas and Propagation Magazine, 57(3), 23–37. doi: 10.1109/MAP.2015.2437274

Davoudi, A., Malhotra, K. R., Shickel, B., Siegel, S., Williams, S., Ruppert, M., Bihorac, E., Ozrazgat-Baslanti, T., Tighe, P. J., Bihorac, A., and Rashidi, P. (2019). Intelligent ICU for autonomous patient monitoring using pervasive sensing and deep learning. Scientific Reports, 9(1), 8020. doi: 10.1038/s41598-019-44004-w

Dhamodaran, M., Praveen, T., Rishaban, G. S., and Vasanth, R. (2020). Implementation of body movement detection system for coma patients using WSN. International Journal of Grid and Distributed Computing, 13(1), 161–166.

Dhinakaran, M., Phasinam, K., Alanya-Beltran, J., Srivastava, K., Babu, D. V., and Singh, S. K. (2022). A system of remote patients' monitoring and alerting using the machine learning technique. Journal of Food Quality, 2022. doi: 10.1155/2022/6274092

Jnnn. (n.d.). Young woman coughing in hospital bed. iStock. https://www.istockphoto.com/video/young-woman-coughing-in-hospital-bed-gm1204004422-346269374

Malche, T., Tharewal, S., Tiwari, P. K., Jabarulla, M. Y., Alnuaim, A. A., Hatamleh, W. A., and Ullah, M. A. (2022). Artificial intelligence of Things- (AIoT-) based patient activity tracking system for remote patient monitoring. Journal of Healthcare Engineering, 2022. doi: 10.1155/2022/8732213

Panini, L., and Cucchiara, R. (2003). A machine learning approach for human posture detection in domotics applications. 12th International Conference on Image Analysis and Processing, 2003. Proceedings., 103–108. doi: 10.1109/ICIAP.2003.1234034

Pardeshi, V., Sagar, S., Murmurwar, S., and Hage, P. (2017). Health monitoring systems using IoT and Raspberry Pi — A review. 2017 International Conference on Innovative Mechanisms for Industry Applications (ICIMIA), 134–137. doi: 10.1109/ICIMIA.2017.7975587

Tuli, S., Gill, S. S., Xu, M., Garraghan, P., Bahsoon, R., Dustdar, S., Sakellariou, R., Rana, O., Buyya, R., Casale, G., and Jennings,

N. R. (2022). HUNTER: AI based holistic resource management for sustainable cloud computing. Journal of Systems and Software, 184, 111124. doi: 10.1016/j.jss.2021.111124

ViralHog. (2020, August 28). Bad way to wake up ‖ ViralHog [Video]. YouTube. https://www.youtube.com/watch?v=CqFxsM5nLGE

Wu, Y., Wang, D. H., Lu, X. T., Yang, F., Yao, M., Dong, W. S., Shi, J. B., and Li, G. Q. (2022). Efficient visual recognition: A survey on recent advances and brain-inspired methodologies. Machine Intelligence Research, 19(5), 366–411. doi: 10.1007/s11633-022-1340-5

24 Innovative technologies and application in libraries

Vijay M. Deshmukh

Librarian, Central Library, St. Vincent Pallotti College of Engineering and Technology, Nagpur, India

Abstract

Library services and facilities are very important for the users and providing them updated services and facilities is one of the biggest problems of libraries. However, better services and facilities can be provided with the use of technology. Application of technology will help libraries to provide advanced facilities to their users where they will get the latest information on their subject. This will create technological environment which will motivate the users towards the library and also the libraries will improve their quality by providing good services and facilities. However, the available technology and its uses, library employees must be technically competent and able to use it as technology has the potential to make the library more productive and efficient as per the need of the users. In this paper, the impact of technology on libraries and users, application of latest technology with their various tools is discussed.

Keywords: Technology application, technology, libraries, library technologies, innovative technology

1. Introduction

The application of technologies in the library provides new trends and opportunities for improving processes, services and capabilities. Technology improves libraries' processes and makes it more predictable about library development, as well as new services, facilities, automation, and quality. Modern technologies are now gradually replacing the traditional functions and services of libraries. Technology is especially important and necessary for people with disabilities to solve their problem of accessing libraries and have a significant impact on the library services of such people.

When a crisis like pandemic arises, technology has the power to overcome that crisis. In those days services were provided remotely. Information and content also delivered with innovative technologies, examples of such technologies are drones, virtual meetings, e-books, e-delivery, etc. The future is more optimistic with technology that provides home based services rather than typical library base services. Due to such technology, time of library people, space and cost of libraries can be saved.

With the development of technologies there are a lot of potential services and facilities possible in libraries. Services such has delivery of documents requested by the user can also be sent to the users' location. Also, people who are dependent due to their physical disability will now get services through assistive technology and will be

vmd0308@yahoo.com

DOI: 10.1201/9781003567653-24

enabled to be less dependent on others for support. Thanks to technology, such people can do their daily work. To overcome the barriers and limitations of technology, library people need to have the vision and training to handle the technology. We are living in the smart technology era where the library can provide services and facilities at the click of a button and voice and does not need to move to provide services.

2. Objectives of Study

- To study new technology in libraries
- To study identify technology products for use in the library.

3. Innovative Technologies in Libraries

3.1. Drones

Drone is a type of robot, which flies from one place to another. It has arrangements for aeronautics. The size of drones is not fixed, they can be big or small in size as per the requirement and operation. They can be operated remotely or independently and are called UAVs (Unmanned Aerial Vehicles). In this system a pilot is not required to fly it.

3.1.1. Application in Library

- If the library area is large then drone can be used to carry books or materials from one place to another place.
- Library books can also be delivered to the required location if the user is unable to come to the library due to physical problems.
- With the help of drones, the scope of access to library reading material can be increased.
- With the help of drones, library staff can locate and monitor the library area.

3.2. Artificial Intelligence

It's also called AI in short. Artificial Intelligence is a technology where machine can think like humans. It is the science of machines where machines can judge and make decision. With the help of AI technology, human can do so many things smartly. AI technology is used with help of computers. In today's technology, AI is such a smart technology that makes human life even easier.

3.2.1. Application in Library

- With help of AI, libraries can improve their facilities and services like search and retrieval process.

Benefits
- Transporation
- Rescue Operation
- Search Operation
- Agriculturation Operation
- Civilian base operation etc.

Features
- GPS operated
- Software opearated
- Remote Operated
- Unmanned Flying Vehicle

Classification By DGCA
- Small drone
- Mediuam drone
- Large Drone
- Micro Drone

Figure 24.1: Drone technology.

Source: https://www.cio.com/article/222238/drones-in-the-supply-chain-the-evolving-drone-landscape-and-ecosystem.html.

Benefits
- 24 x7 availablity
- Decision without biases
- Less time and cost
- Improves Efficiency
- Avoid human errors
- Free up manual processes

Features
- Intelligence
- Solve complex problems
- Helps in automation
- Uses of multiple technologies

Classification
- Not able to learn actions
- Improve responses with memory
- Understand requirements
- Self Awareness

Figure 24.2: AI technology.

Source: https://www.acadecraft.com/blog/role-of-artificial-intelligence-in-education-and-learning/.

- Access to information becomes easier in terms of search and availability of information.
- To solve the queries of the users and provide accurate information and reduce human errors in the information.
- AI technology helps libraries increase their productivity by simplifying their issue process and record maintenance.
- Traditional library work can be replaced with advance library work which helps in reducing operating costs.
- AI helps to increases effectiveness in library services by providing location of reading materials, reference services, cataloguing, indexing.
- AI facilitates theft detection in libraries.

- With the help of AI technology, libraries can provide services and facilities to physically disabled users.

3.3. Augmented Reality

It is called AR in short. Augmented reality is a technology that provides understanding and interpretation of what we see and notice with interactive experiences. AR technology is used with software, hardware, and apps to provide perceptual information with computers. This technique enables the individual to interpret and become aware of something while also developing the senses without judging distances.

Benefits
- 24 x7 availablity
- Collaboration and Team work
- Creat interest
- Enchance understanding
- Practical Learning
- Space Saving

Features
- incorporating with smartphones
- Cost cutting Technology
- Interative learning
- High resolution information materials

Classification
- Marker based
- Location based
- Dynamic AR
- Complex AR

Figure 24.3: AR technology.

Source: xrlabs.com/blog/augmented-reality-can-be-used-in-libraries/.

3.3.1. Application in Library

- With the help of AR, physical libraries can be transformed into virtual libraries. This will help in locating resources and providing available resources at one place.
- AR techniques can provide information about the layout of the library at any location without going inside the library.
- Libraries also provide AR based reading material, which will help users to understand the subject in an interactive manner and they can develop their skills and knowledge.
- It helps people with disabilities access library resources.
- It helps users to get more access to library and books at one place.

- With the help of AR, libraries can provide virtual book to the users so that they can access one book at a time and discuss any topic.

3.4. Library Automation

Library automation means carrying out the traditional activities of libraries with the help of computers and other app based applications. Library automation is also known as library management system. This is the proper management of library housekeeping. It helps to manage all the activities, records, information, services etc. of libraries in an easy way.

Benefits
- 24 x7 availablity
- Save time
- Better management
- Better reporting
- Cost cutting
- Enhance productivity and efficiency

Features
- Space saving technology
- Barcode enabled
- Easy search interface
- Different categories records
- Ready to serve information

Classification
- Fully Automated
- Partial Automated
- Under Automation

Figure 24.4: Library automation.

Source: https://islmblogblog. wordpress. com/2016/05/09/library-automation/.

3.4.1. Application of Library Automation

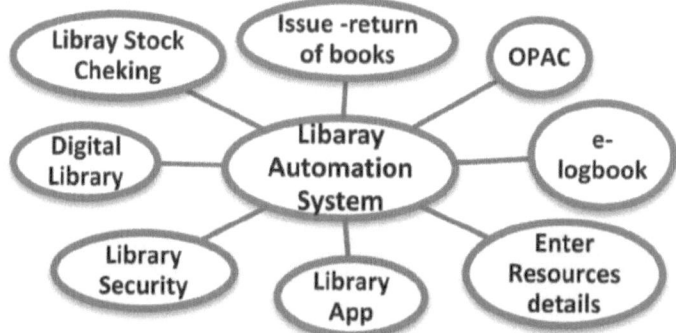

Figure 24.5: Application of library automation.

Source: Author.

4. Conclusion

This paper discusses the importance of technology and focuses on how it can be implemented in libraries. New technology makes life easier regarding information collection through various devices, especially when users have difficulties. All the technological inventions at global level can also be implemented in libraries, which will make the library more effective and productive. This paper emphasizes on the use of technology in library and to develop libraries technologically to serve the users in a better way. With the use of technology the libraries should improve the quality of services, automate process and improve the facilities in the library as per the need of its users. Library communities should take into account the positive aspects of technology to better serve their users. The application of technology in libraries is a step to move forward towards developing their services and facilities for the users. This will meet the needs of the users and solve the problems. The application of technology in libraries focuses on providing good services and facilities to its users and meeting their needs as per their requirement.

References

Akerkar, R. (2005). Introduction to Artificial Intelligence. Prentice Hall of India.

Asole, A. (2014). Augmented reality. St. Vincent Pallotti College of Engineering & Technology, Information Technology.

Evans, N. D. (2018). Drones in the supply chain: The evolving drones landscape and ecosystem. https://www.cio.com/article/222238/drones-in-the-supply-chain-the-evolving-drone-landscape-and-ecosystem.html. (Accessed April 5, 2024)(Figure 1)

IXR Labs. (n.d.). From xrlabs.com/blog/augmented-reality-can-be-used-in-libraries/ (Accessed February 22, 2024). (Figure 3)

Khemani, D. (2017). First course in artificial intelligence. MGH.

Mealy, P. (2018). Virtual and augmented reality. Wiley India (P) Ltd.

NewISLMBlog. (2016). From ISLM Blog Website. https://islmblogblog.wordpress.com/2016/05/09/library-automation/ (Accessed February 22, 2024). (Figure 4)

Santos, W. P. (Ed.). (2023). Frontiers. https://www.frontiersin.org/journals/public-health/articles/10.3389/fpubh.2022.1030656/full (Accessed February 22, 2024).

Shetty, C. G. (2020). Augmented reality: Theory, design and development. MGH.

Smith, A. (2023). The role of artificial intelligence in education and learning. https://www.acadecraft.com/blog/role-of-artificial-intelligence-in-education-and-learning/. (Accessed April 5, 2024) (Figure 2)

Termanini, R. (2023). Drone delivery method and how to communicate with the destination. In R. Termanini (Ed.), Biomedical defense principles to counter DNA deep hacking (pp. 205–219). Elsevier Inc.

25 Learning English via E-Portfolios: a collaborative effort

Shakuntala Vishnu Yadav

Department of Engineering Science and Humanities, St. Vincent Pallotti College of Engineering and Technology Nagpur, India

Abstract

As the word states Innovation, it simply refers to what creativity we can have in ourselves to make the learners learn English effectively and with interest. With the help of Podcast Audio this study aimed to find new tools to learn English. Based on the Lord Gautama Buddha's and Mr Bean's Audio this paper developed understanding and Listening Skills among the students. For innovative ideas, we need to have an accurate, stable, and growing mind set aiding to get more creativity in terms of teaching students. We the pedagogues need to think being in students' shoes whether this sort of teaching is going to make the pupils learn actively, or the pupil's learning is going on letting the information pass without being focused on what has been taught. The present paper is going to enlighten the readers and the pupils about how effectively they can be more observant and more accurate in understanding with an apt structure of sentences and non-verbal communication.

Keywords: Blogging, cartoon films, word games, Podcast, Vlog

1. Introduction

English is the language which has been discerned as the Window to the world via which we can see the entire world. The aspirants who want to have their higher studies overseas and also who are job seekers in foreign countries need to have the apt pronunciation to make the residents understand what we want to convey and also to enable ourselves to comprehend their accent. There are varied techniques we can apply to keep the learners learn English getting entertained. Yuldasheva Dilshoda Musaevna (2020), states that for learning English we need first enrich the vocabulary. Nowadays most of us want to be socially oriented and updated about what is going on around the world,

we can choose the method which we think to be more fascinated to make us learn. The best way to do this is to choose E-portfolios and technology for self-assessment. Taleb Bilal Eli (2021), states that in today's era, students are immersed in technology whether for learning purposes or for getting entertained. So technology can be made to teach them effectually. Let us see what tips and techniques would aid in today's generation to learn English well.

2. Methodology: Survey-based Method

The students were given the audio to listen of Lord Gautam Buddha preaching to the

syadav@stvincentngp.edu.in

DOI: 10.1201/9781003567653-25

people also one video was shared wherein Mr. Bean the comedy character tries to convey something without any dialogue. It was shared via Google form where they were asked questions and were given options to answer. It was a quantitative research design wherein the data of responses was collected. The purpose of sharing the video was to make students realise that nonverbal communication is equally essential to convey a message to people and it enhances our comprehension capability. As per Cristina Georgeta Pielmus (2018), technology is the greatest asset that can be used to improve students' English Listening, Speaking, Reading, and Writing Skills. Survey method and Audio-Visual method have been used in this research work. Some of the methods can also be used like Edublogs which help the pupil to publish their own sets of work, assignments, and material related to their study topics as per the statement of Anggia Susanti et al. (2022). Abdikhalikovna and Termez (2017) avers that online games aid in improving children's communication skills when and only when we select the instructions in the English language. Dwi Astuti Wahyu Nurhayati (2019) states that in today's era of digitalisation, pupils have to have a liberal ambiance to learn the second language.

3. Literature Review

Susanti et al. (2022) states that the use of media and technology empowers the student's vocabulary and structuring sentences in the contemporary era. The paper states about the technique the teacher can use to teach the pupils making them take keen interest in learning English. Ahmet Erdost Yastibas and Gülsah Cinar Yastibas, (2015) focus on e-portfolios which can enhance students' self-assessment skills and enable them to be self- synchronised and self-regulated in terms of completing the portfolio on time. (Goldsmith, 2007) points that E-portfolios also aid in saving the valuable time of pupils and teachers. Via Portfolio they can save the study material well and

utilise them to study whenever they need. Yastıbaş (2013) has shed light on the usage of the portfolio which deals with the help of electronics. Self-judgment is the best way for the students to come up with their weaknesses and strengths carried out a theory on the use of e-portfolios in speaking classes as an assessment tool. The results of the study show that e-portfolio assessment improved students' self-assessment skills as they could monitor their learning process, understand their strengths and weaknesses, and try to overcome their weaknesses (Yastıbaş, 2013). According to Abdi Khalikovna, and Termez (2017), vocabulary is the most essential part of learning a second language which can be enhanced with the latest way of learning. Musaevna, Yuldasheva Dilshoda (2021) put focus on honing the components of vocabulary whilst speaking and writing. Bong and Skaalvik (2003) state that self-efficiency is one of the most significant factors to assess our proficiency. Zarei and Hatami (2012) supported that students must be helped with the supporting tools to learn the second language and incorporate them in honing their skills. Dwi Astuti Wahyu Nurhayati (2019) gives their perception of the era completely involving the learners with keen interest.

4. Data Analysis

Below are figures of the questions given for the podcast containing a video of Mr. Bean and Lord Gautama Buddha. This has been shown to have the clarity of the questions shared with the students and the responses to the questions received. Most of the responses to the question showing the highest percentage seemed to be correct which shows their interest in learning through video emerging technologies. Link (Sources) for Lord Gutama Buddha Video https://www.youtube.com/watch?v=63ZhIYSHbvw. (Source(Link) of the video of Mr Bean(Creepy Bean)the funny video https://www.youtube.com/watch?v=l7A6jNOewFA.

Link of the responses from students https://docs.google.com/spreadsheets/d/1H

0zDPOXwalrmlOZv08EzHvnmRyNVVi
8G1cfOrKFLc6Y/edit?gid=1839736059#
gid=1839736059

The students were asked to go through the videos of Lord Gautama Buddha preaching a man and the video of Mr. Bean (Creepy

Figure 25.1: Question 1 by Lord Gautam Buddha Podcast.

Source: Author.

Figure 25.2: Question 2 by Lord Gautam Buddha Podcast.

Source: Author.

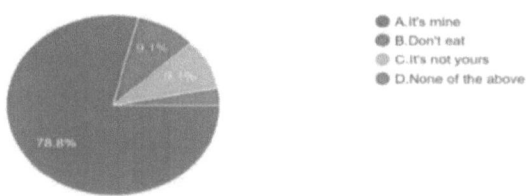

Figure 25.3: Question 1 Mr. Bean's video.

Source: Author.

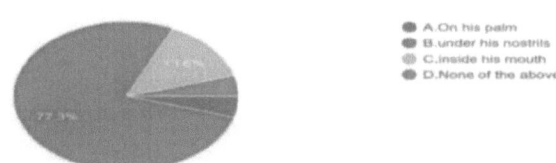

Figure 25.4: Question 2 Mr. Bean's video.

Source: Author.

Bean) the funny video. Figure 25.1 consist the question asked by the man to Lord Gautama Buddha. Figure 25.2 states that what was the thing Gautama Buddha gave to man and asked him to do. Figure 25.3 shows the question Mr. Bean asks the woman while taking popcorn. Figure 25.4 consists the question regarding the place where the popcorn has been put. There were some questions shared on the Podcast of Lord Gautam Buddha's preaching a man about life skills and Mr. Bean including the tenth question as feedback.

5. Discussion and Findings

There were in all sixty-six responses. The details contain the questionnaires based on the funny video of Mr. Bean. The students observed the video well and answered the questions. The best part of this video is that it does not have the dialogues much; rather it contains the non-verbal expression of Mr. Bean. Link of the responses fromstudentshttps://docs.google.com/spreadsheets/d/1H0zDPOXwalrmlOZv08EzHvnmRyNVVi8G1cfOrKFLc6Y/edit?gid=1839736059#gid=1839736059. Based on the responses the analysis has been shown in the table form.

Table 25.1 shows the analysis of the assessment done. We can see that the study enhances their skills of observation and understanding both via these kinds of videos.

There were almost forty-two students listened to the podcast of Lord Gautama Buddha and answered the questions. When the students were asked which mode of learning they prefer, they answered that the audio video methods containing different

Table 25.1: Assessment of the skills

Sr. No.	Enhancement of Skills	Assessment (%)
1	Observation skills	29.20
2	Understanding Skills	9.20
3	Both Observation and Understanding	61.50

Source: Author.

podcasts or videos based on chapters on any topic help them to understand them well and generate curiosity and interest in learning too. The students gave the title to the listening stuff, there were different titles given by them as per their understanding, as it was open-ended. This video helped the teacher to assess the student's comprehension and listening skills.

Table 25.2 shows the analysis of learning. We can see that the students are interested in learning English by Audio Video Method. The students get immersed in video games, online games, and Cartoon films. If this can be done using English audio by switching on the captions to be in English, then they will read all instructions in English. Link of the responses from studentshttps://docs.google.com/spreadsheets/d/1H0zDPOXwalrmlOZv08EzHvnmRyNVVi8G1cfOrKFLc6Y/edit?gid=1839736059#gid=1839736059. Based on the responses the analysis has been shown in the table form.

Table 25.2: Method of learning

Sr. No.	Method of Learning	Preference (%)
1	Audio Video Method	78.50
2	Chalk Board Method	3.00
3	Text Book Method	18.50

Source: Author.

6. Conclusion and Implication

Using the strategies consisting the emerging technologies will help the students and the learners to learn English. Everyone wants 'The Change', hence to avoid monotonous teaching there should be some spice to make the learning more fascinating. Here the emerging technologies motivates the learners to learn in the newer ground. Nowadays everyone wants to speak in English without putting much effort and time and also being entertained. The most renowned and upcoming method for learning English has been found as vlogging. Mentioned emerging technologies in the present paper would help in enriching our vocabulary. The more we speak, the more thirst for words, proverbs, idioms, and phrases will be there. This thirst for enriching vocabulary will be quenched by using the Earning and learning method of vlogging. These Innovative methods are going to be beneficial for especially the new generation and also those who want to make themselves being in learning whilst earning.

References

Abdikhalikovna, T. (2017). Innovative techniques of teaching vocabulary at the intermediate level Khayitova, Cyberleninka.ru, Feruza State University.

Bong, M., and Skaalvik, E. M. (2003). Academic self-concept and self-efficacy: How different are they really? Educational Psychology Review, 15, 1–40

Goldsmith, D. J. (2007). Enhancing learning and assessment through e-portfolios: A collaborative effort in Connecticut. New Directions for Student Services, 119, 31–42. doi: 10.1002/ss.247

Nurhayati, D. A. W. (2019). Students' perspective on innovative teaching model using Edmodo in Teaching English phonology: A virtual class development, Dinamika Ilmu, 19(1). P-ISSN: 1411 3031; E-ISSN: 2442-9651

Pielmus, C. G. (2018). Innovation in teaching English for law enforcement: A technology-integrated approach society. Integration Education, 3, 566–579.

Susanti, A., Kasim, U., Achmad, D., Burhansyah, and Nasir, C. (2022). The use of media in innovative learning to improve students' Achievement in Research in English and Education (READ), 7 (2), 85–90. E-ISSN 2528-746X.

Yastıbaş, A. E. (2013). The application of E-portfolio in speaking assessment and its contribution to students' attitudes towards speaking. MA Thesis, Çağ University, Institute of Social Sciences, Mersin.

Yastibas, A. E., and Yastibas, G. C. (2015). The use of e-portfolio-based assessment to develop students' self-regulated in English language teaching Social and Behavioral Sciences.

Zarei, A. A., and Hatami, G. (2012). On the relationship between self-regulated learning components and L2 vocabulary knowledge and reading comprehension. Theory and Theory and Practice in Language Studies, 2(9), 1939–1944.

26 Fine-tuning of Large Language Model (LLM)

Nilesh Dhannaseth[a], Sameer Walthare[b], Siddharth Supekar[c], Prerna Sorte[d], Kshitija Bais[e], and Sayali Bhagwatkar[f]

Department of Information Technology, St. Vincent Pallotti College of Engineering and Technology, Nagpur, India

Abstract

Fine-Tuning the Llama-2 Language Model for Enhanced Natural Language Understanding Abstract: Natural Language Processing (NLP) has witnessed significant advancements with the advent of large pre-trained language models (LLMs). Llama-2, a state-of-the-art LLM, has demonstrated remarkable capabilities in various NLP tasks. This research paper explores the fine-tuning of the Llama-2 language model to further enhance its performance in specific domains and tasks. The study begins by providing an overview of the Llama-2 architecture and its pre- training methodology. It then delves into the process of fine-tuning, outlining the steps taken to adapt the model to specific datasets and tasks. Special attention is given to avoiding overfitting and optimising hyperparameters during the fine-tuning process. To evaluate the effectiveness of the fine-tuned Llama-2 model, a series of experiments are conducted across diverse NLP benchmarks, including sentiment analysis, named entity recognition, and question answering. The results demonstrate improvements in accuracy and efficiency, showcasing the model's adaptability and versatility. Furthermore, this paper addresses the ethical considerations associated with LLMs, such as potential biases and unintended consequences. Strategies for mitigating biases are discussed, emphasising the importance of responsible AI development. In conclusion, the fine-tuning of the Llama-2 language model proves to be a valuable approach for tailoring its capabilities to specific NLP tasks. The findings of this research contribute to the ongoing discourse on optimising LLMs for improved natural language understanding while emphasising the importance of ethical considerations in AI development.

Keywords: Large Language Model, Fine-tuning, NLP, machine learning, Language Translation

1. Introduction

In recent years, the advent of Large Language Models (LLMs) has revolutionised the landscape of natural language processing, pushing the boundaries of what machines can achieve in understanding and generating human-like text. Among the forefront of these transformative models is the OpenAI GPT (Generative Pre-trained Transformer) series, exemplifying the remarkable capabilities of unsupervised pre-training on vast textual corpora. As these models continue to evolve, researchers and practitioners like Ni Xuanfan and Li Piji (2023) are increasingly turning their attention to the

[a]ndhannaseth@stvincentngp.edu.in, [b]sameerwalthare.21d@stvincentngp.edu.in, [c]siddharthsupekar.fy20@stvincentngp.edu.in, [d]prernasorte.fy20@stvincentngp.edu.in, [e]kshitijabais.fy20@stvincentngp.edu.in, [f]Sayalibhagwatkar.fy20@stvincentngp.edu.in

DOI: 10.1201/9781003567653-26

powerful technique of fine-tuning, recognising its potential to tailor these colossal language models to specific tasks, domains, or applications. This research paper delves into the intricate realm of fine-tuning LLMs, exploring the methodologies, challenges, and implications associated with this crucial process. Fine-tuning serves as a pivotal bridge between the general language understanding capabilities acquired during pre-training and the specialised requirements of diverse real-world applications. Understanding how to effectively fine-tune LLMs is paramount for harnessing their full potential across an array of tasks, including sentiment analysis, named entity recognition, language translation, and more.

2. Overview of Large Language Models

Fine-tuning of Large Language Models (LLMs) is a crucial process that tailors pre-trained models to specific tasks, domains, or applications, enhancing their performance on targeted objectives. The journey typically begins with pre-training, where the LLM is exposed to vast amounts of unlabelled data to grasp general language patterns and structures. Fine-tuning, the subsequent phase, involves training the model on a smaller, task-specific dataset with labelled examples. This allows the LLM to adapt its pre-trained knowledge to the nuances of the specific task, bridging the gap between general language understanding and task-specific requirements. Edward J Hu, et al. (2021).

3. Methodology and Model Specifications

Fine-tuning a large language model (LLM) like GPT-3.5 for financial tasks follows a systematic approach to leverage its natural language understanding capabilities effectively. The process begins with data collection, where a diverse range of financial texts is gathered, including news articles, market reports, company filings, and economic analyses. This data is then pre-processed to clean and tokenise the text handle special characters, and ensure a standardised format across the dataset Fujii et al. (2023). The selected pre-trained LLM, such as GPT-3.5, serves as the base model for fine-tuning. Task-specific layers are added on top of the pre-trained model to adapt it to financial applications. For instance, dense layers may be added for classification tasks like sentiment analysis, while regression layers might be used for numeric predictions such as stock price forecasting. Hyperparameter tuning plays a crucial role in optimising the fine-tuned model's performance. Techniques like grid search, random search, or Bayesian optimisation are employed to experiment with different hyperparameters such as learning rate, batch size, sequence length, and dropout rate. This optimisation process aims to enhance model convergence speed, prevent overfitting, and improve generalisation. Evaluation of the fine-tuned LLM is conducted using appropriate metrics based on the specific financial task. For example, accuracy, F1 score, mean squared error, or custom evaluation metrics may be used to assess model performance on a separate test set. Error analysis helps identify areas where the model struggles and potential biases in the training data, guiding further refinements. Once the fine-tuned LLM demonstrates satisfactory performance, it is deployed in a production environment. Considerations such as latency, scalability, and security are taken into account during deployment. Monitoring tools are implemented to track model performance over time, detect concept drift, and facilitate periodic retraining with new data to ensure continued relevance and accuracy in financial applications (Klein et al., 2020). Process of fineturing LLM's is explained in Figure 26.1.

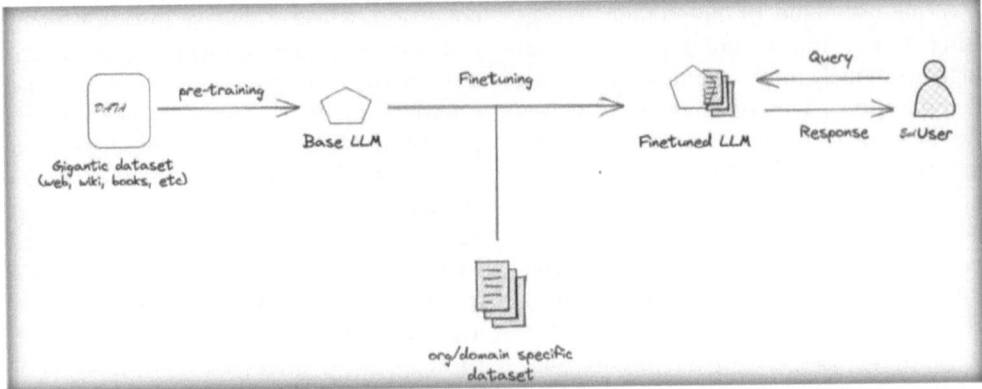

Figure 26.1: Process of finetuning LLM's.

Source: Author.

Note. Adapted from "RAG vs Finetuning — Which Is the Best Tool to Boost Your LLM Application? By Heiko Hotz (2023), Published in Towards Data Science." (https://towardsdatascience.com/rag-vs-finetuning-which-is-the-best-tool-to-boost-your-llm-application-94654b1eaba7)

4. Guidelines for Effective Use of Large Language Models

Effective deployment of Large Language Models (LLMs) involves a systematic process, beginning with task identification. Understanding the specific goal, whether text classification, sentiment analysis, question answering, or text generation, lays the groundwork. Model selection follows, where pre-trained LLMs like GPT- 3, BERT, and RoBERTa are considered based on criteria such as model design, performance, and computational requirements. Wenhao Zhu, et al. (2023). Fine-tuning is crucial, using task-specific data to optimise parameters like learning rate and batch size. The next steps include model analysis, assessing performance through metrics like recall, accuracy, precision, and F1 score, ensuring it meets intended standards before deployment. Arora et al. (2023). Implementation through an API or user interface seamlessly integrates the model into the system, emphasising compatibility with existing infrastructure. Continuous monitoring post-deployment is vital, evaluating performance regularly and retraining as needed with updated data or parameters. User feedback becomes integral for continuous improvement, shaping model updates to address evolving demands and challenges. This iterative approach ensures the model remains relevant and effective in real-world scenarios. Robust deployment guidelines encompass these stages, promoting informed decisions, performance optimisation, and sustained efficacy, contributing to the dynamic landscape of natural language processing applications.

4.1. Ethical Guidelines

The principles outlined prioritise ethical considerations in the development and application of Large Language Models (LLMs). User privacy must be preserved through practices like data minimisation and encryption. Addressing biases inherited in training data is crucial for just and equitable model outputs. Acknowledging the ethical implications, promoting transparency through explainability strategies, and ensuring accountability are paramount to mitigate risks, especially in activities like deepfakes and misinformation dissemination. Encouraging competition prevents monopolisation, requiring collaboration among academia, industry, and government to promote responsible LLM development. Chan et al. (2023). Mitigating environmental impact involves prioritising energy-efficient models and exploring alternative training techniques. The ethical consequences of optimisation in AI and LLM technologies demand careful consideration to prevent exacerbating injustices. Striking a balance between ethical and optimisation factors is essential for responsible and fair use of learning load monitors.

5. Challenges and Limitations of Large Language Models

Large Language Models (LLMs) like GPT-4, despite their potential as forerunners of Artificial General Intelligence (AGI), face significant challenges. Constraints include biases in training data, surface-level pattern dependence, and limitations in common sense and feedback interpretation. LLMs' efficacy is hindered by their high data and computational requirements, as well as issues with interpretability, generalisability, and handling uncommon vocabulary or syntax. Schramowski et al. (2022) Deployment difficulties encompass context-dependent language, emotion and sentiment assessment, multilingual capabilities, memory limitations. Schuurmans (2023), and real-time processing. Maintenance costs, causality, scalability, multimodal inputs, attention span, transfer learning, and understanding beyond textual domains present additional obstacles. Specific challenges include the need for extensive training data, issues with tokenisation introducing biases, high computing resource demands, and difficulties in fine-tuning LLMs. Inference latency, limited context length, information updates, foundation model risks, and potential biases in LLMs are also significant concerns. These models struggle with explaining decisions, reasoning errors, susceptibility to adversarial attacks, modifications in behaviour over time, and errors in spelling and counting.

Addressing these challenges requires collaborative research and development efforts, emphasising transparency, ethical governance, and responsible deployment of LLMs across diverse applications.

6. Applications of Large Language Model

Large Language Models (LLMs), such as GPT-3, have revolutionised artificial intelligence across industries. In healthcare, they analyse medical literature, aiding in information extraction and report generation. LLMs enhance customer support by accurately responding to queries and automate content creation in marketing and blogs. They assist developers in code generation and play a crucial role in translation services. In finance, LLMs analyse market data for decision-making. Education benefits from LLMs in tutoring and grading. In scientific research, LLMs expedite literature review by summarising papers and extracting insights. Overall, LLMs empower sectors with enhanced language processing for efficiency and innovation.

7. Conclusions

Large Language Models (LLMs), exemplified by Llama-2, have transformed Natural Language Processing (NLP). The OpenAI GPT series showcases their prowess in understanding and generating human-like text through unsupervised pre-training. Fine-tuning, a crucial step, tailors LLMs to specific tasks, avoiding overfitting and optimising hyperparameters. The paper evaluates Llama-2's effectiveness in diverse NLP benchmarks, highlighting its adaptability in sentiment analysis, named entity recognition, question answering, and more. The taxonomy of LLM tasks emphasises versatility in activities like language translation, text classification, summarisation, virtual assistance, information extraction, semantic search, and speech recognition. The future holds promise for improved reasoning, explainability, and integration into virtual assistants, chatbots, and speech recognition systems, revolutionising user interactions. LLMs' role in information extraction, knowledge management, and creating structured knowledge graphs is vital for accurate and efficient processes in natural language comprehension and beyond.

References

Arora, S., Yang, B., Eyuboglu, S., Narayan, A., Hojel, A., Trummer, I., and Ré, C. (2023). Language models enable simple systems for generating structured views of heterogeneous data lakes. arXiv preprint arXiv:2304.09433

Chan, A., Bradley, H., and Rajkumar, N. (2023). Reclaiming the digital commons: A public data trust for training data. arXiv preprint arXiv:2303.09001

Edward J. Hu, Yelong Shen, Phillip Wallis, Zeyuan Allen-Zhu, Yuanzhi Li, Shean Wang, Lu Wang, and Weizhu Chen. (2021). Lora: Low-rank adaptation of large language models. arXiv preprint arXiv:2106.09685

Fujii, T., Shibata, K., Yamaguchi, A., Morishita, T., and Sogawa, Y. (2023). How do different tokenizers perform on downstream tasks in scriptio continua languages?: A case study in japanese. arXiv preprint arXiv:2306.09572

Heiko Hotz (2023). RAG vs Finetuning — Which Is the Best Tool to Boost Your LLM Application? Published in Towards Data Science.

Hendy, A., Abdelrehim, M., Sharaf, A., Raunak, V., Gabr, M., Matsushita, H., Kim, Y. J., Afify, M., and Awadalla, H. H. (2023). How good are GPT models at machine translation? A comprehensive evaluation.

Klein, G., Zhang, D., Chouteau, C., Crego, J., and Senellart, J. (2020). Efficient and high-quality neural machine translation with Open-NMT. Proceedings of the Fourth Workshop on Neural Generation and Translation, Stroudsburg, PA, USA. Association for Computational Linguistics, pp. 211–217.

Ni Xuanfan, and Li Piji. (2023). A systematic evaluation of large language models for natural. Proceedings of the 22nd Chinese National Conference on Computational Linguistics (Volume 2: Frontier Forum). Harbin, China. Chinese Information Processing Society of China, pp. 40–56.

Schramowski, P., Turan, C., Andersen, N., Rothkopf, C. A., and Kersting, K. (2022). Large pre-trained language models contain human-like biases of what is right and wrong to do. NMI, 4(3), 258–268.

Schuurmans, D. (2023). Memory augmented large language models are computationally universal. arXiv preprint arXiv:2301.04589

Wenhao Zhu, Hongyi Liu, et al. (2023). Multilingual machine translation with large language models: Empirical results and analysis. arXiv preprint arXiv:2304.04675

27 Design improvement of automobile radiator using evaporative cooling system: a literature review

P. D. Kamble[a], Saurabh Dupare[b], Avadhut Shrimalwar[c], Kshitij Lonkar[d], Rushikesh Jambewar[e], V. M. Korde[f], and S. B. Sahare[g]

Department of Mechanical Engineering, Yeshwantrao Chavan College of Engineering, Nagpur, Maharashtra, India

Abstract

The evolution of automobile radiator systems, aiming to improve cooling efficiency while addressing ecological concerns. The objective is to develop an innovative radiator design that is when hot coolant come though inlet, It passes via an enormous metal plate with several rows of slender metal fins to assist dissipate the heat from the entering hot coolant, on which sprinkling of water using water sprinklers on radiators fins to lower coolants temperature using evaporating water concept, which can be helpful to increase its efficiency and effectiveness for heavy duty vehicles which can be very helpful to increase radiators life expectancy as well as the increase in the performance in the high speed vehicles whereas thermal cooling problems occurred can be resolved not only enhances heat dissipation performance but also contributes to the overall sustainability of automotive systems.

Keywords: Automobile radiator, evaporative cooling system, design, improvement

1. Introduction

An engine cooling system utilizing radiator evaporative cooling combines traditional radiator cooling with evaporative cooling techniques. In this system, radiator is still used to transfer heat from engine coolant to surrounding air, but an additional evaporative cooling element is incorporated. This additional element typically involves a reservoir of water or coolant located near the radiator. As air passes over the radiator, some of it is directed to flow over the reservoir or through channels containing the coolant. The heat from the radiator causes the water or coolant to evaporate, absorbing additional heat from the system. The evaporative cooling effect enhances the overall cooling efficiency of the system, especially in hot climates or under heavy load conditions

[a]drpdkamble@gmail.com, [b]saurabhdupare29@gmail.com, [c]avadhutshrimalwar2002@gmail.com, [d]kshitijlonnkar10@gmail.com, [e]jambewarrushikesh@gmail.com, [f]vmkorde@gmail.com, [g]mrsspkamble@gmail.com

DOI: 10.1201/9781003567653-27

where traditional radiator cooling alone may not be sufficient. However, this system requires regular maintenance to ensure an adequate supply of water or coolant, and it may not be suitable for all engine setups or environments. Figure 27.1 shows the different parts of radiator.

2. Literature Survey

The fifty literatures from renounced journals/patent are considered for making the review. Following are the findings. Sirsikar et al. (2020). This article examines the concept, design and evolution of car radiators, from Ford's use of aluminum in 1983 to its use of carbon fiber between 2000 and 2013. Mohd Zulhelmie Bin Deraman et al. (n.d.). The goal of the project described in this proposal is to monitor and prevent engine overheating through the use of an automobile heat engine management system. By offering alerts, more time to choose a secure stopping spot, and automatic emergency signs, the goal is to improve user safety. Sharma, et al. (n.d.). The application of the Nitinol alloy's shape memory property—which allows it to restore its shape at temperatures between 60 and 80 degrees Celsius—is examined in this study. In order to create a self-driven pulley using the alloy's shape recovery while submerged in hot water, the study focuses on a Nitinol Heat device that uses the alloy as a wiring loop around two wheels Marshall, et al. (1998). This study examined thermal injuries resulting in hospitalization or death among New Zealanders (aged 15 years and over) between 1978 and 1988. According to the study, there were 644 hospitalizations in 1988 and 493 deaths from thermal injuries; Keerthi, Dastagiri, et al. (n.d.). This essay discusses the safety issues raised by opulent car amenities that raise the possibility of fire mishaps. The transmission and receiver sections of the system are introduced in this project. When a fire is detected, the transmission section's smoke sensor notifies the microcontroller and zigbee transceiver Su, Wang, et al. (n.d.). The problems with heat transmission in fuel cell cars during ascents and hot weather are discussed in this study

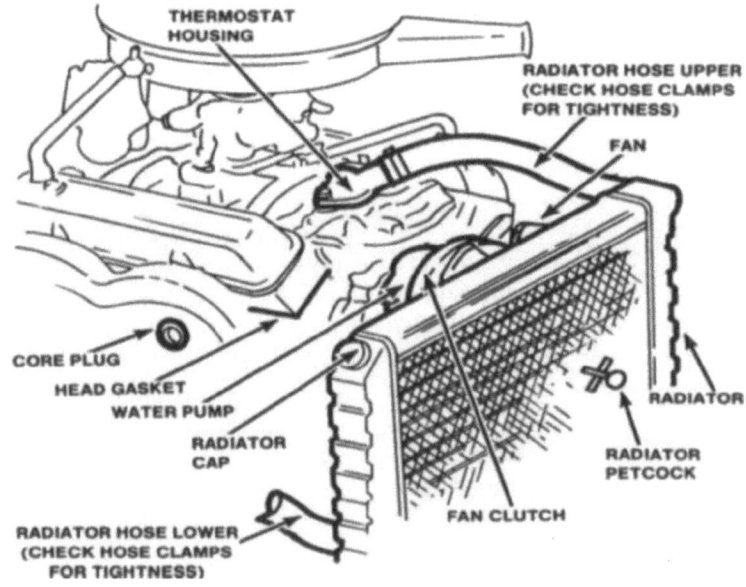

Figure 27.1: Radiator.

Source: Su et al., 2022.

report. The fuel cell's produced water is to be used in the study's suggested spray cooling technique. Mohan, Kuriakose, et al. (2022). This study goes beyond traditional methods such as fin addition to increase surface area and investigates alternate coolants to improve radiator performance. Water, water combined with ethylene glycol or propylene glycol, and, more recently, nano fluids are examples of traditional coolants. Ukueje, Abam, et al. (2017). The innovative heat transfer fluid known as hybrid nanofluids, which combine various nanoparticles to improve thermal conductivity, is the subject of this review. Farahani, et al. (2022). This study presents the findings of an investigation into a two-stage cooling system made up of a cooling coil, an indirect evaporative cooler, and a nighttime radiative unit. Chlorine-free water produced by nighttime radiative cooling is kept in storage for usage throughout the day. Saunders, et al. (2010). This essay charts development of automotive radiators, emphasizing innovations from antiquated water-pipe models to contemporary styles. It highlights the significance of bigger engines, faster speeds, and quieter operation and links inadequate water-jacket design to cooling-system problems. Farmahini-Farahani, et al. (1936). This study investigates the viability and potential of a multi-step system that combines two- stage evaporative cooling with nocturnal radiative cooling in four cities with varying climates. Water runs through radiative panels on hot evenings, transferring heat to the night sky and cooling the system. During the day, a two-stage evaporative cooler filters hot outdoor air before using stored cold water from a tank to power a cooling coil unit. Leshner, et al. (2012). This essay focuses on the advantages of higher coolant temperatures for automobile engines in terms of fuel economy. Maintaining ideal coolant temperatures, avoiding overcooling, and guaranteeing temperature homogeneity throughout the engine are the basic goals of evaporative engine cooling. France, Smith, et al. (1983) a unique hybrid radiator-cooling system that modifies the amount of cooling that the vehicle's coolant radiator can provide. By integrating traditional air-side finned surface cooling with active evaporative water cooling, the system maximizes heat transport under varying driving conditions. Active evaporative cooling is only employed in extreme situations; for most settings, the air-side finned surface is sufficient. This hybrid method allows for a smaller finned surface than standard radiators, which results in a more efficient system. Kubozuka, Ogawa, et al. (2013). It investigates a improve engine cooling system that uses radiative heat transfer via a condenser and boiling in the water jacket. Three main features are demonstrated by experimental results from a real engine and a model boiler: a more even temperature distribution across the engine body, efficient heat transfer up to $10 \cdot kW/m^2$ with minimal coolant flow, and an efficient variable temperature control system that lessens knocking and increases power. Jafari, Dunne, et al. (1987). This analysis examines evaporative cooling technologies that have been proposed for 100 years in the context of vehicle engine thermal control. The analysis's goals are to draw attention to system flaws and identify open research issues that are vital for industry adoption. Harrison, et al. (2017). The author of this research questions the term "steam cooling," suggesting that "evaporative cooling" is a better description of the technique. The author proposes a method that uses boiling water in an internal combustion engine, based on seven years of experiments. Nagarjun Vinukonda et al. (1926). The assessment highlights the significance of developing and maintaining effective engine cooling systems, with a focus on radiators, given their substantial influence on the overall performance of the engine. Fell, Brandon et al. (2007–2012). In this study, we discuss how the development of improved nanofluids with superior thermal conduction and convection capabilities presents a fresh opportunity to design a lightweight, extremely energy-efficient car radiator. Carl, Guy, Leyendecker, et al. (2012). This looks into the process of heat transfer in car radiators using both experimental and theoretical analysis. Internal fluid flow via noncircular tubes is first investigated in order to get the convective heat transfer coefficients for water. Subsequently, the external fluid flow between radiator tubes and fins is analyzed to determine the convective heat transfer coefficient for air. Trivedi, & Vasava, et al. (2012). In this paper we study challenge of insufficient heat dissipation in car radiators—a consequence of requiring larger engines in less spaces—is discussed in this study. It talks on the shortcomings of the design approaches used today, which are based on empirical evidence and trial-and- error testing. *Deshpande, et al. (2016).* To reduce size and increase heat transfer efficiency, the usage of heat pipes in automobile radiators is examined in this research. The radiator's function in controlling temperature and preventing engine overheating is emphasized. Heat pipes are suggested as a way to improve the radiator's compactness or heat transmission capacity because of their high thermal conductivity. Mallick, Das, Banik, et al. (n.d.). In order to guarantee smooth engine running and avoid overheating, the article focuses on improving engine cooling efficiency. It recognizes that considerable heat is produced when engines run and that efficient heat dissipation is crucial to preserving engine performance. Arunpandiyan, et al. (2016). The difficulty the automotive industry faces in producing engines that are both powerful and efficient for today's cars is covered in the paper. It highlights how crucial the cooling system is to improving heat transfer, fuel efficiency, and engine performance. Jafari, DunneF., Langari, et al. (2017). The study looks at evaporative cooling system innovations for engine temperature control in automotive applications over the last 100 years. Its goal is to pinpoint research topics and system weaknesses that

automakers might use. It begins by restating the advantages of evaporative cooling systems, including decreased emissions and increased engine efficiency. Rahman, Mahmud, et al. (2023). In this paper, A quick overview of the engine cooling system's essential role is given in preserving the ideal operating temperature of contemporary internal combustion engines is highlighted. The significance of avoiding overheating is emphasized in order to guarantee effective engine running, optimal fuel usage, and prolonged engine life. Pang, Kalam, et al. (Pang et al., 2012). The importance of the engine cooling system in preserving ideal engine operating temperatures is covered in the article. Yadav, & Singh et al. (Yadav & Singh, 2011). This research presents a comprehensive numerical parametric study of car radiators using the thermal resistance concept and the finite difference approach. Radiator performance is evaluated by adjusting test setup factors such as coolant mass flow rate and input coolant temperature. Furthermore, a comparison of water and a water-propylene glycol mixture (40:60 ratio) is carried out as coolants. According to the study, Although dissolved salts and water's corrosive nature limit its long-term usefulness, water is still the finest coolant. Ismael, Yun, et al. (2020). In order to evaluate heat dissipation, the article performs parametric experiments on automobile radiators. Maheswari, Reddy et al. (2020). Given the scarcity of water these days, one efficient method for thermal power plants to conserve it is the dry cooling system. Pang, Kalam, et al. (2012) apparatus, this paper presents an experimental investigation of the heat transfer capabilities of a tube-and-fin radiator for automobiles. Figure 27.2 shows the assembly of radiator.

3. Research Gap

The research gap between traditional radiators and radiator evaporative cooling technology lies in the need for comprehensive

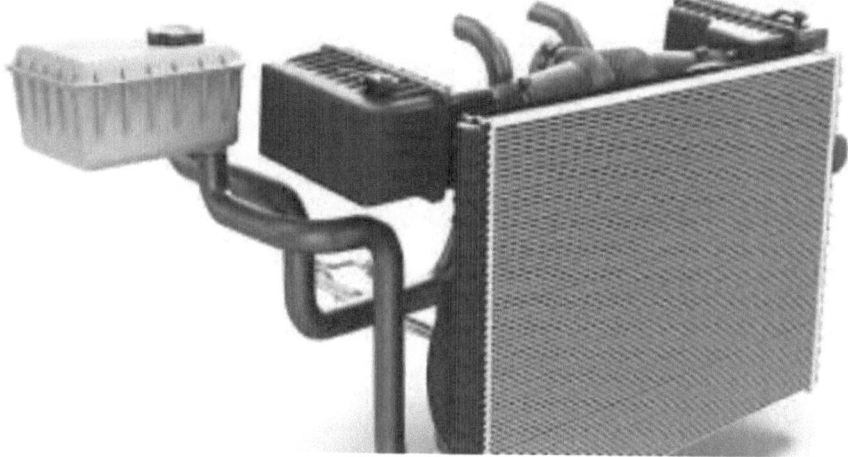

Figure 27.2: Radiator assembly.

Source: Farmahini-Farahani & Heidarinejad, 2012.

studies evaluating the effectiveness, energy efficiency, and practicality of the latter. Further research should explore real-world applications, potential limitations, and the environmental impact of adopting evaporative cooling in automotive and industrial radiator systems. Additionally, a comparative analysis of performance metrics, such as cooling efficiency and maintenance requirements, would contribute to filling this research gap. In the design of radiators using evaporative cooling technology lies in optimizing the integration of this technology to enhance heat dissipation efficiency while addressing potential challenges such as water consumption, system durability, and overall cost- effectiveness. Further investigation is needed to develop innovative materials, advanced control mechanisms, and sustainable practices for an efficient and environmentally friendly evaporative cooling radiator system.

4. Future Scope

The future of projects focused on improving automotive radiator design using evaporator cooling technology has the potential for growth in automotive thermal management.

Here are some statements about the future of the project:

1. Increase Efficiency and Performance Learn how to improve your evaporator cooling system to improve heat and make electricity more powerful. Discover new materials and designs that improve the overall performance of the engine, help transfer heat well and reduce engine temperature.
2. Compact and lightweight design investigating ways to reduce the overall size and weight of air conditioning systems can help improve fuel efficiency and vehicle performance. Explore state- of-the-art manufacturing techniques such as additive manufacturing to create a process and create a lightweight thermal design that maintains process integrity and thermal efficiency.

5. Methodology

5.1. Define Objectives

The objectives that your literature review aims to address. For example, you might want to understand the existing technologies, problems, future scope and advancements related to automobile radiator cooling system.

5.2. Search Strategy

Develop a systematic search strategy to identify relevant literature review. Utilize academic databases, journals, conference proceedings, and reputable online repositories. Keywords might include "Cooling technology," "Radiator enhancement," "Sprinkling of water," and "Temperature drop".

5.3. Selection Criteria

Define inclusion and exclusion criteria for the literature to ensure that the selected studies align with the scope of your review. Criteria might include publication date, relevance, and the quality of the research.

5.4. Literature Search

Conduct a comprehensive literature search using academic databases like IEEE Xplore, and Google Scholar. Record and organize the search results for later analysis.

5.5. Review and Selection

Screen the identified literature based on the selection criteria. Initially, review titles and abstracts to narrow down the relevant studies. Obtain and read full texts for potentially eligible papers.

5.6. Data Procurement

Systematically extraction of relevant information from each selected study. This may include key findings, methodologies employed, technologies discussed, Challenges identified, And any theoretical frameworks used.

5.7. Identification of Research Gaps and Future Scope

Identify gaps in the existing literature. What aspects of radiator cooling technology

methods other than the standard one? Use this information to suggest potential areas for future research.

References

Arunpandiyan, D. (2016). A Review of Automotive Radiator Performance. IJIRST – International Journal for Innovative Research in Science & Technologyl, 2(10).

Carl, M., Guy, D., Leyendecker, B., Miller, A., & Fan, X. (2012). The theoretical and experimental investigation of the heat transfer process of an automobile radiator. In ASEE Gulf Southwest Annual Conference, 1(128), 1–12.

Deshpande, A., Patil, V., & Patil, R. (2016). Theoretical design of radiator using heat pipes. Int J Eng Res Technol, 5(11).

Farahani, M. F., Heidarinejad, G., & Delfani, S. (2010). A two-stage system of nocturnal radiative and indirect evaporative cooling for conditions in Tehran. Energy and Buildings, 42(11), 2131–2138.

Farmahini-Farahani, M., & Heidarinejad, G. (2012). Increasing effectiveness of evaporative cooling by pre-cooling using nocturnally stored water. Applied Thermal Engineering, 38, 117–123

Fell, Brandon; Janowiak, Scott; Kazanis, Alexander; Martinez, Jeffrey 2007–12.

France, D., Smith, D., & Yu, W. (2013). Efficient, Active Radiator-Cooling System. SAE International Journal of Commercial Vehicles, 6(2013-01-9017), 249–256

Harrison, H. C. (1926). Evaporative cooling. SAE Transactions, 314–337.

Jafari, S., Dunne, J. F., Langari, M., Yang, Z., Pirault, J. P., Long, C. A., & Thalackottore Jose, J. (2017). A review of evaporative cooling system concepts for engine thermal management in motor vehicles. Proceedings of the Institution of Mechanical Engineers, Part D: Journal of Automobile Engineering, 231(8), 1126–1143.

Jafari, S., Dunne, J. F., Langari, M., Yang, Z., Pirault, J. P., Long, C. A., & Thalackottore Jose, J. (2017). A review of evaporative cooling system concepts for engine thermal management in motor vehicles. Proceedings of the Institution of Mechanical Engineers, Part D: Journal of Automobile Engineering, 231(8), 1126–1143.

Keerthi, N., Dastagiri, M., & Deepthi, N. International Journal of Engineering Sciences & Research.

Kubozuka, T., Ogawa, N., Hirano, Y., & Hayashi, Y. (1987). The development of engine evaporative cooling system. SAE transactions, 96–106.

Leshner, M. D. (1983). Evaporative Engine Cooling for Fuel Economy, 1983.

Maheswari, C., REDDY, R. M., & Meenakshi, R. (2020). Enhancement of the heat transfer of the dry cooling system using a rotating radiator. Int. J. Mech. And Production Engineering Research and Development, 10(3), 669–678.

Maheswari, C., REDDY, R. M., & Meenakshi, R. (2020). Enhancement of the heat transfer of the dry cooling system using a rotating radiator. Int. J. Mech. And Production Engineering. Research and Development, 10(3), 669–678.

Mallick, B., Das, R. K., Banik, S. C., Noor, M. F. B., & Habib, A. Performance Enhancement of an Automobile Radiator by Using a Nozzle Arrangement.

Mohan, S. V., Kuriakose, J., Nidhin, A. R., Varghese, B., & Fidus, F. (2017). Experimental Comparison of Heat Transfer in a Down Flow Louvered Fin Auto Radiator using Water, Water and Ethylene Glycol Mixture & Water, Ethylene Glycol and Sugarcane Juice Mixture as Coolants. Glob. Res. Dev. J. Eng, 2(5), 215–223.

Mohd zulhelmie bin deraman, m. Z. (2011). Automotive engine iieat management system shms-smart heat engine management.

Pang, S. C., Kalam, M. A., Masjuki, H. H., & Hazrat, M. A. (2012). A review on air flow and coolant flow circuit in vehicles' cooling system. International Journal of Heat and Mass Transfer, 55(23–24), 6295–6306.

Pang, S. C., Kalam, M. A., Masjuki, H. H., & Hazrat, M. A. (2012). A review on air flow and coolant flow circuit in vehicles' cooling system. International Journal of Heat and Mass Transfer, 55(23–24), 6295–6306.

Rahman, M. R., Mahmud, M. R., Hasan, M.L., Shamim, M., & Ali, M. S. (2023). Engine Water Cooling System of Automobile Engineering (Doctoral dissertation, Sonargoan University (SU)).

Saunders, L. P. (1936). Radiator development and car cooling. SAE Transactions, 496–516.

Sharma, S., Mattoo, A., Singh, A. P., & Mishra, R. Applications of Nitinol Propulsion Device.

Sirsikar, S., Mehta, V., Chafekar, K., Sonawane, G., & Dandekar, Y. (2020). Review paper of engine cooling system. Int. Res. J. Eng. Technol, 7, 4538- 4541.

Su, C. Q., Wang, S., Liu, X., Tao, Q., & Wang, Y. P. (2022). Experimental and numerical investigation on spray cooling of radiator in fuel cell vehicle. Energy Reports, 8, 1283–1294.

Trivedi, P. K., & Vasava, N. B. (2012). Effect of variation in pitch of tube on heat transfer rate in automobile radiator by CFD analysis. International Journal of Engineering and Advanced Technology, 1(6), 180–3.

Technology Minimization of the Fire Accidents in Automobiles by Automatic Spraying of Fire Retards and Utilization of a Fire Fighting Robot.

Ukueje, W. E., Abam, F. I., & Obi, A. (2022). A perspective review on thermal conductivity of hybrid nanofluids and their application in automobile radiator cooling. Journal of Nanotechnology, 2022.

Waller, A. E., Marshall, S. W., & Langley, J. D. (1998). Adult thermal injuries in New Zealand resulting in death and hospitalization. Burns, 24(3), 245–251.

Yadav, J. P., & Singh, B. R. (2011). Study on performance evaluation of automotive radiator. SAMRIDDHI: A Journal of Physical Sciences, Engineering and Technology, 2(02), 47–56.

28 Harvesting sustainability: the solar-powered drip irrigation solution for agriculture

Shiv Sankar Das[a], Ronismita Mishra[b], and Rajani Agrawalla[c]

School of Management, Centurion University of Technology and Management, Bhubaneswar, Odisha, India

Abstract

As global agriculture faces challenges related to climate change, water scarcity, and sustainable resource management, there is a growing need for innovative and eco-friendly irrigation solutions. The present study emphasizes on a solar-operated drip irrigation system as a sustainable and efficient approach to address these challenges. The system integrates solar energy, advanced sensor technology, and drip irrigation techniques to optimize water usage in agricultural practices. The solar-operated drip irrigation system offers several advantages like it relies on clean and renewable solar energy, resulting in lower operating costs and reduced reliance on non-renewable resources. The solar-operated drip irrigation system presents a promising solution for sustainable and resource-efficient agriculture. By harnessing solar energy and incorporating intelligent irrigation practices, this system contributes to environmental conservation, cost savings, and improved agricultural productivity. The study aims to establish a framework and is applied in the field as a descriptive case for examining the usability of the system by rural farmers.

Keywords: Solar drip irrigation system, sustainable development, agricultural productivity

1. Introduction

In today's ever-evolving world, sustainability is becoming a top priority, especially when it comes to agriculture. Farmers worldwide are seeking innovative solutions to conserve water and minimize their environmental impact. A solar-powered drip irrigation, a game-changing solution that combines renewable energy and precise water delivery to maximize crop yields while minimizing resource usage. Harnessing the power of the sun, solar panels generate energy to pump water through a drip irrigation system. This system utilizes strategically placed drippers to deliver controlled amounts of water directly to the plant roots, ensuring optimal hydration and minimizing wastage. By minimizing evaporation and runoff, this eco-friendly solution not only saves water but also reduces the need for chemical fertilizers, ultimately enriching the soil and enhancing crop quality (Energy, 2023).

[a]shivsankar.das@cutm.ac.in, [b]ronismita.mishra@cutm.ac.in, [c]rajani.agrawalla@cutm.ac.in

DOI: 10.1201/9781003567653-28

2. The Importance of Sustainable Agriculture

Sustainable agriculture is crucial for ensuring the long-term viability of our food production systems. As the world's population continues to grow, so does the demand for food. However, traditional farming practices often place a heavy burden on our natural resources, leading to water scarcity, soil degradation, and environmental pollution. To address these challenges, sustainable agriculture focuses on minimizing the negative impact of farming while maximizing productivity and efficiency (Vr, 2023). By adopting sustainable practices, farmers can reduce their reliance on non-renewable resources, conserve water, and protect biodiversity.

2.1. Understanding Drip Irrigation and its Benefits

Drip irrigation is a widely recognized irrigation method that delivers water directly to the plant roots through a series of tubes and emitters. Drip irrigation provides controlled and targeted hydration, minimizing water waste and maximizing efficiency. The key components of a drip irrigation system include a water source, a filtration system, distribution pipes, and emitters. Water is pumped from the source, filtered to remove impurities, and then distributed through the pipes. Emitters, also known as drippers, are strategically placed near the plant roots to deliver water in controlled amounts. This ensures that each plant receives the right amount of water, reducing water stress and promoting healthy growth. Drip irrigation allows for precise water delivery, ensuring that each plant receives the necessary amount of hydration. This promotes optimal plant growth and minimizes water stress, leading to higher crop yields and improved quality. It helps conserve water by minimizing evaporation and runoff (Ashraf & Jamil, 2022). It reduces the need for chemical fertilizers (Rosa, 2022). Lastly, it promotes soil health and reduces erosion (Admin, 2023).

3. Solar-Powered Drip Irrigation: An Innovative Solution

Solar-powered drip irrigation combines the benefits of both renewable energy and precision irrigation. By harnessing the power of the sun, solar panels generate electricity to power the irrigation system, eliminating the need for grid electricity or diesel generators. This makes it a cost-effective and sustainable solution for farmers, particularly in remote areas with limited access to electricity. The solar-powered drip irrigation system consists of solar panels, a pump, a controller, and a drip irrigation network. The solar panels convert sunlight into electricity, which is used to power the pump. The pump draws water from a water source, such as a well or a reservoir, and delivers it to the crops through the drip irrigation network (Grant et al., 2022). Solar drip irrigation system reduces dependency on fossil fuels, minimizing greenhouse gas emissions and contributing to a cleaner environment. It saves on electricity costs, making it more affordable for farmers in the long run. It provides a reliable and consistent water supply, even in areas with unreliable grid electricity or limited access to water sources.

3.1. Framework for Adoption of Solar Drip Irrigation System in Rural Areas of India

From Figure 28.1 various stakeholders are identified for adoption of solar drip irrigation system in rural areas. Stakeholders include government, NGO, research institutions, financial institutions, technology providers and farmers.

Here farmers act as end-users and are the primary beneficiaries of the solar drip irrigation system. They are the essential stakeholders who directly interacts with and benefit from the implemented

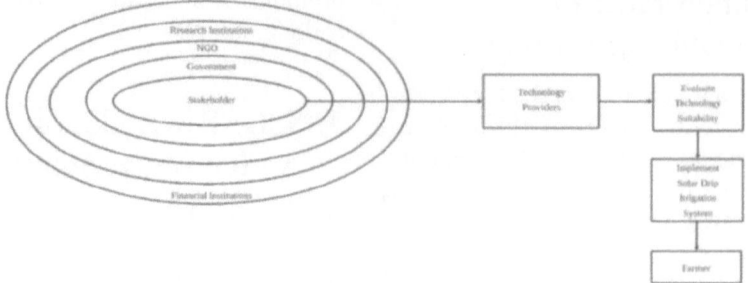

Figure 28.1: Framework for adoption of solar drip irrigation system in rural areas of India.

Source: Author.

irrigation system. Their needs, preferences, and feedback are crucial for the success of the adoption process. Government act as a regulatory authority and provider of support and incentives. Government involvement includes policy-making, providing financial incentives, and ensuring a supportive regulatory environment. Non-Governmental Organizations (NGOs) support and advocacy for sustainable agriculture. They contribute by providing expertise, support, and advocacy for sustainable agricultural practices. They need to assist in conducting awareness campaigns, training programs, and connecting farmers with relevant resources. Financial Institutions are the providers of financial resources and support. Financial institutions, such as banks or funding agencies, business houses play a critical role in providing financial resources for the implementation of solar drip irrigation systems. They will offer loans, grants, or financial assistance to farmers and other stakeholders. Research institutions are the providers of research, innovation, and expertise. Research institutions contribute by providing relevant research, innovation, and expertise in agricultural practices. They will conduct studies on the effectiveness of solar drip irrigation and offer insights to improve the adoption process. Technology Providers are the suppliers of solar drip irrigation technology. Technology providers are essential stakeholders who will supply the necessary equipment and technology for solar drip irrigation. They will offer guidance on system selection, installation, and maintenance. Each stakeholder has a unique role and contribution to the overall ecosystem, emphasizing the collaborative effort required for the successful adoption of solar drip irrigation in a rural context. The flowchart visualizes the relationships and interactions among these stakeholders and the key processes involved in the adoption journey.

4. Case Studies: Success Stories of Solar-Powered Drip Irrigation

4.1. Case Study: Mr. Dilip Majhi's Success with Solar-Powered Drip Irrigation

One success story worth mentioning is Dilip Majhi, who implemented a solar-powered drip irrigation system on their vegetable farm in a village in South Odisha, India. Before adopting this technology, he struggled with inconsistent crop yields due to water scarcity in their region. However, by harnessing the power of the sun and carefully delivering water to each plant, he saw remarkable results. With solar-powered drip irrigation, he was able to increase their vegetable yield by 50%. The controlled application of water ensured that each plant received the necessary amount, resulting in healthier crops and higher market value. The reduced water usage also allowed him to sustain their farm even during periods of drought, securing their income and livelihood. Furthermore, it had significantly reduced their carbon footprint, contributing to a greener and more sustainable future. The success of Dilip's farm served as an inspiration to neighbouring farmers, who also started adopting this technology.

4.2. *Case Study: Ms. Pranati Barik's Success with Solar-Powered Drip Irrigation*

Ms. Pranati Barik's success story showcases the potential of solar-powered drip irrigation in transforming arid lands into productive farms, located in a region in South Odisha, India with limited access to water, Pranati used the drip irrigation system to ensure that each plant received the necessary water, allowing her to grow high-value crops in a previously barren land. The increased crop yields significantly improved Pranati's income and provided opportunities for economic growth in the community. Furthermore, solar-powered drip irrigation allowed her to optimize their farm's resource usage. The precise delivery of water reduced the need for fertilizers, as nutrients could be directly applied to the roots. This not only reduced costs but also minimized the risk of water pollution and soil degradation. Her success story inspired other farmers in the region to adopt solar-powered drip irrigation, leading to a positive transformation of the local agricultural landscape.

4.3. *Challenges and Limitations of Solar-Powered Drip Irrigation*

While solar-powered drip irrigation offers significant advantages, it is not without its challenges and limitations. Some of the notable ones include, the initial cost of installing a solar-powered drip irrigation system can be a barrier for some farmers, particularly those with limited financial resources. However, long-term cost savings and increased crop yields often outweigh the upfront investment. Proper installation and maintenance of solar-powered drip irrigation systems require technical expertise. Farmers may need assistance from trained professionals or organizations to ensure the system operates efficiently. Solar-powered systems are dependent on sunlight availability. During cloudy or rainy periods, the efficiency of the system may be affected, leading to reduced water and energy savings. Scaling up solar-powered drip irrigation systems to larger agricultural operations can present logistical challenges. Adequate planning and coordination are required to ensure the system can meet the water demands of a larger area. While these challenges exist, it can be addressed through proper planning, training, and support from agricultural extension services and organizations. The cost of implementing a solar-powered drip irrigation system varies depending on several factors, including the size of the farming operation, the chosen technology, and the local market conditions. The initial investment includes the cost of solar panels, pumps, controllers, drip irrigation components, installation, and training. The cost of solar panels has been decreasing in recent years, making them more affordable for farmers.

5. Conclusion

Solar-powered drip irrigation is revolutionizing the agricultural landscape by offering an innovative and sustainable solution for water-efficient and energy-efficient irrigation. By harnessing the power of the sun, this technology enables farmers to conserve water, reduce their carbon footprint, and enhance crop yields. The pressing need for sustainable agriculture, coupled with the increasing demand for food, calls for the adoption of innovative technologies like solar-powered drip irrigation. By combining renewable energy with precision irrigation, farmers can meet the challenges of water scarcity, climate change, and resource conservation. Solar-powered drip irrigation offers a promising path towards a greener and more efficient future. By investing in this technology, farmers can contribute to a sustainable food system, protect natural resources, and secure a prosperous future for generations to come.

References

Admin. (2023). Revolutionizing Agricultural Irrigation with Solar-Powered Water Pumps. Harvest, Solar Pump, Harvest Solar Pump, Harvest Solar Water Pump. https://www.harvestpump.com/revolutionizing-agricultural-irrigation-with-solar-powered-water-pumps/

Ashraf, A., and Jamil, K. (2022). Solar-powered irrigation system as a nature-based solution for sustaining agricultural water management in the Upper Indus Basin. Nature-Based Solutions, 2, 100026. doi: 10.1016/j.nbsj.2022.100026

Energy, E. C. (2023). Powering the Fields solar energy role in modern agriculture. Energy. https://energy5.com/powering-the-fields-solar-energy-role-in-modern-agriculture

Grant, F., Sheline, C., Sokol, J., Amrose, S., Brownell, E., Nangia, V., and Winter, A. G. (2022). Creating a solar-powered drip irrigation optimal performance model (SDrOP) to lower the cost of drip irrigation systems for smallholder farmers. Applied Energy, 323, 119563. doi: 10.1016/j.apenergy.2022.119563

Rosa, L. (2022). Adapting agriculture to climate change via sustainable irrigation: biophysical potentials and feedbacks. Environmental Research Letters, 17(6), 063008. doi: 10.1088/1748-9326/ac7408

Vr, A. (2023). Harvest the benefits of solar energy in agriculture. Republic of Solar. https://arka360.com/ros/solar-energy-in-agriculture-and-irrigation/

29 Breast cancer detection in its early stages with spider monkey optimization using the MIAS dataset

Kanchan Warkar[1,a], Sandhya Dhage[1,b], Ashish Dandekar[2,c], and Bhakti Thakre[3,d]

[1]Department of Computer Engineering, SVPCET, Nagpur, India
[2]Department of Data Science, SVPCET, Nagpur, India
[3]Department of Computer Science and Engineering (Cyber Security), SVPCET, Nagpur, India

Abstract

Millions of women around the world have a higher chance of survival thanks to early and accurate breast cancer diagnosis. A key component of detection is computer-aided diagnosis, which relieves medical experts of the load of precise detection. A number of machine learning and deep learning algorithms were used in the computer-aided diagnosis of breast cancer. This research suggests a feature extraction and optimization approach for the identification of breast cancer. The suggested methodology used stationary wavelet transform techniques to extract features. An optimization algorithm called spider-monkey was used to extract the features. A dynamic population-based metaheuristic function called "spider-monkey optimization" can improve the feature optimization of medical pictures. We suggested a neural network (NN)-based classifier encoder deep learning for the identification of breast cancer. Breast cancer diagnosis accuracy is increased by the suggested deep learning-based classifier based on a neural network encoder. MATLAB 2018R software was used to simulate the suggested technique. The MIAS dataset was used by the algorithm for the valuation. The effectiveness of the suggested method in comparison to other breast cancer algorithms, including CNN, Ransom-RF, SVM, and NN.

Keywords: Breast cancer, optimization, SMO, SWT, NN, MIAS

1. Introduction

For the society of medical and biomedical engineering, the survival rate of breast cancer patients represents a major problem. The lives of breast cancer patients are saved by early identification of the disease. To extract features from cancer images, utilize the transform function. The textural trait is the most noticeable and potent characteristic of breast cancer. For the processing of images related to breast cancer, the wavelet

[a]kwarkar@stvincentngp.edu.in, [b]sdhage@stvincentngp.edu.in, [c]adandekar@stvincentngp.edu.in, [d]bthakre@stvincentngp.edu.in

DOI: 10.1201/9781003567653-29

transform function offers significant advantages over the feature extractor. Here we employed SMO; for a long time, the development of optimization algorithms has benefited from the study of social animals' foraging behaviour (Kwegyir et al., 2021). The global optimization algorithm Spider Monkey Optimization (SMO) was modeled after the Fission-Fusion Social (FFS) structure of spider monkeys during their foraging behavior. SMO provides a great illustration of two fundamental concepts in swarm intelligence: division of labor and self-organization. Swarm Message Organizer (SMO) is a swarm intelligence-based algorithm that has gained popularity recently and is now used to resolve a lot of engineering optimization problems. The Mammographic Image Analysis Society, or MIAS. The original MIAS (Mammographic Image Analysis Society) dataset has been preprocessed to create this dataset. 1,679 photos with the following designations are included in it: normal (0), benign (1), and malignant(2).

2. Related Work

This section examines current methods for detecting breast cancer through the use of deep learning and machine learning algorithms. Here is a literature survey on the earlier detection of breast cancer presented.

Authors, "Narayanan et al. (2022) proposed that the development of a hybrid deep learning-based assist system for breast cancer detection and classification from mammogram images could potentially lead to significant improvements in breast cancer detection accuracy." (Narayanan et al., 2022)

Authors "Kwegyir, Daniel, Emmanuel Asuming Frimpong (2021) suggest that the integration of local leader phase Spider Monkey Optimization in neural network training could lead to enhanced convergence rates and improved accuracy in feedforward neural network training."

Authors "Chouhan, Naveed, Asifullah Khan, and Jehan Zeb Shah (2021) proposed that the utilization of a deep convolutional neural network combined with emotional learning techniques could potentially advance breast cancer detection accuracy, particularly when applied to preprocessing digital mammography images."

Authors "Ahmed and Muhammad Shahzad Younis suggest that the implementation of transfer learning and wavelet transform techniques holds promise for improving breast cancer detection accuracy, especially through the preprocessing of mammography images using wavelet transform."(2021)

Authors Yala et al. (2019) proposed that the application of artificial intelligence to mammography offers significant potential for improving breast cancer detection. Their research focused on developing an AI

Type of images	Sample image 1	Sample image 2
Benign	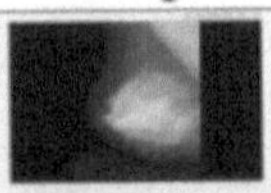	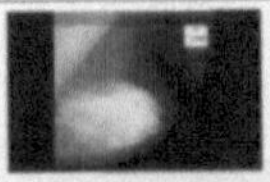
Normal	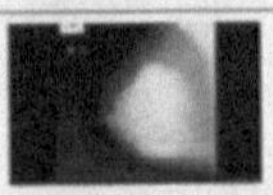	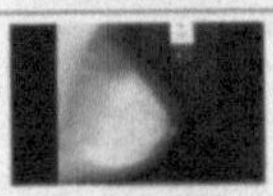
Malignant	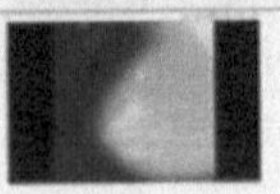	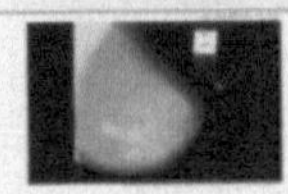

Figure 29.1: MIAS Mammogram samples.

Source: Reference [9].

algorithm with high sensitivity and specificity for early breast cancer detection using mammograms. By leveraging advanced machine learning techniques, such as deep learning and convolutional neural networks, their proposed approach aims to enhance diagnostic accuracy and enable earlier detection of breast cancer, ultimately leading to better patient outcomes.

3. Proposed Methodology

Three sections examine the suggested methods for detecting breast cancer. The process of feature extraction is explained in the first part. The procedure of feature optimization is examined in the second stage. A deep learning-based detection technique using neural network encoders is examined in the third section.

3.1. Section I

In a number of studies, wavelet transform has been utilized to help radiologists comprehend picture scans. It improves the image scan's characteristics, which help neural networks learn and categorize images more successfully. Wavelets are mostly utilized in medical imaging as a pre-processing method for extracting features from images. Wavelet transforms are useful for highlighting key features in breast scan images, such as the boundaries of tumors. SWT applied recursive dilated filters in spite of sampling. The filter dilated by adding is the transform's scaling factor. The bandwidth from one level to another is decreased during the transformation process.

$$F1_k^{(j)} = \begin{cases} \frac{F1_k}{2^j}, & k = 2^j m \, if \, m \epsilon Z \\ 0 & else \end{cases} \quad (1)$$

$$F2_k^{(j)} = \begin{cases} \frac{F2_k}{2^j}, & k = 2^j m \, if \, m \epsilon Z \\ 0 \, else \end{cases} \quad (2)$$

3.2. Section II

Spider monkey optimization (SMO) is a relatively new optimization strategy among swarm intelligence-based algorithms.

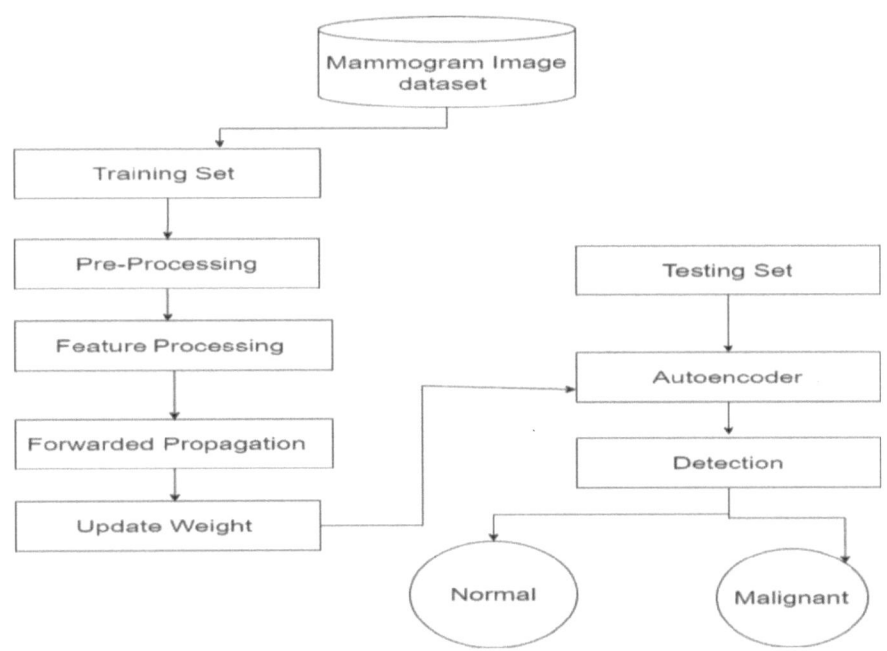

Figure 29.2: Proposed methodology.

Source: Author.

The update equations are derived using Euclidean distances between potential solutions.

The optimization procedure of the spider monkey algorithm consists of the following six phases:

1. The local leader phase, in which group members and the local leader exchange information, causing subgroup members to reposition themselves.
2. Phase of the global leader: Members of every subgroup are rearranging their positions in response to information from the global leader and the local leader of the subgroup.
3. Learning phase of the global leader: To choose the spider monkey closest to a food supply, the global leader used a greedy selection technique on all members, switching positions. Subsequently, the spider monkey assumed the role of the world leader.
4. Local leader learning happens in the subterritory and is comparable to phase 3. organization's
5. Local leader decision phase: In order to avoid stagnation, all members of the subgroup will update their positions in response to information from the global leader and the local leader, provided that the local leader modifies her stance within a set amount of time.
6. Global leader choice phase: in addition, in the event that the global leader changes her mind within a set amount of time, the group is divided into two subgroups in the first iteration, three subgroups in the second, and so forth, up to the maximum number of groups that are allowed. The global leader merges all of the sub-groups into one group when the spider monkey swarm has been split up into the maximum number of groups and she has been transferred. (Selvakumar & Aravindan, 2019)

3.3. Section III

After the process of feature optimization, process employed neural network encoder deep learning-based classification approach for the detection of being or malignant cells in women breast image datasets (Frazer et al., 2021).

4. Experimental Analysis

To evaluate the performance of the proposed algorithm for breast cancer detection and the existing algorithm for breast cancer detection, use MATLAB tools with version 2018R.

$$Accuracy = \frac{TP + TN}{TP + TN + FP + FN} \times 100$$

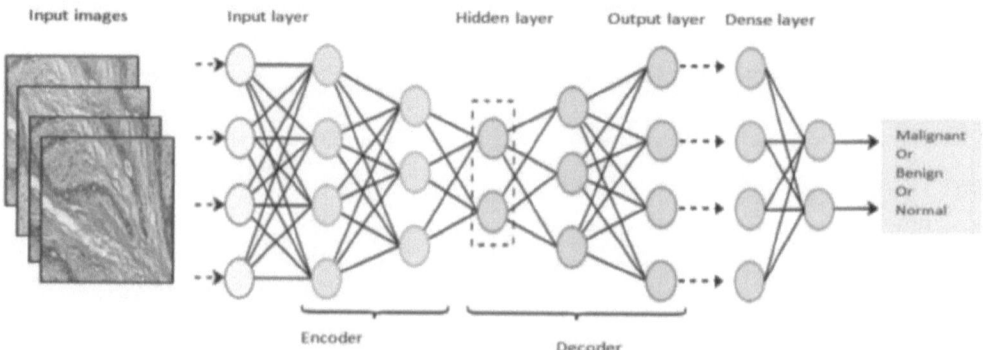

Figure 29.3: Detection using neural network encoder deep learning.

Source: Reference [18].

$$Sensitivity = \frac{TP}{TP + FN} \times 100$$

$$Specificity = \frac{TN}{TN + FP} \times 100$$

$$F1 = 2 \times \frac{Pricision \times Recall}{Pricision + Recall} \times 100$$

TP: True Positive, TN: True Negative, FP: False Positive, FN: False Negative

5. Conclusion

In order to diagnose breast cancer, we used our main technique, which is based on deep learning neural network encoding. The suggested approach combines feature optimization with feature extraction. A stationary wavelet transform was used in the feature extraction procedure; this transform gets beyond the drawbacks of conventional transform functions. A spider-monkey optimization technique was used in the feature optimization procedure. The optimization algorithm known as spider-monkey lowers the noise content. Enhancing the precision of identification is the cancer MIAS dataset's best feature. The suggested method is compared to other breast cancer algorithms that are currently in use, including Random-RF, SVM, CNN, and NN. According to our research, proposed NNs can outperform CNNs on small datasets, despite the slight variation in performance. The suggested model achieved 99.2% classification accuracy using NN, and 98.5% classification accuracy using CNN.

Table 29.1: Performance analysis of accuracy using method proposed method, random-RF, SVM, NN, and CNN

Perf. Analysis	No. of Attributes				
	5	*10*	*15*	*20*	*25*
CNN	98.9	97	96.9	96	97
NN	94.9	97.8	95.7	96	95.9
SVM	97.8	97	97	97.7	97
Random-RF	97.8	97	97.8	97.7	98
Proposed	98.9	98.8	99.5	96.4	100
	Accuracy				

Source: Author.

Table 29.2: Performance analysis of sensitivity using method proposed method, random-RF, SVM, NN, and CNN

Perf. Analysis	No. of Attributes										
	20	22	24	26	28	30	32	34	36	38	40
CNN	96.5	97	97.5	98	97.2	97	96.8	96.2	96.1	97	98
NN	97	96.4	95.6	95.5	95.3	95	95.3	95.5	95.7	96.1	96.5
SVM	95.5	96	96.3	96.5	97	97.1	96.9	96.6	96.5	96	95.5
Random-RF	97	96.7	96.8	96.4	96	95.5	96.2	96.5	97.3	97.1	96.1
Proposed	97.9	98.3	98.4	98.4	98.3	98.4	98.6	98.9	98.8	98.7	98.6
	Sensitivity										

Source: Author.

References

Chouhan, N., and Khan, A., Shah, J. Z., and Hussnain M. (2021). Deep convolutional neural network and emotional learning based breast cancer detection using digital mammography. Computers in Biology and Medicine, 132 104318.

Frazer, H. M. L., Qin, A. K., Hong Pan, and Brotchie, P. (2021). Evaluation of deep learning-based artificial intelligence techniques for breast cancer detection on mammograms: Results from a retrospective study using a breast screen Victoria dataset. Journal of Medical Imaging and Radiation Oncology, 65(5): 529–537.

Kwegyir, D., Frimpong, E. A., and Opoku, D. (2021). Optimization of feedforward neural network training using modified local leader phase spider monkey optimization.

Narayanan, K. L., Santhana Krishnan, R., and Harold Robinson, Y. (2022). A hybrid deep learning based assist system for detection and classification of breast cancer from mammogram images. International Arab Journal of Information Technology 19(6).

Rasheed, A., Younis, M. S., Qadir, J., and Bilal, M. (2021). Use of transfer learning and wavelet transform for breast cancer detection. arXiv preprint arXiv:2103.03602

Selvakumar, G., and Aravindan, P. (2019). Spider monkey optimization algorithm for feature selection and parameters optimization in bioinformatics problems. Proceedings of the 2019 3rd International Conference on Computer Science and Artificial Intelligence (CSAI), Sanya, China, pp. 259–263. doi: 10.1145/3368567.3368593

Yala, A., Lehman, C., Schuster, T., Portnoi, T., and Barzilay, R. (2019). A deep learning mammography-based model for improved breast cancer risk prediction. Radiology, 292(1), 1–270.

30 Navigating security measures in "WhatsApp Banking"

Priti Golar[a], Prajwal Behalpande[b], Sahil Dohate[c], Vikas Yadav[d], Rupal Bethriya[e], Shruti Damdu[f], and Janhavi Belsare[g]

Department of Information Technology, St. Vincent Pallotti College of Engineering and Technology, Nagpur, India

Abstract

This research explores the integration of WhatsApp into banking, focusing on information availability, trust, and perceived usefulness. Utilizing advanced tools like PLS-SEM, it analyzes 400 active banking customers, revealing intricate relationships and positioning WhatsApp as a transformative catalyst in customer-bank interactions. The paper serves as a guide, highlighting the synergy between WhatsApp and banking, contributing to discussions on effective customer engagement and the transformative potential of modern communication platforms in finance.

Keywords: WhatsApp, banking services, trust, PLS-SEM (Partial Least Squares Structural Equation Modelling)

1. Introduction

Within the dynamic landscape of the swiftly evolving banking sector, the significance of technological advancements cannot be overstated, as they assume a paramount role in the metamorphosis of customer interactions. This particular segment takes a profound dive into the metamorphic influence wielded by WhatsApp on communication channels entrenched within the labyrinthine structure of financial institutions. It endeavours to unravel the intricate tapestry of how WhatsApp, as a ubiquitous messaging platform, seamlessly aligns itself with the perpetually evolving expectations of consumers. Furthermore, it elucidates the strategic deployment of cutting-edge technological innovations as a fulcrum for revolutionizing the overarching customer experience within the financial realm. This exploration traverses the synergistic confluence of digital prowess and customer-centric strategies, delving into the nuanced mechanisms through which WhatsApp orchestrates a paradigm shift in the symbiotic relationship between financial entities and their clientele.

2. Literature Review

The literature review on WhatsApp Banking reveals its transformative potential within the financial landscape, aligning with the evolving expectations of consumers

[a]pgolar@stvincentngp.edu.in, [b]prajwalb.it20@stvincentngp.edu.in, [c]sahildohate.fy20@stvincentngp.edu.in, [d]vikasyadav.fy20@stvincentngp.edu.in, [e]rupalbethriya.fy20@stvincentngp.edu.in, [f]shrutidamdu.fy20@stvincentngp.edu.in, [g]janhavib.it20@stvincentngp.edu.in

DOI: 10.1201/9781003567653-30

(Venkatesh et al., 2003). Conventional communication channels in banking, characterized by sluggishness and a lack of adaptability, face inherent limitations in meeting the dynamic demands of modern banking (Czajkowski et al., 2001). WhatsApp Banking emerges as a responsive, agile, and immediate communication conduit that addresses these challenges (Acar and Murthy, 2018). This innovative approach introduces a symbiotic relationship between technology and financial services, offering real-time and seamless communication between customers and financial institutions (Bank for International Settlements, 2020). The integration of cutting-edge chatbot technology within WhatsApp Banking further underscores its adaptability, providing personalized customer support experiences (Acar and Murthy, 2018). The literature review contributes to understanding the responsiveness, agility, and symbiotic integration of technology within WhatsApp Banking, positioning it as a catalyst for redefining the customer-bank relationship (Venkatesh et al., 2003).

The identification of maximally homologous subsequence among sets of long sequences is an important problem in molecular sequence analysis. The problem is straightforward only if one restricts consideration to contiguous subsequence (segments) containing no internal deletions or insertions. The more general problem has its solution in an extension of sequence metrics Smith and Waterman (1981). In the broader context of technological integration, the literature points to the significance of digital channels in sculpting the future trajectory of banking services (V. World Bank, 2017). WhatsApp Banking, as an exemplar of this digital evolution, transcends mere utilitarian aspects and delves into nuanced dynamics, enriching customer experiences and fortifying engagement bonds (Venkatesh et al., 2003). This synthesis of factors signifies a holistic approach where the digital platform augments the efficiency and effectiveness of financial services, addressing the

shortcomings of traditional systems (Bank for International Settlements, 2020). The references Venkatesh et al. (2003), Acar and Murthy (2018) and Bank for International Settlements (2020) validate the literature's credibility, providing a robust foundation for understanding WhatsApp Banking's transformative influence in contemporary banking practices. This review sets the stage for further exploration into the multifaceted nature of technological integration and its implications for the future of financial services.

3. Data and Variables

It conducts a rigorous examination of 400 dynamic banking customers actively using WhatsApp, employing Partial Least Squares Structural Equation Modelling (PLS-SEM) as an analytical framework. The sample selection process involves meticulous criteria to ensure representation across demographics, emphasizing randomness and detailed participant profiling. Data collection strategies encompass quantitative and qualitative methods, including surveys, interviews, and transactional data validation, adhering to ethical standards. PLS-SEM is chosen for its ability to unravel intricate relationships, exploring variables influencing WhatsApp adoption, and employing exploratory data analysis. The section underscores precision and depth achievable through PLS-SEM, considering trade-offs for model complexity and validating reliability through robustness checks.

The study's analytical framework further explores direct and latent effects, formulating hypotheses about variable pathways and utilizing advanced statistical techniques within PLS-SEM, such as bootstrapping. The emphasis on achieving a comprehensive understanding involves synthesizing findings from various methodological approaches, considering socio-economic and technological contexts, and validating transferability to diverse banking environments. Reflexivity is encouraged within the research team to address biases and assumptions, ensuring the study's relevance and applicability.

4. Methodology and Model Specifications

In fortifying the study's methodological foundations, a meticulous approach is adopted. The investigation focuses on 400 dynamic banking patrons engaged through WhatsApp, utilizing Partial Least Squares Structural Equation Modelling (PLS-SEM) for an exhaustive exploration of interconnections among variables. PLS-SEM is chosen for its capacity to discern direct and latent effects, providing depth and precision in understanding WhatsApp adoption determinants in banking.

4.1. Sample Selection

- Employ rigorous criteria to identify and recruit a diverse cohort consisting of 400 active and engaged banking customers who utilize WhatsApp for financial interactions.
- Ensure inclusivity across demographic variables such as age, income, and geographical location to represent a broad spectrum of the target population.

4.2. Data Collection

- Implement a comprehensive data collection strategy, combining quantitative and qualitative methods to capture the multifaceted aspects of customer interactions on WhatsApp.
- Utilize surveys, interviews, and transactional data to compile a rich dataset, ensuring the inclusion of both explicit and implicit user feedback.

4.3. Analytical Framework

- Optional Practical Training (OPT) for Partial Least Squares Structural Equation Modelling (PLS-SEM) as the

analytical framework due to its versatility and applicability in handling complex relationships.
- Validate the appropriateness of PLS-SEM for the specific research objectives and characteristics of the dataset.

4.4. Variable Exploration

- Identify key variables influencing WhatsApp adoption in the banking sector, drawing from existing literature and preliminary data analysis.
- Categorize variables into independent, dependent, and mediator variables to establish a clear conceptual framework.

4.5. Precision and Depth

- Emphasize the precision and depth of analysis achievable through PLS-SEM, particularly in capturing complex relationships that may be overlooked by traditional methods.
- Leverage the ability of PLS-SEM to handle multiple dependent and independent variables simultaneously, allowing for a more comprehensive examination.

4.6. Direct and Latent Effects

- Investigate not only the direct effects of variables on WhatsApp adoption but also the latent and indirect effects that may manifest through mediating factors.
- Formulate hypotheses regarding the specific pathways through which variables exert their influence, considering both linear and non-linear relationships.

4.7. Richness of Insights

- Prioritize the generation of rich and nuanced insights by triangulating findings from multiple data sources and analytical methods.
- Integrate qualitative data, such as customer feedback and perceptions, with

quantitative results to provide a holistic understanding.

4.8. *Comprehensive Understanding*

- Strive for a comprehensive understanding of the intricate dynamics influencing WhatsApp adoption in the banking sector by synthesizing findings from various methodological approaches.
- Consider the broader socio-economic and technological context in which the study is situated, acknowledging external factors that may impact the research outcomes.

Conclude the methodology section by summarizing the steps taken to guarantee the robustness, validity, and reliability of the research approach.

5. Working of System

A detailed exposition on the operational dynamics of the WhatsApp Banking System is presented. It explores the secure interfaces embedded within the platform, focusing on end-to-end encryption protocols and the amalgamation of advanced algorithms. The system's functionality transcends mere transmission of information, embodying a fusion of technological prowess, security fortification, and user-centric design. The methodology and working of proposed system as shown in Figures 30.1 and 30.2. These figures are inspired by the work of Ascar and Murthy (2018), titled "WhatsApp for teaching and learning: A case study."

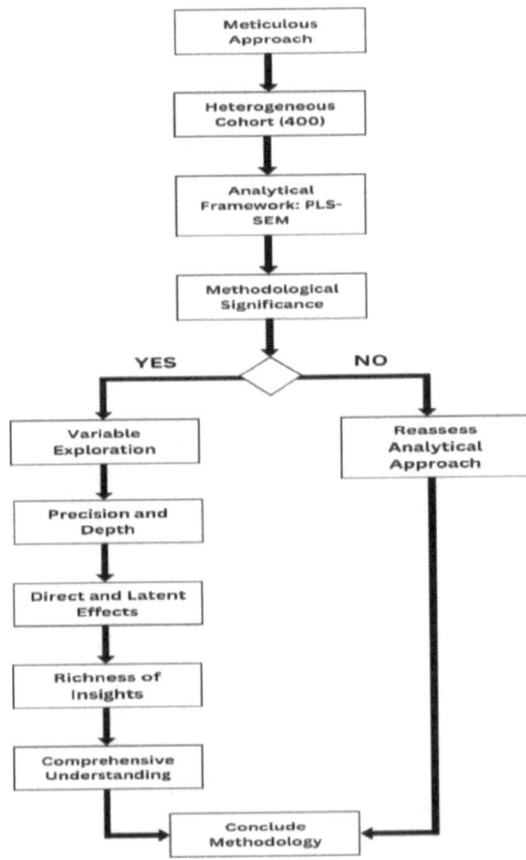

Figure 30.1: Flow chart of methodology.

Source: Author.

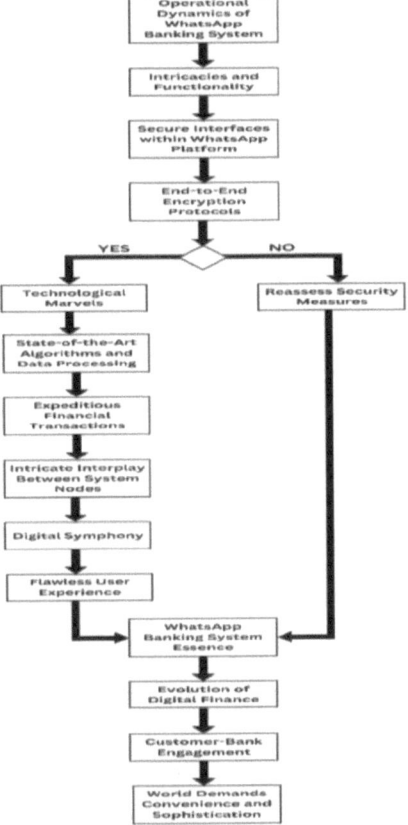

Figure 30.2: Flow chart of working.

Source: Author.

6. Empirical Results

This analytical approach serves to validate hypotheses concerning information availability, trust, and perceived benefits, with a particular focus on their impact on sustained WhatsApp usage in the context of banking services. The intricacies of the relationships among these variables are meticulously scrutinized, providing a comprehensive and nuanced understanding of the factors influencing customer engagement within the WhatsApp banking landscape. The result of Login in WhatsApp Banking, balance enquiry and account as shown in Figure 30.3, which illustrates the system's output, is inspired by the findings from our own project conducted during a study of the research paper "Navigating security measures in WhatsApp Banking".

7. Conclusion

In conclusion, this research highlights the transformative impact of WhatsApp integration on customer experience and engagement within the banking sector. It emphasizes the profound influence of real-time information accessibility, the establishment of trust, and the perception of tangible benefits, portraying a mosaic of factors that synergistically contribute to the sustained adoption of WhatsApp in banking interactions. The conclusion further underscores the symbiotic relationship among these

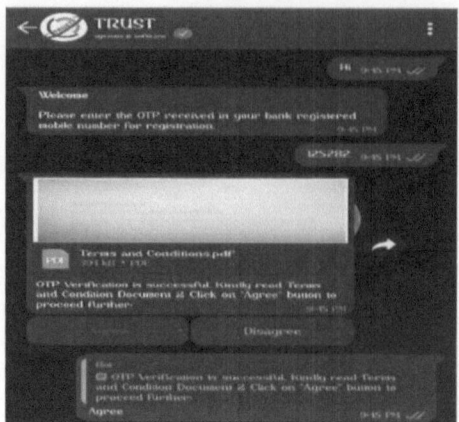

i. Login in WhatsApp Banking.

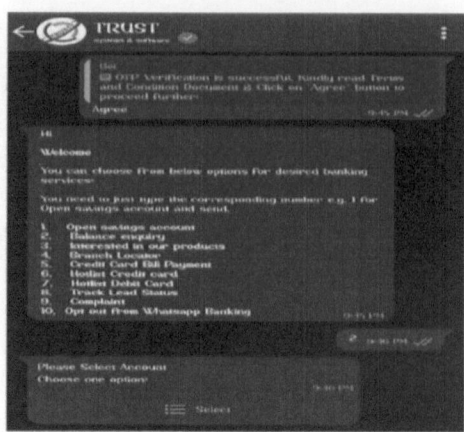

ii. Selecting any option, like; balance enquiry.

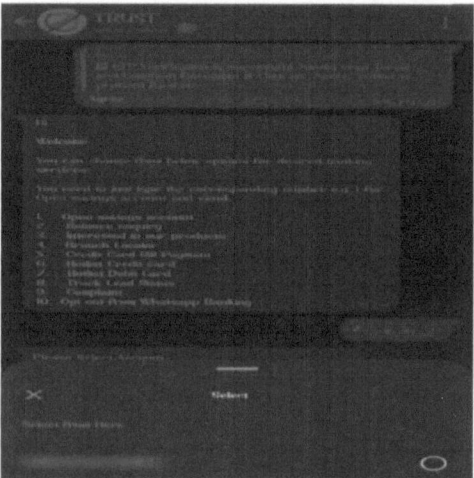

iii. Selecting account if you linked multiple.

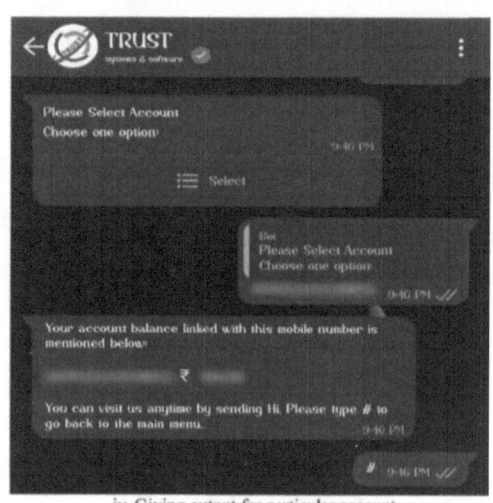

iv. Giving output for particular account

Figure 30.3: Results.

Source: Author.

elements, enriching customer experience and fortifying engagement bonds. Beyond utilitarian considerations, the holistic view explores the nuanced dynamics shaping the evolution of contemporary banking practices, serving as a testament to the multifaceted nature of technological integration.

In light of these revelations, the conclusion advocates the pivotal role of digital channels in shaping the future trajectory of banking services. It calls for a continuous commitment to exploration and optimization, urging stakeholders to adapt strategies and frameworks to the ever-evolving landscape of digital interactions. This conclusion extends beyond a mere summary, transforming into a call to action for the continuous pursuit of innovation and refinement in the banking realm to stay abreast of the dynamic currents of the digital age.

References

Acar, A. Z., and Murthy, D. (2018). WhatsApp for teaching and learning: A case study. Interactive Learning Environments, 26(7), 865–877.

Bank for International Settlements. (2020). Central Bank Digital Currencies. BIS Papers, No. 107.

Czajkowski, K., Fitzgerald, S., Foster, I., and Kesselman, C. (2001). Grid information services for distributed resource sharing. 10th IEEE International Symposium on High-Performance Distributed Computing, IEEE Press, New York, pp. 181–184.

Foster, I., Kesselman, C., Nick, J., and Tuecke, S. (2002). The physiology of the grid: An open grid services architecture for distributed systems integration. Technical report, Global Grid Forum.

May, P., Ehrlich, H. C., and Steinke, T. (2006). ZIB structure prediction pipeline: Composing a complex biological workflow through web services. In W. E. Nagel, W. V. Walter, and W. Lehner (Eds.). (2006). Euro-Par 2006. LNCS, vol. 4128, pp. 1148–1158. Springer, Heidelberg.

Nunnally, J. C. (1978). Psychometric Theory. McGraw-Hill.

Smith, T. F., and Waterman, M. S. (1981). Identification of common molecular subsequence. Journal of Molecular Biolog, 147, 195–197.

V. World Bank. (2017). The global findex database 2017: Measuring financial inclusion and the fintech revolution.

Venkatesh, V., Morris, M. G., Davis, G. B., and Davis, F. D. (2003). User acceptance of information technology: Toward a unified view. MIS Quarterly, 27(3), 425–478.

31 Design and analysis of multi-purpose cartesian co-ordinate robot

Mohammad Athar Hayat , Raju Sahu^a, Pratik Raulkar, Harsh Sahare, Saket Tirpude, and Rahul Dhoke

Department of Mechanical Engineering, St. Vincent Pallotti College of Engineering and Technology Nagpur, India

Abstract

A Cartesian coordinate robot comprises three interconnected transmission mechanisms. The first transmission mechanism is composed of a first guide rail, a first slider that smoothly moves along the first guide rail, and a first driving module responsible for propelling the first slider. Connecting at a right angle to the first guide rail, the second transmission mechanism consists of a second guide rail, a second slider that slides along it, and a second driving module propelling the second slider. Perpendicular to both the first and second guide rails, the third transmission mechanism includes a third guide rail, a third slider moving along it, and a third driving module propelling the third slider. The main goal of designing this robot is to increase the affordability for the low scale manufacturers and to make it flexible for both vertical milling and MIG welding.

Keywords: Cartesian co-ordinate, vertical milling, MIG welding, ANSYS, CATIA, SolidWorks

1. Introduction

The design and analysis of a multipurpose Cartesian coordinate robot involve a system with three linear joints operating in the Cartesian coordinate system (X, Y, and Z). These robots move along orthogonal axes, providing linear motion in different directions based on Cartesian geometry. They are easily programmable and controllable, moving to specific points in space according to coordinates along the three axes. Cartesian robots are versatile and find applications in various industrial sectors due to their cost-efficiency, compact size with scalability, straight-line movement capabilities, high precision,

adaptability, strength to carry light to heavy loads over long distances, stability against vibrations, and ease of programming. These robots can be configured as stages attached in an x, y, and z-axis setup using joining brackets. The design typically involves linear stages with linear actuators supported by linear bearings to handle moment loads.

The robot is designed for vertical milling and MIG welding by simply changing its end effector.

The following designed components are used in this project:

1. **Base Fixture (Motor):** This fixture is used to provide support to the linkages and

^{a}rajusahu82296@gmail.com

DOI: 10.1201/9781003567653-31

components responsible for the motion. Additionally, it provides an extension for the overhanging part of the motor and supports the vertical linkages of the robot.

2. **Base Fixtures (Free End):** This fixture provides support to the free end by holding the bearings and the lead screw, allowing the motion of the lead screw to provide the necessary motion.

3. **Side Fixtures:** The following fixture is used to sync the movement of the X-axis by providing support to the Y-axis and also providing support to the additional components of the Y-axis.

4. **Z-axis Fixture:** This fixture holds all the components of the Z-axis and is used to sync the Z-axis and Y-axis to provide the expected output motion according to the given inputs.

2. Literature Review

Cartesian coordinate robots, also known as gantry robots or XYZ robots, have garnered significant attention in the field of industrial automation due to their versatility, precision, and ease of programming. This literature review aims to provide a comprehensive overview of the state-of-the-art research in Cartesian coordinate robotics, focusing on key areas such as kinematics, dynamics, control algorithms, applications, and advancements in design.

Kinematics is a fundamental aspect of Cartesian robots, governing their motion planning and trajectory generation. Furthermore (Jin et al., 2019) proposed a hybrid kinematic model combining traditional analytical methods with machine learning algorithms to enhance the accuracy and adaptability of Cartesian robot motion control.

3. Material Selection

In the material selection the materials considered for this project are selected on the basis of its properties like Tensile strength, Ultimate tensile strength, etc. To ensure that the fixtures can withstand the required static and dynamic loads. The purpose of material selection is to keep the product cost effective with necessary structural strength.

3.1. PLA

Polylactic acid (PLA) is a biodegradable as well as recyclable polyester made from renewable feedstock. It uses lactic acid as a raw material produced by fermentation of glucose and sucrose.

3.2. ABS Plastic

Acrylonitrile butadiene styrene (ABS) plastic is a common thermoplastic polymer it is made from a blend of two plastics and one rubber: acrylonitrile, polystyrene and butadiene.

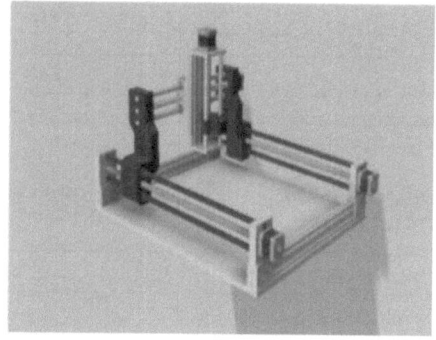

Figure 31.1: CAD assembly of model.

Source: SolidWorks assembly.

Figure 31.2: Working model.

Source: Assembled model.

Table 31.1: Properties of PLA

Properties of PLA	
Tensile Strength (MPa)	35.6
Young's Modulus (MPa)	3600
Density (g/cm³)	1.25
Impact Strength (kJ/m²)	16.5
Compressive Strength (MPa)	48.2

Source: https://www.matweb.com/search/DataSheet.aspx?MatGUID=ab96a4c0655c4018a8785ac4031b9278.

Table 31.2: Properties of ABS plastic

Properties of Alluminium6061-T6	
Tensile Strength (MPa)	240
Young's Modulus (MPa)	6890
Density (g/cm³)	2.7
Impact Strength (kJ/m²)	520
Compressive Strength (MPa)	250

Source: https://www.matweb.com/search/DataSheet.aspx?MatGUID=3a8afcddac864d4b8f58d40570d2e5aa.

Table 31.3: Properties of Aluminium 6061-T6

Properties of ABS	
Tensile Strength (MPa)	43.8
Young's Modulus (MPa)	1470
Density (g/cm³)	1.05
Impact Strength (kJ/m²)	58
Compressive Strength (MPa)	36.5

Source: https://www.matweb.com/search/DataSheet.aspx?MatGUID=b8d536e0b9b54bd7b69e4124d8f1d20a.

3.3. *Aluminium 6061-T6*

Aluminium 6061-T6 is an aluminium alloy which is known for its strength, corrosion resistance and finishing. It's also easy to weld and has similar strength to mild steel. The "T6" represents the tempering process grade where the metal is heated at specific temperature to improve its characteristics.

From the above comparison we selected PLA plastic as material for the manufacturing and fabrication of the fixtures for our model, the PLA plastic satisfies all the necessary requirements for strength. The availability of PLA and its cost effectiveness also makes it easier to procure. Hence, we selected PLA as material for fabricating the fixtures for model.

4. Load Calculations

For the analysis of the fixtures we first need to calculate maximum loads acting on each fixture, by calculating the forces we can apply boundary conditions for analysis. The calculations below are considered for analysis of the fixtures:

Components load at maximum
2motors = 2.4kg
Slide fixtures = 1.4kg
End effector = 0.5kg
Z axis fixture = 1kg
Lead screw = 2kg
Smooth bar = 2.5kg
Milling motor = 1.5 kg
Cutting tool attachment = 0.2

Total weight maximum weight 11.5kg = 115N, **F = 115N**
Required torque to propel the weight

$$T = F \times r$$

Where r is considered to be the radius of lead screw

$$T = 115 \times 12 \text{ mm}, T = 1380 \text{ N mm}$$

Torque provided by motor

$T = 25$ kg.cm, $T = 2500$ N.mm
$P = V \times I$, $P = T/t$, $P = T/1$, Maximum power output by motor, **P = 2.5 Watt**

So, from the above calculations the boundary conditions were applied in the Static structural model with the help of ANSYS software for the analysis and real time simulations.

The CAD model for this analysis was made with the help of SolidWorks and CATIA software. First the CAD models were designed in the above softwares and then the CAD model was imported in the analysis with the help of ANSYS software. We got satisfactory results by applying the above calculated boundary conditions.

Hence, the complete analysis was carried out through the above process.

5. CAD and Analysis

Design and Analysis is a decision making process that involve various analytical methods, tools and software's to develop a product model. It's crucial step in the manufacturing process that ensures the final product is safe, high quality and meets the intended purpose.

5.1. CAD Design

The CAD design for this product model was created with the help of SolidWorks and

CATIA by keeping the major aspects such as "material properties" and "ease of assembly" in consideration the models were designed. The design of the fixtures is made to sustain the various loads on product model and to keep it sturdy and safe while operating.

The following image is a rendered image of product model, the CAD for all the parts of product model was designed and assembled in the SolidWorks. After that the model was imported in CATIA and was rendered.

5.2. Analysis

The parts of following cad model was imported separately and with the help of the ANSYS software the analysis was performed on these parts. Below mentioned are the results of analysis of each part:

6. Conclusions

The goal for this project is to design and analyse the product model of Multi-Purpose Cartesian Co-ordinate Robot while the

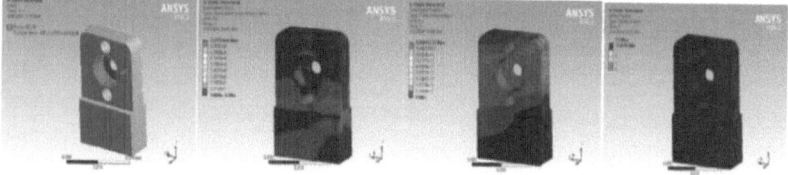

Figure 31.3: Analysis of base fixture.

Source: Author.

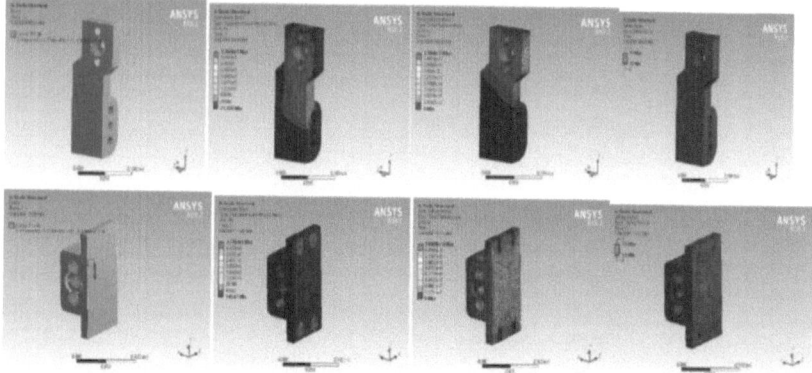

Figure 31.4: Analysis of side fixture.

Source: Author.

major focus was to optimize the Design. This project is designed to be beneficial to the low scale manufacturers of furniture design and other low scale manufacturers as well. With lighter, compact design while keeping it cost effective this project is best for low scale manufacturers.

References

Cen, L., and Melkote, S. (2017). Effect of robot dynamics on the machining forces in robotic milling. Procedia Manufacturing, 10, 486–496. doi: 10.1016/j.promfg.2017.07.034

Denkena, B., Bergmann, B., and Lepper, T. (2017). Design and optimization of a machining robot. Procedia Manufacturing, 14, 89–96. doi: 10.1016/j.promfg.2017.11.010.9

Garnier, S., Subrin, K., and Waiyagan, K. (2017). Modelling of robotic drilling. Procedia CIRP, 58, 416–421. doi: 10.1016/j.procir.2017.03.246

Jin, M., Liu, Q., Wang, B., and Liu, H. (2019). An efficient and accurate inverse **kinematics** for 7-DOF redundant manipulators based on a **hybrid** of analytical and numerical method. Ieee Access, 8, 16316–16330.

Lehmann, C., Halbauer, M., Euhus, D., and Overbeck, D. (2012). Milling with industrial robots: Strategies to reduce and compensate process force induced accuracy influences. doi: 10.1109/ETFA.2012.6489741

Möller, C., Schmidt, H., Koch, P., Böhlmann, C., Kothe, S., Wollnack, J., and Hintze, W. (2017). Machining of large scaled CFRP-Parts with mobile CNC-based robotic system in aerospace industry. Procedia Manufacturing, 14, 17–29. doi: 10.1016/j.promfg.2017.

Mousavi, S., Gagnol, V., Bouzgarrou, B. C., and Ray, P. (2017). Stability optimization in robotic milling through the control of functional redundancies. Robotics and Computer-Integrated Manufacturing, 50, doi: 10.1016/j.rcim.2017.09.004

Pan, Z., Zhang, H., Zhu, Z., and Wang, J. (2006). Chatter analysis of robotic machining process. Journal of Materials Processing Technology, 173, 301–309.

Wayland, M., and Landgraf, M. (2018). A cartesian coordinate robot for dispensing fruit fly food. Journal of Open Hardware, 2. doi: 10.5334/joh.9

Zhang, Hui., and Pan, Zengxi. (2008). Robotic machining: Material removal rate control with a flexible manipulator. 2008 IEEE International Conference on Robotics, Automation and Mechatronics, RAM 2008, 30–35. doi: 10.1109/RAMECH.2008.4690881

32 Deep learning-powered traffic analysis: YOLOv7 and DeepSORT Integration in Smart Vision Traffic

Jagdish Pimple[1,a], Mohit Agarwal[2,b], Dipmala Kathale[1,c], Kajal Matte[1,d], Rutuja Hande[1,e], and Shruti Jiwane[1,f]

[1]Department of Information Technology, St.Vincent Pallotti College of Engineering and Technology, Nagpur, India
[2]School of Computer Science Engineering and Technology, Bennett University, Greater Noida, India

Abstract

A unique vehicle division system designed for dynamic metropolitan areas is introduced in this research. By utilising state-of-the-art technology such as computer vision, deep learning, and real-time warning systems, it seeks to improve road safety, lessen traffic, and efficiently handle unforeseen events. Using the YOLO approach, a robust vehicle instance segmentation model that can identify automobiles in real-time from several sources is initially created. The vehicles are counted and a preset threshold is set up in the following phase. The technology will detect and suppose that traffic is clogged if the number of vehicles in the video or image exceeds a predetermined threshold. Users receive updates on anomaly detection and congestion analysis through user-friendly interfaces during the final phase. This article intends to provide a responsive, safe, and efficient urban transportation experience by combining state-of-the-art technologies like machine learning and comprehensive approaches to modernise traffic control.

Keywords: YOLOv7, DeepSORT, OpenCV, object detection, multiclass segmentation, object counting

1. Introduction

The purpose of this study is to design an enhanced vehicle segmentation system that integrates state-of-the-art technology in response to the problems presented by dynamic urban areas. The first step focuses on developing a robust vehicle instance segmentation model for real-time vehicle recognition in pictures or video streams using the YOLO and DeepSORT approaches. The study uses multi-class segmentation in addition to vehicle classification to provide a thorough analysis of the road infrastructure. The investigation of traffic congestion, the identification of bottlenecks using

[a]jpimple@stvincentngp.edu.in, [b]26.mohit@gmail.com, [c]dipmalaakathale@gmail.com, [d]kajalmatte02@gmail.com, [e]rnhande02@gmail.com, [f]shrutijiwane914@gmail.com

DOI: 10.1201/9781003567653-32

optical flow and density estimates, and the use of deep learning algorithms for anomaly detection are the next goals. During the final stage, users will receive real-time traffic notifications using user-friendly interfaces that give segmentation data, congestion analysis, and abnormality detection.

2. Problem Statement

The problem statement for this paper can be specified as

2.1. Traffic Management Complexity

Traffic management is a complex and tedious task in India, requiring effective solutions to handle various challenges.

2.2. Accidents on Roads

Road accidents are a common occurrence, and managing traffic around accident sites is a significant challenge for authorities.

2.3. VIP Movement

The movement of VIPs through specific roads can disrupt regular traffic flow, leading to congestion and delays.

3. Proposed System

The proposed intelligent traffic monitoring system (Figure 32.1) involves deploying high-resolution cameras at traffic signals for real-time video capture. Advanced computer vision and machine learning algorithms classify vehicles, emphasising three, four, and five-wheelers. Continuous real-time density calculations inform a deep learning-based traffic jam detection system. Alerts are triggered when predefined thresholds are surpassed. The Work flow of proposed system (Figure 32.2) elaborated the stepwise execution of the proposed system.

4. Methodology Used

1. For the real-time video monitoring and alert system in traffic control, the research paper's technique combined DeepSORT (Deep Simple Online and Realtime Tracking) with YOLOv7 (You Only Look Once version 7).
2. Modern object identification algorithm YOLOv7 was used because of its effectiveness in identifying and categorising items in every frame of the real-time video streams.
3. Furthermore, DeepSORT was used to track and preserve the identity of objects recognised across a series of frames.

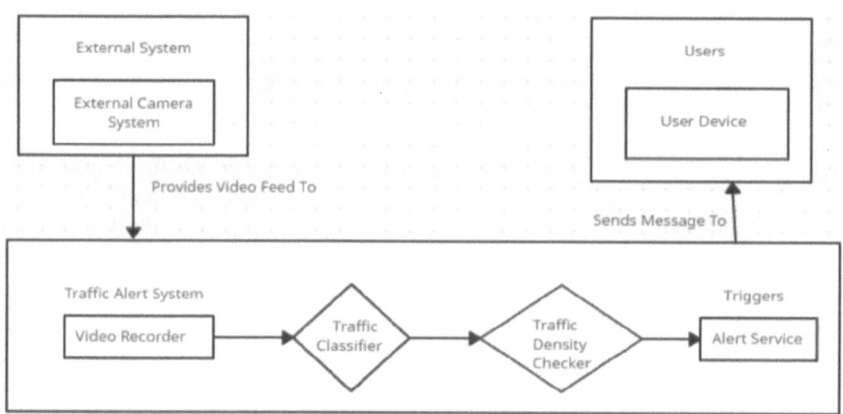

Figure 32.1: Data flow of proposed system.

Source: Author.

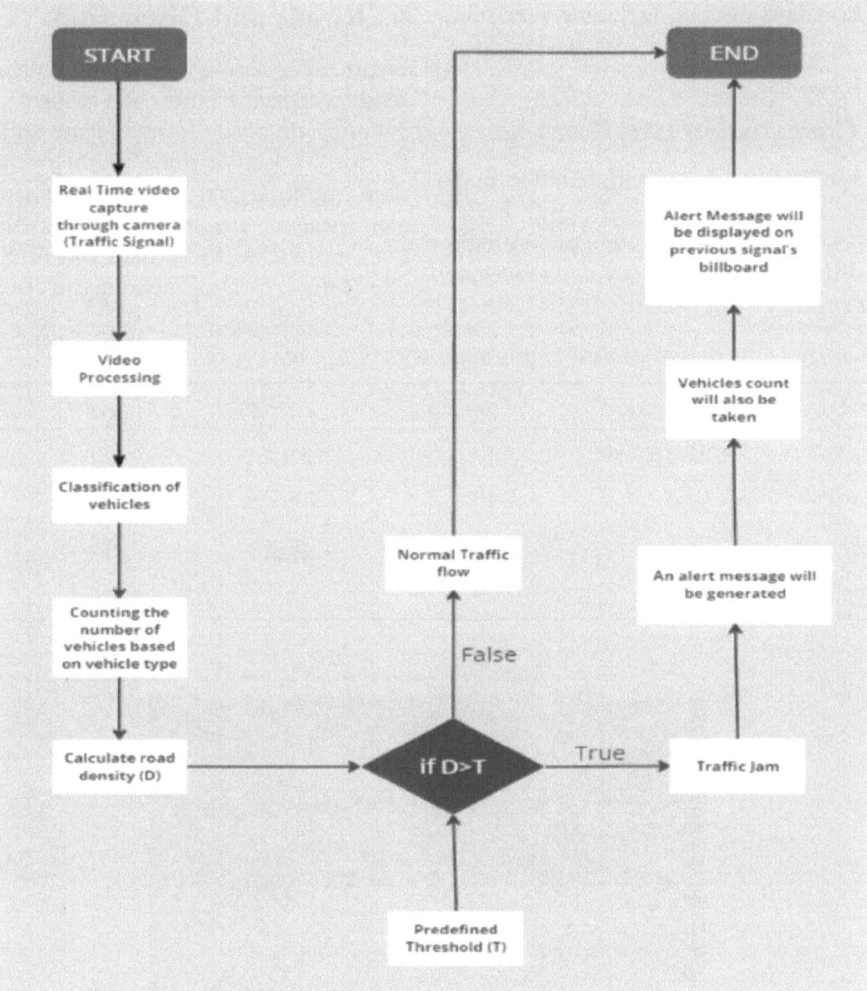

Figure 32.2: Work flow of proposed system.

Source: Author.

5. Implementation Related Work

In the proposed system, we utilise the YOLOv7 algorithm for real-time video object detection. The algorithm identifies various objects such as cars, trucks, and pedestrians. After detection, the system counts the instances of each detected object class and displays the counts on the screen. The system provides valuable traffic management and analysis information by accurately counting different types of vehicles and individuals.

5.1. Model Architecture and Versions

We utilised YOLOv7 as our primary object detection framework due to its real-time capabilities and high accuracy. Additionally, we integrated the DeepSORT algorithm with YOLOv7 for object tracking.

5.2. Model Evaluation

We evaluated the performance of our YOLOv7 + DeepSORT implementation using standard metrics, including precision, recall, and F1-score for object detection.

5.3. *Comparisons and Baselines*

We compared our model with baseline models, including YOLOv4, SSD, and Faster RCNN (Table 32.1). The comparison highlighted the superior accuracy and efficiency of our proposed approach.

6. Result and Discussion

Figure 32.3 shows object counting on the road whereas Figure 32.4 shows vehicle counting on road. To investigate and assess the use of deep learning techniques in real-time traffic signal video monitoring and alert systems, the research paper "Innovative Approaches to Traffic Control: Systematic Review of Deep Learning in Real-Time

Table 32.1: Comparison of different models with YOLOv7

Models	Accuracy (%)	Frame Size	AP^{val} (%)	AP^{val}_{50} (%)
YOLOv7	92.86	640	51.2	69.7
YOLOv4	43.5	640	49.7	68.2
SSD	79.8	512	28.8	48.5
Faster RCNN	83.92	600	39.8	59.2

Source: Author.

Figure 32.3: Object counting.

Source: Author.

Figure 32.4: Vehicle counting.

Source: Author.

Signal Video Monitoring & Alerts" was written. The study provides a thorough overview of the developments, difficulties, and possible future directions in this field. All things considered, the systematic review offers insightful information about the state of deep learning applications in traffic control today, laying the groundwork for further study and advancement in this quickly developing area.

7. Conclusion

The development of intelligent transportation systems has been made with the deployment of the YOLO algorithm-based vehicle recognition and traffic monitoring system. A thorough grasp of road conditions is provided by the precise classification and counting of vehicles, as well as the determination of road density and real-time traffic congestion identification. With the timely information our real-time alert system provides, drivers and pedestrians may make more educated decisions about their travel plans and routes, which eventually helps to enhance traffic management and lessen congestion.

References

Ahmed, S. (2023). Object tracking with Deep-SORT. ResearchGate 368646461.

Dhar, S., and Kundu, M. (2021). Accurate multi-class image segmentation using weak continuity constraints and neutrosophic set. ISI Digital Commons. doi: 10.1016/j.asoc.2021.107759

George, S., Sudhan, A., Shajan, J., Sreelakshmy, A. R., and Joy, J. R. (2022). Traffic sign board recognition and voice alert system. IJCRT. 3, ISSN: 2320 2882.

Hengyuan Zhang, Suyao Zhao, Ruiheng Liu, Wenlong Wang, Yixin Hong, and Runjiu Hu (2022). Automatic traffic anomaly detection on the road network with spatial-temporal graph neural network representation

learning. Wireless Communications and Mobile Computing, 2022, 4222827.

Jeon, Hyunjeong, Jincheol Lee, and Keemin Sohn. (2018). Artificial intelligence for traffic signal control based solely on video images. *Journal of Intelligent Transportation Systems* 22(5), 433–445.

Jiang Xin, Wang Yu, and Zheng Rui (2022). YOLO based multi-objective vehicle detection and tracking. Research Square. doi: 10.21203/rs.3.rs-2006332/v1

Kumar, M., Kumar, K., and Das, P. (2021). Study on road traffic congestion: A review. Recent Trends in Communication and Electronics (1st ed.). Taylor and Francis group.

Lukman, A., and Kolo, J. (2020). In-vehicle traffic accident detection and alerting system using distance-time based parameters and radar range algorithm. 2020 IEEE PES/IAS PowerAfrica. IEEE.

Pujara, A., and Bhamare, M. (2022). Deep-SORT: Real time & multi-object detection and tracking with YOLO and TensorFlow. 2022 International Conference on Augmented Intelligence and Sustainable Systems (ICAISS) in Trichy, India. IEEE.

Rajguru, Nikhil S. (2021). Implementation paper of Traffic Signal Detection and Recognition using deep learning. *Turkish Journal of Computer and Mathematics Education (TURCOMAT)*, 12(1S), 212–219.

Ravi, A., Nandhini, R., Bhuvaneshwari, K., Divya, J., and Janani, K. (2021). Traffic management using machine learning algorithm. International Journal of Innovative Research in Technology (IJIRT), 7(11).

Shepelev, V., Gushkov, A., Gritsenko, A. V., Dmitry, and Nevolin (2022). Assessing the Traffic capacity of urban road intersections. Frontiers in Built Environment, 8, 968846.

Yunhao Du, Zhicheng Zhao, Yang Song, Yanyun Zhao, Fei Su, Tao Gong, and Hongying Meng. (2022). StrongSORT: Make Deep-SORT great again. Accepted by IEEE Transactions on Multimedia 2023. IEEE.

Yusuf, S., Aldawasari, A., and Souissi, R. (2022). Automotive parts assessment: applying real-time instance-segmentation models to identify vehicle parts. arXiv:2202.00884

33 Prediction of early strength of concrete using integrated thermo-sensors

Suyog U. Dhote[1,a], Yugandh Bangale[1], Geetshri Padole[1], Samiksha Bhandwalkar[1], Gopichand Thaware[1], Sanket Thombare[1], Ayush Agnihotri[1], and Nakul Pathrabe[2]

[1]Civil Engineering Department, St. Vincent Pallotti College of Engineering and Technology, Nagpur, India
[2]MSc in Civil Engineering, Technical University of Munich, München, Germany

Abstract

At present, quality standards for the use of concrete became essential as well as more demanding in our country. India is categorised in developing country in which concrete has major demand. The concrete is tested for compressive strength. The formwork removal is dependent on concrete compressive strength. This will lead to formwork cost and cost benefits for any construction projects. The paper proposes an economical method to monitor early strength of concrete during the curing period. The setup consists of thermos-couples, temperature sensors integrated in concrete cubes. M25 mix design was used for experimentation. The concrete strength with temperature relationship was established. A concrete cube prototype with an embedded temperature sensor were cast in laboratory. Data collected from the temperature sensor were used to predict the strength of the cube. The early removal of formwork saves the substantial cost and also helps in early completion of activities on a construction project.

Keywords: Concrete, concrete mix design, thermocouples, temperature sensors, etc.

1. Introduction

The setting and hardening of concrete influence the form removal time. Present advances in construction industry have expanded with super-high-rise buildings and long-span bridges. The demand requires huge quantities of concrete within the shortest time span. Construction sector is one of the prominent sectors which requires continuous evaluation. With the evaluation, monitoring for structural stability and reliability is having utmost priorities. Concrete strength at early stages monitoring can reduce structural failures as well as fatal accidents. The appropriate time for the form removal is important to reduce the construction duration and costs.

Concrete compressive strength is the important parameter to ensure safety at construction sites. The given approach is helpful in predicting the appropriate time of removal of scaffoldings. The maturity method shows the relationship between the maturity index,

[a]sdhote@stvincentngp.edu.in

DOI: 10.1201/9781003567653-33

concrete temperature, and concrete strength (Carino & Lew, 2001). In general, the maturity method has been used for monitoring and estimation of concrete strength since the 1950. Maturity method is one of the nondestructive methods, can be used for predicting the early age strength of concrete. The heat of hydration is the key feature in evaluating the early strength of concrete.

The thermos-sensor was inserted inside the fresh concrete. The real-time temperature variation collected as output. The sensors used in the study have a wide range of temperature, but the poor workmanship can create inappropriate results. Different mix proportions vary from M25 to M40 were adopted for the study. The Figure 33.1 below shows the experimentation setup of M25 concrete cube with thermos sensor.

2. Methodology

The complete experimental work was divided into seven subheads i.e. selection of cement, concrete mix design, water cement ratio, use of temperature sensors, early strength results, comparison and conclusion. It was proposed to cast concrete cubes of having dimensions as prescribed in Indian standard. The cement is the utmost important ingredient in concrete. The Portland pozzolana cement was used for the study.

The standard concrete mix design for M25 was used to monitor real time temperature variation during early age. The thermosensor was used to record real time concrete temperature variation. Two set of three cubes each were casted. The experiment was done with one set was kept in curing tank and real time monitoring of temperature were recorded. Another set of three cubes were exposed to atmosphere and monitored the temperature variation. The temperature variation received from the temperature sensors for two sets of cubes were recorded and compared. The conclusion based on the comparison of temperature variation has been incorporated in this study.

The reference temperature is taken as 27°C for the present study. The model developed in the laboratory with and without curing. The Figures 33.2 and 33.3 shows the cubes with and without curing respectively.

Figure 33.2: Concrete cube setup in set 01.

Source: Laboratory of the institute during the research work.

Figure 33.3: Concrete cube setup in set 02.

Source: Laboratory of the institute during the research work.

Figure 33.1: Experimental setup with thermo-sensor.

Source: Institute laboratory.

3. Results

The temperature variation taken by integrated temperature sensors have been recorded for the cube not considered for curing shown in Figures 33.4 and 33.5.

In the similar manner, the cube considered for curing and temperature recorded with temperature sensor is shown in Figures 33.6 and 33.7.

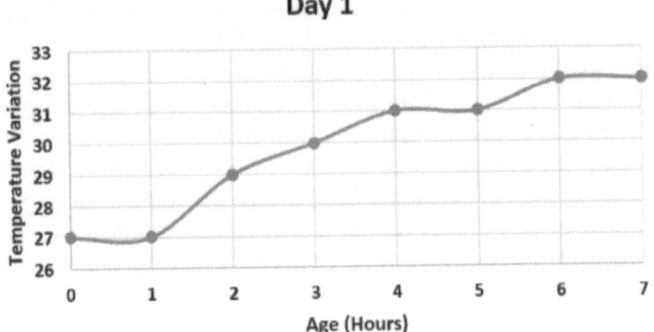

Figure 33.4: Temperature variation Day 1.

Source: Author.

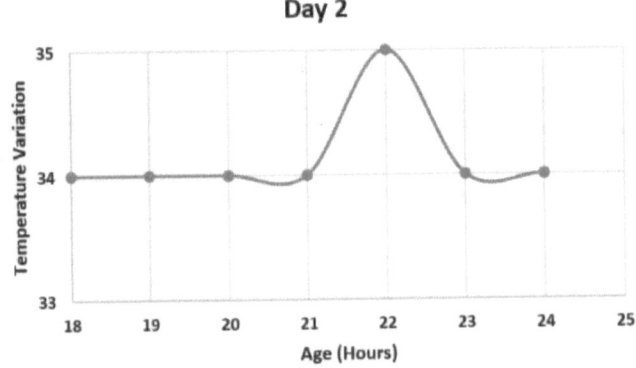

Figure 33.5: Temperature variation Day 2.

Source: Author.

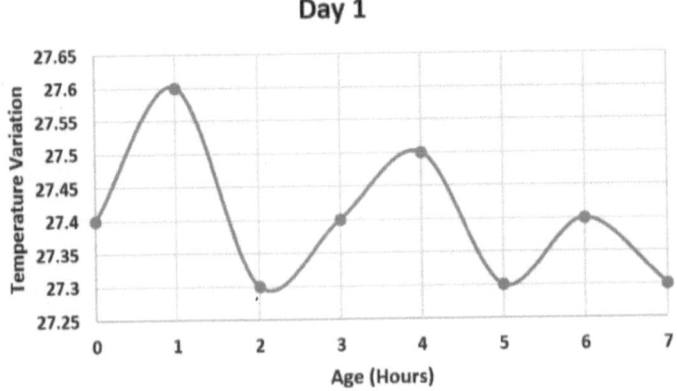

Figure 33.6: Temperature variation Day 1.

Source: Author.

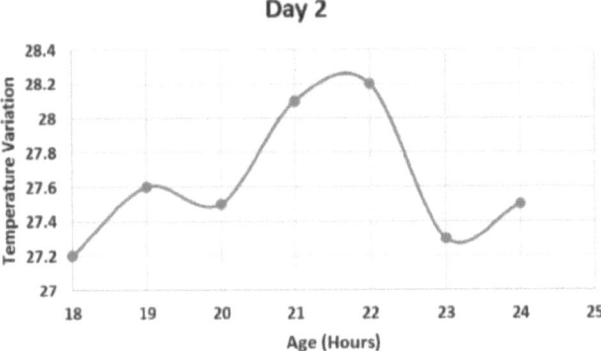

Figure 33.7: Temperature variation Day 2.

Source: Author.

4. Conclusion

Following conclusion drawn on the basis of maturity approach.

1. The present study shows the relationship between strength and thermos-sensor model. The maturity can vary with variety of variables. Includes type of aggregates, chemical the compositions of cement and the curing temperature.
2. The response of maturity model is also dependent on workmanship. Also, the way in which the temperature sensors integrated in the concrete cube.
3. For predicting compressive strength, a maturity method might be implemented.
4. Considering groups of maturity models can assess the maturity and strength of concrete at early stages.
5. The exposure conditions can vary the model results.

References

Barroca, N., Borges, L. M., Velez, F. J., Monteiro, F., Górski, M., and Castro Gomes, J. (2013). Wireless sensor networks for temperature and humidity monitoring within concrete structures. Construction and Building Materials, 40, 1156–1166. doi: 10.1016/j.conbuildm t.2012.11.087

Bloem, D. L. (1954). Effect of curing condition on compressive strength of concrete test specimens. European Chemical Bulletin, 12(issue 1), 5225–5241

Carino, N. J., and Lew, H. S. (2001). The maturity method: From theory to application, Structures 2001: A structural engineering Odyssey. doi: 10.1061/40558(2001)17

Fink, G. J. (1944). The effects certain variations in consistency and curing conditions on the compressive strengths of cement-lime mortars, ASTM Proc, 44, 780–792

Gardner, N. J. (1990). Effect of temperature on the early-age properties of Type I, Type II, and Type III/fly ash concrete with temperature. Mater J, 87(1), 68–78

Helal, J., Sofi, M., and Mendis, P. (2015). Non-destructive testing of concrete A review of methods. Electronic Journal of Structural Engineering, 14(1), 97–105.

Hulshiz, A. J. A. J., and Em, A. R. M. A. (1984). Implementation of concrete strength-maturity concept yields construction economies, ASCE-1984 Spring Convention, Atlanta, Georgia.

Hulshizer, A. J., Edgar, M. A., Daniels, R. E., Suminsby, J. D., and Myer, G. E. (1984). Maturity concept proves effective in reducing form removal time and winter curing cost. ACI Symposium Paper Publication, 82, 351–376 Doi: 10.14359/6563

John, S. T., Roy, B. K., Sarkar, P., and Davis, R. (2020). IoT enabled real-time monitoring system for early-age compressive strength of concrete. Journal of Construction Engineering and Management, 146(2), 05019020. doi: 10.1061/(ASCE)co.1943-7862.0001754

Kumar, B. S., and Raju, P. M. 2023). Early age compressive strength concrete using maturity models. European Chemical Bulletin, 12(issue 1), 5225–5241.

Pinto, R. C., and Hover, K. C. (1999). Application of maturity approach t setting times. Mater J, 96(6), 686–691

Rudeli, N., Santilli, A., and Arrambide, F. (2015). Striking of vertical concrete elements: an analysis using the maturity method. Engineering Structures, 95, 40–48. doi: 10.1016/j.engstruct.2015.03.021

Sof, M., Mendis, P. A., and Baweja, D. (2012). Estimating early-age in situ strength development of concrete slabs. Construction Building Materials, 29, 659–666.

Soutsos, M., Kanavaris, F., and Hatzitheodorou, A. (2018). Critical analysis of strength estimates from maturity functions. Case Studies in Construction Materials, 9, 1–19. doi: 10.1016/j.cscm.2018.e00183

Sun, B., Noguchi, T., Cai, G., and Chen, Q. (2020). Prediction of early compressive strength of mortars at different curing temperatures and relative humidity by a modified maturity method. Struct Concrete. doi: 10.1002/suco.202000041

34 A Review on microbial fuel cells used for generation of electricity

Abhay G. Hirekhan[1,a], Ayush Gawande[1], Hemant Kondaborla[1], Isha Tembhurne[1], Mukul Khorgade[1], Shruti Borkar[1], Tejas Dhale[1], Tejasvi Ghugal[1], and Ishwar Chandra[2]

[1]Department of Civil Engineering, St. Vincent Pallotti College of Engineering and Technology, Nagpur, India
[2]Department of Civil Engineering, Technical University of Munich, Munich, Germany

Abstract

Urbanisation and global population growth have accelerated energy demand, primarily in cities, leaving rural regions underserved. When fossil fuels like coal, oil, and natural gas are burned, they release greenhouse gases and other hazardous pollutants into the environment. This has raised the global temperature and has a number of negative impacts on the environment. It is therefore more essential than ever to switch to renewable energy sources, which are considerably less damaging to the environment and can lessen the effects of climate change. Examples of these sources are wind, solar, and natural gas. Microbial fuel cells (MFCs) have emerged as a promising sustainable energy solution. MFCs use biomass and bacteria to generate electricity, with electrode materials playing a crucial role in power generation and reducing the system costs. This review highlights the potential of MFC technology in generating electricity from diverse substrates, including alternative feedstocks like cow dung and biodegradable kitchen waste. The study emphasises the importance of efficient and cost-effective materials in MFC design and operation, offering insights into their future applications in sustainable energy and solid-waste management.

Keywords: Microbial fuel cells, electricity, fossil fuel, MFC, bacteria

1. Introduction

The need for energy rises in unison with population growth. The urban area has an increased demand for energy than the rural one. In order to meet demand, energy is diverted to urban areas, leaving rural communities in the dark. Fossil fuels, hydrocarbon-based energy sources derived from thousands of years of decomposing plants and animals, are the primary source of energy. These fuels are found in the carbon and hydrogen-rich Earth's crust. Coal, Crude Oil (Petroleum), and Natural Gas are the primary types of fossil fuels. These fossil fuels can be burned to produce energy (Dey et al. 2022; Mittal et al., n.d.).

[a]ahirekhan@stvincentngp.edu.in

DOI: 10.1201/9781003567653-34

However, there are a lot of adverse effects of using fossil fuels to generate energy. When fossil fuels are burned, they emit a lot of different pollutants into the atmosphere, including carbon dioxide (CO_2), sulphur dioxide (SO_2), nitrogen oxides (NO), carbon monoxide (CO), and other particulates. These pollutants which are released in the atmosphere produce unfavourable health conditions for humans and wildlife. By far this release of gases has shown many harmful consequences, which include an increase in temperature, a rise in sea level, and severe weather events (Mittal et al., n.d.; Varma et al., 2013).

The data in Figure 34.1 shows that India generated 1294.92 billion units of electricity through thermal power plants in FY2023.

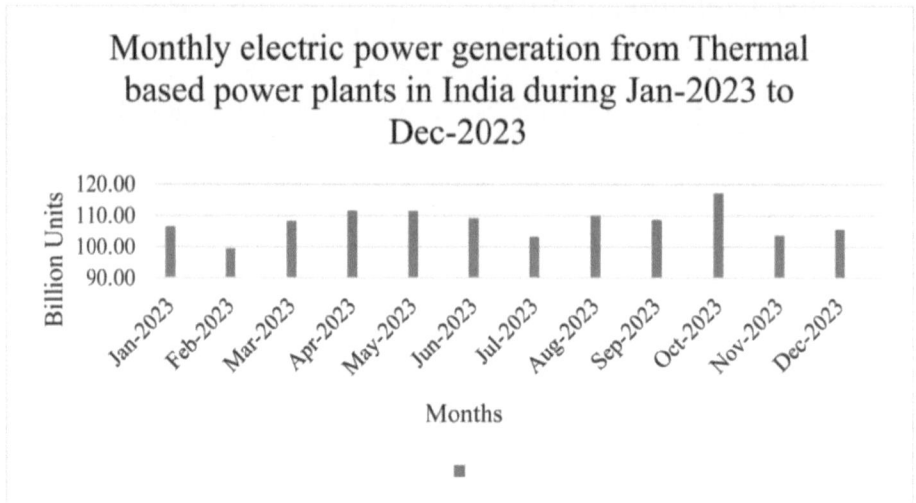

Figure 34.1: All India power generation from thermal power plant for FY2023.

Source: Central Electricity Authority, Ministry of Power, Government of India, 2023–24.

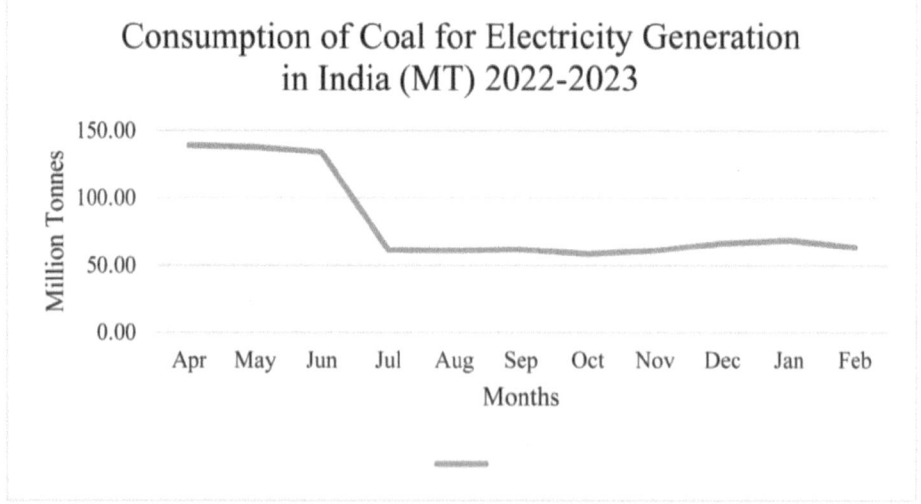

Figure 34.2: Consumption of coal in thermal power plant for FY2022–23.

Source: Central Electricity Authority, Ministry of Power, Government of India, 2023–24.

Coal consumption for the financial year 2022–23 is estimated to be around 6688.12 metric tons as shown in Figure 34.2. The annual growth rate in power generation using fossil fuels between FY2023–24 is 9.91%, indicating an increasing energy demand. However, this also means a higher dependence on fossil fuels and a significant release of CO_2 (Ministry of Power, Government of India, 2023–2024; Central Electricity Authority, Ministry of Power, Government of India, 2023–24).

The world's energy landscape has undergone a paradigm shift in favour of alternative and renewable sources in response to the serious concerns posed by the widespread usage of fossil fuels. The limited amount of fossil fuel reserves and the environmental effects of extracting and burning them are the main causes of the urgency. Among the prominent alternatives, wind, solar, and hydroelectric energy have emerged as pivotal contributors to a sustainable and resilient energy future (Dey et al., 2022).

The data that is shown in Figure 34.3 makes it clear that the production of power from renewable energy sources is progressing remarkably. Specifically, in FY2022–23, solar energy generated FY2023 96993.9

MU, wind energy generated 184995.9 MU, and other renewable sources generated 20591.1 MU, showcasing a significant shift towards sustainable and environmentally friendly energy sources. The annual growth rate stands at an impressive 19.10%, indicating the promising future of renewable energy with the advancement of technology and infrastructure. However, it is crucial to acknowledge that these sources have limitations and drawbacks, with the biggest being their dependency on weather conditions, such as wind speed and daylight hours (Ministry of Power, Government of India, 2023–2024; Central Electricity Authority, Ministry of Power, Government of India, 2023–24).

According to the data presented in Figure 34.3, the electricity generated by wind energy sources experienced a gradual increase from May 2023 to August 2023. However, it then decreased until October 2023, which was attributed to the changes in wind speed (Ministry of Power, Government of India, 2023–2024) (Central Electricity Authority, Ministry of Power, Government of India, 2023–24). On the other hand, solar energy demonstrated a consistent output throughout the period. Furthermore, there

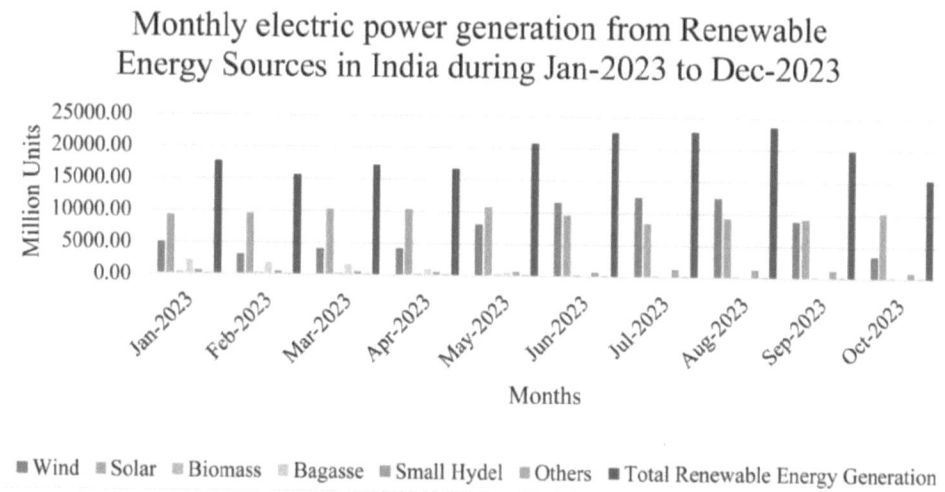

Figure 34.3: Renewable Energy Generation in FY2023.

Source: Central Electricity Authority, Ministry of Power, Government of India, 2023–24.

was a significant increase in electricity generation by solar energy sources during the months of March to May 2023, which was a result of favourable weather conditions. Both wind and solar energy are valuable sources of renewable energy. However, they are both subject to weather conditions, which can affect their output levels (Ministry of Power, Government of India, 2023–2024; Central Electricity Authority, Ministry of Power, Government of India, 2023–24).

Investigating cutting-edge technology has emerged as a critical undertaking in the search for environmentally friendly energy sources and ways to reduce environmental effects. Microbial fuel cells (MFCs) are a potential new development in the field of power generation in the pursuit of sustainable energy solutions.

2. Microbial Fuel Cell

A Microbial Fuel Cell is a device that generates electricity using biomass and bacteria. The process involves the decomposition of biomass by anaerobic bacteria that produce electrons. These electrons are then used to generate electricity. The MFC comprises two cells, namely an anode chamber and a cathode chamber, separated by a cation exchange membrane, also known as a proton exchange membrane (PEM). This membrane allows the flow of protons but not electrons (Akatah et al., 2019; Li, 2013; Thulasinathan et al., n.d.).

In the anode chamber, the bacteria feed on the substrate (Biomass) to produce hydrogen ions (Protons) and Electrons in an anaerobic environment. The electrons flow through the electrode (Anode), where the bacteria attach themselves. On the other hand, the protons flow through the PEM (Akatah et al., 2019; Li, 2013; Thulasinathan et al., n.d.).

The overall reaction in the anode chamber is as follows (Vishwanathan, 2021):

$$C_6H_{12}O_6 + 6H_2O \longrightarrow 6CO_2 + 24e^- + 24H^+ \quad (1)$$

As the electrons pass through the electrode(anode) with a connection wire, they are allowed to flow to the cathode chamber where the electrons and protons react with each other in the presence of oxygen.

The overall reaction in the cathode chamber is as follows (Vishwanathan, 2021):

$$6O_2 + 24H^+ + 24e^- \longrightarrow 12H_2O \quad (2)$$

The overall reaction (Vishwanathan, 2021):

$$C_6H_{12}O_6 + 6O_2 \longrightarrow 6H_2O + 6CO_2 \quad (3)$$

3. Literature Review

Dey S., et al. (Dey et al., 2022): Solar, wind, tidal, geothermal, biomass, hydrogen, and nuclear energy are examples of renewable energy sources in India. India has 33,791.74 MW of total capacity for renewable energy, of which 22,465.03 MW comes from wind energy. Because non-renewable energy sources have negative consequences on both the environment and human life, renewable energy sources are crucial for protecting the environment.

Akatah B. M., et al. (2019): An experiment was conducted to generate electricity from septic wastewater using microbial fuel cells (MFCs). Four MFCs were used, and each MFC produced approximately 3.0V of electricity. The MFCs could produce up to 10.5V when connected in parallel. The experiment was carried out for 12 weeks, and the power output was measured every week.

Jessica Li. (2013): Using microorganisms and organic debris, the microbial fuel cell (MFC) experimental setup investigates the production of power. The proton-exchange membrane, anode chamber, cathode chamber, and external circuit are the four components of the system. Protons and electrons are produced by bacteria during oxidation, and the electrons are carried to the cathode via the external circuit. The electricity generated utilising various organic samples over a 10-day retention period is analysed and discussed.

Thulasinathan B., et al. (n.d.): Emphasise the potential for addressing the issues of MFC systems' poor power production and lack of economic viability by utilising the plentiful waste biomass resources. With the use of nitrogen-rich protein food industry wastewater as the substrate in a dual-chambered MFC, 81 MW/m² of electricity and 95% COD removal were achieved. According to the data, acetate is the most suitable substrate for MFC performance, with highest power density and COD removal rates of 114 MW/m² and 79%, respectively.

Vishwanathan A. S. (2021): Discusses microbial fuel cells (MFCs) in particular and gives a thorough overview of microbial electrochemical systems. Potter originally explained the idea in a 1911 description. Still, until the last few decades, there were few real-world applications. The catalysis of electron release from organic substrates is a key function of microbes in MFCs. Stable, conductive, and biocompatible are the three qualities of an effective electrode. Though they have several uses, MFCs are mostly used for producing power and treating wastewater.

Chaturvedi V., and Verma P. (2016): The oxidation of organic substances at the anode of microbial fuel cells (MFCs) produces electricity. The electrons are transferred to the cathode, where they are used for the reduction of various electron acceptors. Various substances, including hexavalent chromium, agro-wastes, azo dyes, selenite, and nitrate, have been investigated as potential substrates for electricity generation in MFCs.

Tomar A., et al. (2020): This study focused on the composition and microflora of cow dung and identified Bacillus as the bacteria found in the highest amount compared to other bacteria such as Enterococcus, Diplococcus, and Pseudomonas. These findings provide insights into the microbial diversity of cow dung and its potential antimicrobial properties.

Gupta K. K., et al. (2016): Cow dung is a versatile and easily available bioresource that has various potential uses and benefits. It contains microorganisms that can produce different metabolites and is a major source of biogas in India. Cow dung can be used in biogas plants to produce methane gas, which can be used for cooking, lighting, and even generating electricity. Additionally, it has the potential to be used as a source of activated carbon for supercapacitors, which are used for energy storage in electrochemical batteries.

Priyambada Ika Bagus and Wardana Irawan Wisnu (2018): Food waste is the leftover food that's not consumed and waste generated during food preparation, cooking, and serving. It's made up of animal and vegetable waste, with a density of about 1 pound per cubic meter and a moisture content ranging from 50% to 80%. Composting is a process of decomposing organic matter through biological processes with the help of decomposers. Macronutrients consist of nitrogen (N), phosphorus (P), potassium (K), calcium (Ca), magnesium (Mg), and sulphur (S). On the other hand, micronutrients consist of iron (Fe), manganese (Mn), zinc (Zn), copper (Cu), molybdenum (Mo), sodium (Na), and boron (B).

Gupta S. S., et al. (2020): Discuss the issue of solid waste management in India, specifically focusing on food waste management. It introduces a method of addressing this problem through rotary drum composting. The proposed solution uses a rotary drum composter to manage food waste. The key advantage is the acceleration of the composting process. This method offers agitation, aeration, and waste mixing, creating favourable conditions for composting various waste types, including food waste, vegetable waste, dry leaves, and paper waste.

Unuofin F. O. and Siswana M. (2019): Experimented to investigate the impact of different sources of phosphorus (P) on the decomposition of waste mixtures during vermicomposting. The primary focus is on whether P or calcium (Ca) from various sources, including rock phosphate (RP), triple superphosphate (TSP), phosphoric acid (PHA), and calcium chloride (CaCl2), affect the decomposition process.

Peighambardoust S. J. et al. (2010): Provide a thorough introduction to fuel cells, emphasising those with proton exchange membranes (PEMs). Fuel cells are electrochemical devices that, without burning, transform a fuel's chemical energy such as that of hydrogen, methanol, or ethanol—into electrical energy. In a PEM fuel cell, hydrogen undergoes core reactions on the anode electrode where it is converted into hydrogen ions and electrons and on the cathode electrode where it combines with oxygen to form water and heat

Jincheng Wei, et al. (2011): This research discusses microbial fuel cells (MFCs), a technology that generates electricity with bacteria while purifying wastewater. However, large-scale implementation of MFCs has been hindered by low power generation yields and high costs. To overcome these challenges, the design and materials of the electrodes used in MFCs should have good conductivity, chemical stability, mechanical strength, and low cost. The study focuses on cost-effective materials and architectures to scale up MFCs for practical wastewater treatment systems.

Kamau J. M., et al. (2017): Discuss the concept of electrical resistance in the context of microbial fuel cells (MFCs). While MFCs are designed to generate electrical current, they also have internal resistance. When electrons flow within the MFCs, they encounter resistance from the materials within the cell. Internal resistance in MFCs can be further categorised into anodic, cathodic, and electrolyte resistance (including the membrane if present). The study presents an experimental setup for MFCs. Different resistors were used, and the MFC was operated under anaerobic conditions for 9 days.

Jonathan Ramirez-Nava, et al. (2021): Microbial fuel cells (MFCs) depend primarily on electrochemical membranes. Membranes range into two primary categories: porous and nonporous. Bipolar membranes (BPMs), anion-exchange membranes (AEMs), and cation-exchange membranes (CEMs) are the three types of nonporous membranes. In MFCs, a membrane separator's principal objectives are to prevent O_2 from moving from the aerobic to the anaerobic chamber, ionic conduction, and short circuiting.

Schneider G., et al. (2023): Focus on isolating electrogenic bacteria from environmental mud samples, to explore their potential applications in Microbial Fuel Cell (MFC) systems. To achieve this goal, the researchers employed a series of methods and assays to characterise these isolates' electrogenic properties, biofilm-forming capabilities, and enzymatic potential.

KeChrist Obileke A., et al. (2021): Put forward a review of Microbial Fuel Cell, as a renewable energy source for electricity generation. The research is focused on the different components of an MFC. It gives a comprehensive overview of factors responsible for the efficient performance of MFC such as pH level, temperature, material for electrodes, and proton exchange membrane. The author also gives a review of single-chamber MFC and double-chamber MFC.

Jianfei Wang, et al. (2022): Microbial Fuel Cells (MFCs) use microorganisms to turn organic matter into electricity. Selecting and modifying the right microbial strains is crucial for MFCs' performance. Genetic engineering can enhance MFC systems. However, scaling up MFCs and substrate selection are still challenging. Available substrates include defined substrates, wastewater, and lignocellulosic biomass, each with different benefits. The choice of substrate depends on factors such as the specific application, availability, and desired outcomes.

4. Discussion

The research demonstrates the potential of MFCs as a sustainable technology for electricity generation. The studies show electricity generation from various substrates, including septic wastewater and waste biomass. Several studies focus on the environmental application of MFCs, particularly in wastewater treatment. MFCs can simultaneously clean up wastewater while generating electricity, making them a promising solution for sustainable water management. There's a recognition of the importance of electrode materials in MFC design. Efficient and cost-effective materials are crucial for enhancing power generation and reducing the overall cost of MFC systems. While MFCs hold great promise, challenges such as low power output, internal resistance, and cost constraints need to be addressed for broader practical implementation in fields like solid-waste treatment and renewable energy generation. The studies mentioned offer valuable insights and contribute to our understanding of MFCs and related fields, paving the way for future advancements and applications in this exciting area of research.

5. Conclusion

Sewage wastewater can pose several limitations for microbial fuel cells (MFCs) due to its complex mixture of organic and inorganic compounds with varying pH levels. This mixture can have a negative impact on the stability and performance of MFCs. The presence of harmful contaminants like heavy metals, antibiotics, and pathogens can decrease the effectiveness of MFCs, making it necessary to remove them before use. Although sewage contains natural matter, it may lack essential vitamins needed for microbial growth, which can limit the ability of electrogenic micro-organisms to generate energy within the MFC. Additionally, it can cause fouling and scaling on the electrodes and membranes of MFCs, which reduces efficiency and requires frequent

maintenance. To overcome these challenges, Cow Dung and Biodegradable Kitchen waste can be used as they have uniform pH levels and do not contain any chemical or biological contaminants, such as heavy metals, antibiotics, and pathogens.

References

Akatah, B. M., Kalagbor, I. A., and Gwarah, L. S. (2019). Electricity generation from septic waste water using septic tank as microbial fuel cell. Sustainable Energy, 7(1), 1–5, doi: 10.12691/rse-7-1-1.

Central Electricity Authority, Ministry of Power, Government of India, 2023–24.

Chaturvedi, V., and Verma, P. (2016). Microbial fuel cell: A green approach for the utilization of waste for the generation of bioelectricity. Bioresour. Bioprocess, 3, 38. doi: 10.1186/s40643-016-0116-6

Dey, Subhashish, Sreenivasulu, Anduri, Veerendra, G.T.N., Rao, K. V., and Anjaneya Babu, P. S. S. (2022). Renewable energy present status and future potentials in India: An overview. doi: 10.1016/j.igd.2022.100006

Gupta, K.K., Aneja, K.R., and Rana, D. (2016). Current status of cow dung as a bioresource for sustainable development. Bioresour. Bioprocess. 3, 28. doi: 10.1186/s40643-016-0105-9

Gupta, S. S., Bisen, R. D., Badole, S. H., Nandagawali, V., Dewalkar, P., and Marathe, N. (2020). A paper on the decomposition of food waste by rotation drum method. International Research Journal of Engineering and Technology (IRJET), 07(02):2477-2478.

Jessica Li. (2013). An experimental study of microbial fuel cells for electricity generating: Performance characterization and capacity improvement. Journal of Sustainable Bioenergy Systems, 03(03), 171–178. doi: 10.4236/jsbs.2013.33024

Jianfei Wang, Kexin Ren, Yan Zhu, Jiaqi Huang, Jiaqi Huang, and Shijie Liu (2022). A review of recent advances in microbial fuel cells: Preparation, operation, and application. Advances in Environmental. Biotechnology (AEB). BioTech 2022, 11(4), 44. doi: 10.3390/biotech11040044

Jincheng Wei, Peng Liang, and Xia Huang (2011). Recent progress in electrodes for microbial fuel cells. Bioresource Technology, 102(20). doi: 10.1016/j.biortech.2011.07.019

Kamau, J. M., Mbui, D. N., Mwaniki, J. M., Mwaura, F. B., and Kamau, G. N. (2017). Microbial fuel cells: Influence of external resistors on power. Current and Power Density Journal of Thermodynamics & Catalysis, 08(01). doi: 10.4172/2157-7544.1000182

Ministry of Power, Government of India, 2023–2024.

Mittal, M. L., Sharma, C., and Singh, R. Estimates of emissions from coal-fired thermal power plants in India.

Obileke, KeChrist, Onyeaka, H., Meyer, E. L., Nwokolo, N. (2021). Microbial fuel cells, a renewable energy technology for bio-electricity generation: A mini-review. Electrochemistry Communications, 125(5):107003.

Peighambardoust, S. J., Rowshanzamir, S., and Amjadi, M. (2010). Review of the proton exchange membranes for fuel cell applications. International Journal of Hydrogen Energy, 35(17): 9349–9384

Priyambada, I. B., and Wardana, I. W. (2018). Fast decomposition of food waste to produce mature and stable compost. Sustinere Journal of Environment and Sustainability, 2(3), 156–167. doi: 10.22515/sustinere.jes.v2i3.47.

Ramirez-Nava, J., Martínez-Castrejón, M., García-Mesino, R. L., López-Díaz, J. A., Talavera-Mendoza, O., Villagrana, A. S., Rojano, F., and Hernández-Flores, G. (2021). The implications of membranes used as separators in microbial fuel cells. Applied Ion-Exchange Membrane Technologies for Sustainable Energy Production,

Membranes, 11(10), 738. https://doi.org/10.3390/membranes11100738.

Schneider, G., Pásztor, D., Szabó, P., Kőrösi, L., Kishan, N. S., Raju, P. A. R. K., and Kaur Calay, R. (2023). Isolation and characterisation of electrogenic bacteria from mud samples. New Electrogenic Microbes.

Thulasinathan, B., Jayabalan, T., Arumugam, N., Kulanthaisamy, M. R., Woong Kim, Kumar, P., Govarthanan, M., and Alagarsamy, A. (2022). Wastewater substrates in microbial fuel cell systems for carbon-neutral bioelectricity generation: An overview. doi: 10.1016/j.fuel.2022.123369

Tomar, A., Choudhary, S., Kumar, L., Singh, M., Dhillon, N., and Arya, S. (2020). Screening of bacteria present in cow dung. International Journal of Current Microbiology and Applied Sciences. doi: 10.20546/ijcmas.2020.902.073

Unuofin, F. O., and Siswana, M. (2019). Enhancing organic waste decomposition with the addition of phosphorus and calcium through different sources. International Journal of Recycling of Organic Waste in Agriculture, 8, 139–150. soi: 10.1007/s40093-018-0239-1

Varma, S. A. K., Srimurali, M., and Varma, M. K. R. (2013). Ozone pollution in India due to power plant emission. International Journal of Scientific Research, 2(5). doi: 10.15373/22778179/May 2013/55.

Vishwanathan, A.S. (2021). Microbial fuel cells: a comprehensive review for beginners. 3 Biotech, 11, 248. doi: 10.1007/s13205-021-02802-y

35 Supercapacitor-based metro bus transportation

Gaurav Gadge[a], Jyoti Rothe[b], Laxmi Katwale,
Dhanshree Ingole, Yash Dongre, Samay Patil, Jai Kongalwar,
and Vijay Kumar Sahu

Department of Electrical Engineering, SVPCET, Nagpur, India

Abstract

The "Super Capacitor Based Bus" system is a pioneering urban transportation solution designed to optimise energy efficiency, reduce environmental impact, and ensure reliable mass transit. It employs state-of-the-art supercapacitor technology as its primary energy storage, surpassing conventional fossil fuel buses. Integration of advanced components like Arduino Nano, IR sensors, motors, diodes, and LM2595 voltage regulators ensures smooth operation and peak performance. The system architecture features an Arduino Nano microcontroller as the central processing unit, coordinating sensor and actuator communication. IR sensors provide real-time data to prevent collisions, while a supercapacitor-powered motor drives the bus's wheels for smooth movement along routes. Diodes regulate voltage, and LM2595 voltage regulators ensure a stable power supply. Comprehensive testing shows the system's energy efficiency, reliability, and environmental benefits, leading to significant savings in expenses, emissions, and energy usage compared to traditional buses. Its scalability, adaptability, and compatibility with future supercapacitor advancements make it a transformative solution for urban transit challenges, promising cleaner, greener, and more efficient mass transit networks globally.

Keywords: Super capacitor based bus system, energy efficiency, urban transportation challenges

1. Introduction

The "Super Capacitor Based Bus" system revolutionises urban transit by enhancing energy efficiency, reducing environmental impact, and ensuring reliable mass transit (Ang et al., 2022). Utilising supercapacitor technology, it overcomes the limitations of traditional fossil fuel-dependent buses, offering rapid charging, high power density, and longer cycle life (Ortenzi et al., 2018; Ortenzi et al., 2019). Advanced components like Arduino Nano microcontrollers, infrared sensors, and supercapacitor-powered motors ensure seamless functionality, collision avoidance, and smooth navigation (Yu et al., 2018). The system addresses global challenges of energy consumption, traffic congestion, and pollution, demonstrating improved energy efficiency and sustainability. With its scalability, compatibility with future advancements and potential cost savings, it provides a transformative and greener alternative to traditional buses, optimising passenger comfort, safety, and system reliability.

[a]ggadge@stvincentngp.edu.in, [b]jrothe@stvincentngp.edu.in

DOI: 10.1201/9781003567653-35

2. Literature Review

The Super Capacitor Based Bus system offers a solution to urban transportation challenges by using supercapacitors for energy storage, enhancing energy efficiency and sustainability. It leverages rapid charging capabilities suitable for bus applications like regenerative braking. Ortenzi et al. focus on efficient supercapacitors for electric buses, achieving rapid charging in 30 seconds without DC–DC converters (Ortenzi et al., 2019). Sung explores supercapacitors' potential in energy storage, classifying them into electrochemical double layer and pseudo capacitors (Sung & Shin, 2020). Lin *et al.* propose an energy-management technique for supercapacitors, optimising bidirectional energy flow and showcasing rapid response capabilities through simulations (Lin, 2021). Makar et al. advocate for supercapacitors in elevators, capturing regenerative energy with a six-phase interleaved DC/DC converter, demonstrating significant energy-saving potential in simulations (Makar et al., 2022). Momcilovic *et al.* highlight e-buses as zero-emission urban transit options, addressing range anxiety challenges with a simulation framework that considers real-world conditions to improve service quality and ridership (Momcilovic et al., 2023).

3. Working

The Figure 35.1 shows the Block diagram of the working of the Super Capacitor Based Bus system involves several interconnected components and subsystems working together to ensure efficient energy management, propulsion, and operation. The PCB Layout of the Super Capacitor Based Bus system as shown in Figure 35.2. Below are detailed points outlining the working of the entire system:

- **Energy Storage and Management:** The Super Capacitor Based Bus system uses supercapacitors for energy storage.

During regenerative braking, they capture and store energy, converting heat into electrical energy. An advanced algorithm optimises the supercapacitors' charging and discharging, enhancing energy efficiency.

- **Power Conversion and Regulation:** LM2595 converters regulate voltage for system components, stepping down the supercapacitor bank's voltage to match the requirements of sensors, microcontrollers, and actuators. They ensure stable operation under varying loads.

- **Sensor Integration and Collision Avoidance:** Infrared (IR) sensors detect obstacles, aiding collision avoidance, pedestrian detection, and proximity sensing. Detected objects prompt actions like braking or driver alerts.

- **Microcontroller Coordination:** An Arduino Nano microcontroller coordinates sensor and actuator communication, interpreting sensor signals and initiating responses. It monitors system parameters and manages energy flow for seamless operation.

- **Actuator Control and Propulsion:** Motors powered by supercapacitors drive the bus, managing acceleration, deceleration, and route navigation. Controlled by the microcontroller, they adapt power delivery to real-time conditions like speed and braking, optimising energy use and performance.

- **Safety Features and Regulatory Compliance:** The system features overvoltage and overcurrent protection, guarding against spikes and excessive currents. It meets regulatory standards, ensuring system reliability and compatibility with existing infrastructure.

- **Testing, Validation, and Optimisation:** Rigorous testing evaluates energy consumption, efficiency, and reliability. Feedback guides parameter adjustments, addressing issues and enhancing system efficiency. The Super Capacitor Based Bus system aims for sustainable and efficient mass transit.

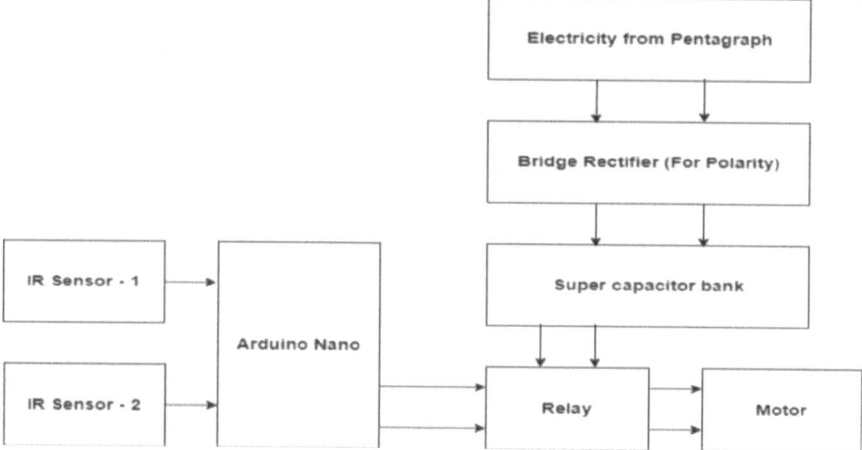

Figure 35.1: Block diagram.

Source: Author.

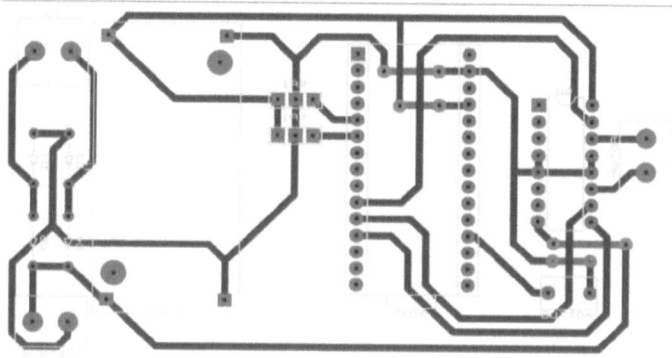

Figure 35.2: PCB layout.

Source: Author.

4. Implementation

The Super Capacitor Based Bus system's implementation as shown in Figure 35.3 starts with analysing requirements, designing hardware and software, and testing for compatibility and reliability. Components are assembled into a prototype, with software algorithms enabling intelligent decision-making. Simulation tools validate system behaviour prior to field trials in urban settings, assessing performance and guiding refinements. This approach signifies a notable advancement in sustainable urban mass transit.

5. Results

The implementation of the Super Capacitor Based Bus system has delivered promising outcomes in energy efficiency and environmental impact. Rigorous testing reveals significant reductions in energy consumption compared to traditional buses, thanks to supercapacitor technology and smart energy management algorithms. Integration of regenerative braking systems further enhances energy efficiency. Additionally, the system improves passenger comfort and safety with infrared sensors for collision avoidance and real-time monitoring,

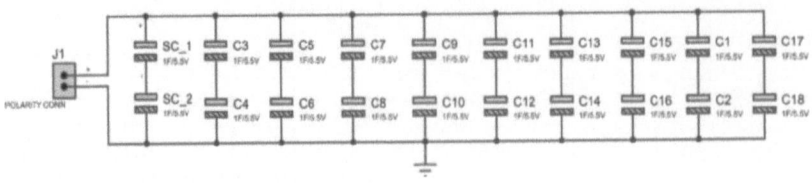

Figure 35.3: Super capacitor circuit.

Source: Author.

demonstrating its potential to transform urban transportation sustainably.

6. Conclusion

The Super Capacitor Based Bus system is a groundbreaking solution for sustainable urban transportation, integrating advanced technologies like supercapacitors and electric motors. Its implementation has led to significant reductions in energy consumption and emissions, enhancing urban environments. Improvements in passenger safety and comfort are evident through the integration of infrared sensors and real-time monitoring. The system's success highlights the importance of innovation and collaboration in addressing urban challenges. As cities tackle urbanisation and climate change, initiatives like this offer hope for building smarter, greener cities.

References

Ang, J. H., Yusup, Y., Zaki, S. A., Salehabadi, A., and Ahmad, M. I. (2022). Comprehensive energy consumption of elevator systems based on hybrid approach of measurement and calculation in low- and high-rise buildings of tropical climate towards energy efficiency. Sustainability, 14, 4779.

Lin, B. (2021). Review of energy management strategy for super capacitor energy storage system based on phase shifted full bridge converter.

Magrini, A., Lentini, G., Cuman, S., Bodrato, A., and Marenco, L. (2020). From Nearly Zero Energy Buildings (NZEB) to Positive Energy Buildings (PEB): The next challenge – The most recent European trends with some notes on the energy analysis of a forerunner PEB example. Developments in the Built Environment, 3, 100019.

Makar, M., Pravica, L., and Kutija, M. (2022). Review of supercapacitor-based energy storage in elevators to improve energy efficiency of buildings.

Momcilovic, V., Dimitrijevic, B., and Stokic, M. (2023). Review of supercapacitor electric bus modeling and simulation framework.

Ortenzi, F., Pasquali, M., and Prosini, P. P. (2019). Review of design and validation of ultra Ortenzi, F., Pede, M. P. G., Lidozzi, A., and di Benedetto, M. (2018). Ultra-fast charging infrastructure for vehicle on-board ultracapacitors in urban public transportation applications. Proceedings of the EVS31— The 31st International Electric Vehicles Symposium & Exhibition & International Electric Vehicle Technology Conference, Kobe, Japan, September 30–October 3.

Sung, J., and Shin, C. (2020). Review of recent studies on supercapacitors with next-generation structures.

Wilberforce, T., Olabi, A.G., Sayed, E.T., Elsaid, K., Maghrabie, H.M., and Abdelkareem, M.A. (2021). A review on zero energy buildings—Pros and Cons. Energy and Built Environment, In press.

Yu, H., Cheli, F., and Castelli-Dezza, F. (2018). Optimal design and control of 4-IWD electric vehicles based on a 14-DOF vehicle model. IEEE Transactions on Vehicular Technology, 67, 10457–10469.

36 GSM and IOT based smart agriculture monitoring system

Kirti Vaidya[1,a], Nitin Dhote[2,b], Jayant Meshram[3,c], Shubhangi Kaikadi[1,d], Sagar Vaidya[1,e], Hrishika Dhakad[1,f], Raj Shripad[1,g], Yuktashri Kurapati[1,h], and Vedant Nandokar[1,i]

[1]Department of Electronics and Telecommunication, St. Vincent Pallotti College of Engineering and Technology, Nagpur, India
[2]Department of Electrical Engineering, St. Vincent Pallotti College of Engineering and Technology, Nagpur, India
[3]Principal Scientist, ICAR-Central Institute of Cotton Research, Nagpur, India

Abstract

In every country, agriculture, considered both a science and an art, has been practiced for ages. As technology advances in our daily lives, it becomes imperative to modernize agricultural practices as well. This is where IoT (Internet of Things) steps in, playing a pivotal role in the realm of smart agriculture. IoT sensors are utilized to gather crucial information about agricultural fields, facilitating efficient monitoring and management. With the advent of IoT, smart agriculture is powered by devices like NodeMCU, incorporating various sensors such as humidity, temperature, moisture, and even a DC motor. This system initiates monitoring of humidity and moisture levels, automatically triggering watering when levels fall below a certain threshold. Moreover, it adjusts irrigation schedules based on changes in temperature, while providing detailed data on humidity, moisture, date, and time. Meanwhile, the Agriculture Monitoring System, employing cutting-edge technology like ESP32, sensors, and GSM modules, offers a comprehensive solution for optimizing agricultural activities. It ensures real-time tracking of environmental parameters like humidity, soil moisture, and temperature. By integrating a relay module and pump, automated irrigation is facilitated based on sensor data, promoting sustainable farming practices, minimizing water wastage, and empowering farmers to make informed decisions. This amalgamation of IoT-driven smart agriculture and advanced monitoring systems signifies a significant leap towards enhancing agricultural productivity and sustainability.

Keywords: Internet of Things, soil moisture sensor, temperature sensors, ESP 32, relay, Wi-Fi module ES, Blynk app, agriculture monitoring system

[a]kvaidya@stvincentngp.edu.in, [b]ndhote@stvicnentngp.edu.in, [c]j.h.meshram@gmail.com, [d]shubhangikaikadi00@gmail.com, [e]sagarvaidya251@gmail.com, [f]hrishikadhakad1@gmail.com, [g]rajshripad0710@gmail.com, [h]yuktasrikurapati0803@gmail.com, [i]vedantnandokar3@gmail.com

DOI: 10.1201/9781003567653-36

1. Introduction

India, renowned for its agricultural heritage, relies heavily on this sector as the primary source of livelihood for millions. However, the burgeoning population necessitates a significant increase in agricultural yield to sustain the growing demand for food. By 2050, (World Population Projected to Reach 9.8 Billion in 2050, and 11.2 Billion in 2100, n.d.) the challenge looms large: food production must double to meet the needs of the expanding populace (Food Production Must Double by 2050 to Meet Demand From World's Growing Population, n.d.).

Traditional farming methods, while effective, are labor-intensive and demand constant physical presence and monitoring from farmers. This not only consumes valuable time but also limits their ability to engage in other activities that could improve their standard of living. Considering these challenges, there is a need for innovative solutions that can optimize agricultural practices and empower farmers to manage their fields remotely.

The emergence of the Internet of Things (IoT) offers a promising avenue for revolutionizing agricultural management. IoT technology enables the seamless combination of actuators, sensors, and communication devices, allowing for real-time data collection, analysis, and control of agricultural processes. This paradigm shift towards smart agriculture presents opportunities for enhanced efficiency, optimal resource management, remote monitoring, and improved crop yields.

In this context, we propose a GPS and IoT-based agriculture monitoring system aimed at addressing the challenges faced by farmers in India and beyond. By leveraging GPS technology for location tracking and IoT devices for data acquisition and transmission, our system offers farmers unprecedented insights into the conditions of their fields. From soil moisture levels to weather patterns, farmers can remotely monitor crucial parameters and make informed decisions to optimize crop production.

Furthermore, the integration of GPS and IoT technologies enables precision agriculture, wherein all resources such as fertilizers, water, pesticides are used as per requirement. This not only minimizes waste but also maximizes resource utilization, leading to higher yields and reduced environmental impact.

In conclusion, the proposed GPS and IoT-based agriculture monitoring system represents a paradigm shift in agricultural management, offering farmers the tools they need to overcome challenges posed by population growth and increasing food demand. By harnessing the power of technology, we aim to empower farmers, enhance productivity, and ensure food security for generations to come.

2. Literature Review

The projected global population growth, as outlined by the United Nations (World Population Projected to Reach 9.8 Billion in 2050, and 11.2 Billion in 2100, n.d.), poses significant challenges for food production and agricultural management. The estimated population is 9.8 Billion by 2050. So there is a pressing need to increase food production to meet the demands of this growing population (Food Production Must Double by 2050 to Meet Demand from World's Growing Population, n.d.).

In response to these challenges, researchers have been exploring innovative technologies to improve agricultural management and enhance food production efficiency. One such technology is the integration of wireless sensors and Internet of Things (IoT) devices into agricultural systems. Navulur and Prasad (2017) highlighted the potential of agricultural management through wireless sensors and IoT, emphasizing its role in optimizing resource utilization and enhancing crop yields.

Several studies have focused on use of internet of things for crop field inspection monitoring and automation of irrigation. Rajalakshmi and Devi Mahalakshmi (2016) presented an IoT-based approach for crop field monitoring and automation of irrigation, demonstrating its effectiveness in optimizing water usage and improving crop health. Similarly, in (Gutierrez et al., 2013) author proposed an irrigation system which is automated using a wireless sensor network. Also it uses GPRS module. This system helps in monitoring field conditions and irrigation processes.

Furthermore, IoT technologies have been applied to develop real-time automation plus monitoring systems for modernized agriculture researchers (Vidya Devi & Meena Kumari, 2013) introduced a real-time automation along with monitoring system designed to enhance agricultural productivity and streamline management practices.

Beyond irrigation and crop monitoring, IoT solutions have been explored for other agricultural applications as well. Researchers in (Basha and Rus, 2007) focused on the design of early warning flood detection systems for developing countries, highlighting the importance of leveraging IoT technologies for disaster management in agricultural regions.

Moreover, IoT-based agriculture systems have been developed using various platforms and methodologies in (Jyostna Vanaja et al., n.d.) proposed an IoT-based agriculture system using NodeMCU, demonstrating the feasibility of integrating low-cost IoT platforms for agricultural applications. Mat et al. (2016) explored the use of wireless moisture sensor networks in precision agriculture, emphasizing the role of IoT in optimizing resource usage and improving crop quality.

Additionally, researchers have developed intelligent management systems for agricultural greenhouses based on IoT principles. In his paper author (Zhao Chan Li et al., 2017) presented the design of an intelligent management system for agricultural greenhouses, leveraging IoT technologies to monitor environmental conditions and optimize greenhouse operations.

In summary, the literature review highlights the growing interest in leveraging IoT technologies for agriculture monitoring systems. These systems offer opportunities to enhance productivity, optimize resource utilization, and mitigate risks, thereby contributing to the sustainable management of agricultural resources in the face of increasing global population and food demand.

3. Working of AMS

In this paper we propose an IOT based agriculture management system (AMS)which will help the farmers to monitor the fields conditions sitting at their homes or from anywhere. The important soil parameters are displayed on cloud using blynk software. Depending on soil moisture motor pump can be made on for certain amount of time.

The Agriculture Monitoring System, depicted in Figure 36.1, represents a sophisticated solution aimed at bolstering the efficiency and productivity of agricultural endeavours. Central to its operation is the ESP32, a robust microcontroller serving as the system's core processing unit.

Equipped with an array of sensors, including the DHT11 sensor for humidity and temperature measurement, the PIR sensor for motion detection, and a soil moisture sensor for continuous monitoring of soil moisture content, this system empowers farmers with actionable data for informed decision-making. Furthermore, incorporating a GSM Module enables seamless remote communication, allowing users to oversee and manage the system's functionality from any location. A DC to DC Converter ensures compatibility among various modules by adjusting voltage levels as needed.

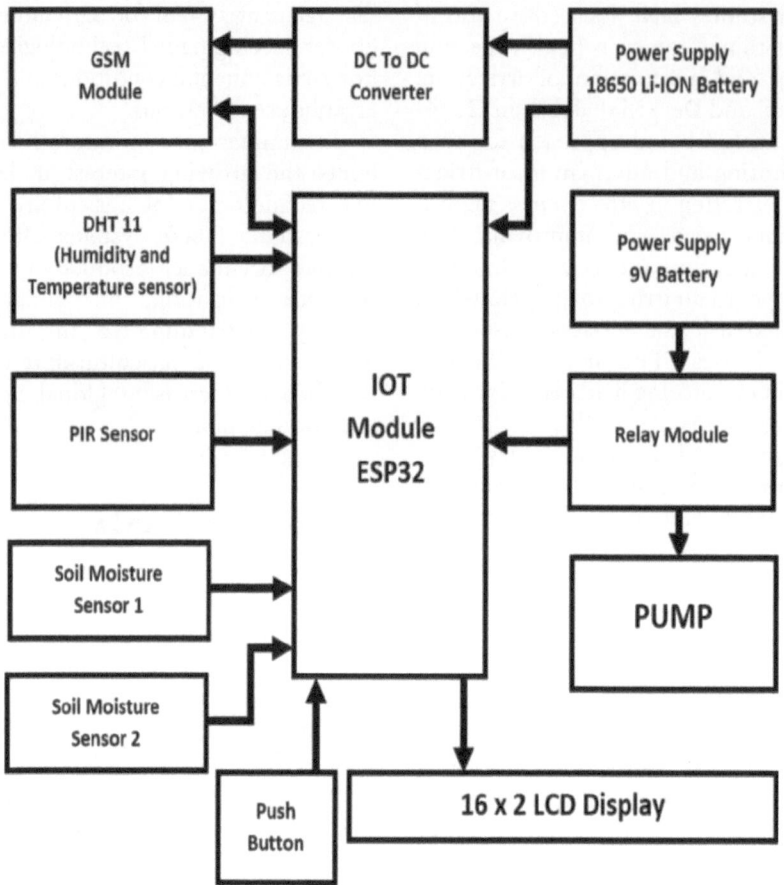

Figure 36.1: Block diagram of agriculture monitoring system.

Source: Author.

With a relay module linked to a pump for automated irrigation, water consumption is optimized based on soil moisture sensor data, ensuring crops receive the appropriate amount of water optimal times while conserving this vital resource. Real-time sensor readings are displayed on a 16 × 2 LCD Display, providing users with immediate insights into system performance.

Recognizing the significance of agriculture as a primary livelihood source in India and its pivotal role in sustaining human life, there's a growing imperative to enhance agricultural productivity, particularly given the challenges posed by seasonal variations and water scarcity. In response, IoT-based smart agriculture systems are being deployed to address these challenges effectively.

Global and regional agricultural monitoring systems leverage IoT technology to offer timely updates on food production, facilitating informed decision-making for farmers. Utilizing sensors for monitoring various field conditions like light, humidity, temperature, and soil moisture, IoT-based smart farming systems enable remote monitoring and enhance efficiency compared to traditional methods.

An example of this is the proposed IoT-based Irrigation System utilizing the ESP32 Module and DHT11 Sensor. This system not only automates irrigation based on soil moisture levels but also transmits data to Blynk Server for land condition tracking. Recent advancements in sensor technology and the evolution of IoT and WSN (Wireless Sensor Networks) have paved the way for automatic irrigation systems, capable of monitoring crucial parameters like water quantity, weather, soil properties, and limited use of fertiliser. These systems offer a holistic approach to smart irrigation, leveraging wireless technologies to optimize agricultural practices for improved productivity and sustainability.

4. Results

This work is done as a part of undergraduate project of final year students of St. Vincent Pallotti college of Engineering and Technology. It is done in association with Central Institute for cotton research Nagpur.

The Figure 36.2 is the overall look of the project consists of microcontroller ESP32, soil moisture sensors which is used. for smart farming in the agriculture field. These sensors are connected to microcontroller ESP32 which is responsible for controlling of these sensors. Threshold values are fixed based on the sensors.

For example if the threshold value of sensor is 1000

Sensor 1- If the (crop 1 = Rice) condition is 400 its for rice cultivation. If it satisfied the condition of 700 then the motor is automatically off. If not then it automatically on.

Sensor 2- If the (crop 2 = sugarcane) condition is 800 its for rice cultivation. If it satisfied the condition of 700 then the motor is automatically off. If not then it automatically on.

It indicates the value of soil moisture sensors.

It indicates temperature and humidity

It also shows motor on or off.

It also show whether the motion is detected or not. If not detected it will show off.

Moisture sensor sense the moisture content of the soil and based on the value automatic water pump will be ON to increase the moisture content. The use of automated monitoring and management system are gaining increasing demand with the technological advancement. In agricultural field loss of yield mainly occurs due to widespread of disease and traditional farming

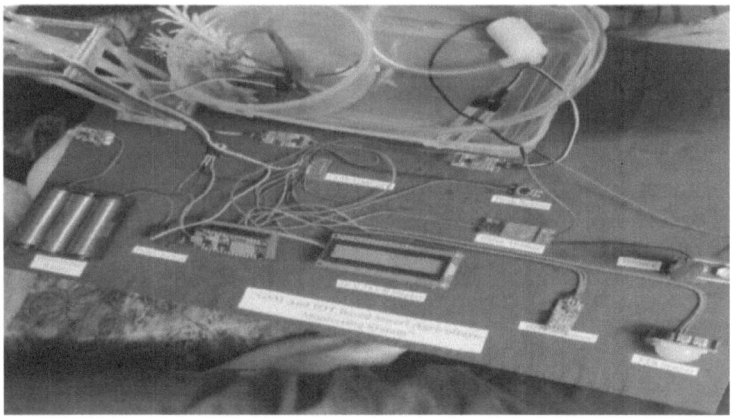

Figure 36.2: Implementation of AMS system.
Source: Author.

techniques. Mostly the detection and identification of the disease is noticed when the disease advances to severe stage.

5. Conclusion

The Agriculture Monitoring System, incorporating components such as the ESP32, various sensors, and a GSM Module, represents a significant advancement in precision agriculture. Offering real-time monitoring and automation of critical parameters, it optimizes resource usage, enhances sustainability, and improves crop yield and quality. With features like remote accessibility and a user-friendly interface facilitated by the Blynk server and Arduino IDE, the system empowers farmers to make informed decisions and maximize productivity. Overall, the Agriculture Monitoring System embodies the potential of technology to revolutionize farming practices, offering efficient, sustainable solutions to address the challenges of modern agriculture.

References

Basha, E., and Rus, D. (2007). Design of early warning flood detection systems for developing countries. International Conference on Information and Communication Technologies and Development.

Food Production Must Double by 2050 to Meet Demand From World's Growing Population. (n.d.). Accessed: Apr. 5, 2019. [Online]. https://www.un.org/press/en/2009/gaef3242.doc.htm

Gutierrez, J., and Villa-Medina, J. F., et al. (2014). Automated irrigation system using a wireless sensor network and GPRS module. IEEE Transactions on Instrumentation and Measurement, 63(1), 166–176.

Jyostsna Vanaja, K., Suresh, A. et al. (n.d.). (2018). IOT based agriculture system using NodeMCU. International Research Journal of Engineering and Technology.

Li, Zhaochan, Jinlong Wang, Russell John Higgs, Li Zhou, and Wenbin Yuan. (2017). Design of an intelligent management system for agricultural greenhouses based on the internet of things. IEEE International Conference on Computational Science and Engineering and IEEE International Conference on Embedded and Ubiquitous Computing, 02(2017), 154–160.

Mat, I., Kasim, M. R. M., Harun, A. N., and Yusuf, I. M. (2016). IOT in precision agriculture applications using wireless moisture sensor network. IEEE Conference on Open Systems (ICOS), October 10-12-2016, Langkaw, Malaysia.

Navulur, S., and Giri Prasad, M. N. (2017). Agricultural management through wireless sensors and Internet of Things. International Journal of Electrical and Computer Engineering, 7(6), 3492–3499.

Rajalakshmi, P., and Devi Mahalakshmi, S. (2016). IOT based crop field monitoring and irrigation automation. 10th International Conference on Intelligent Systems and Control (ISCO).

Vidya Devi, V., and Meena Kumari, G. (2013). Real time automation and monitoring system for modernized agriculture. International Journal of Review and Research in Applied Sciences and Engineering, 3(1), 7–12.

World Population Projected to Reach 9.8 Billion in 2050, and 11.2 Billion in 2100. (n.d.). Accessed: Apr. 18, 2019. [Online]. https://www.un.org/development/desa/en/news/population/world-population-prospects2017.html

37 A survey on design and development of travel and itinerary planner

Shankar Gadhave[a], Shreya Pise[b], Nabha Bhongle[c], Ayushi Kolhe[d], Atharva Palandurkar[e], and Raghav Jugade[f]

Department of Information Technology, SVPCET, Nagpur, India

Abstract

With a view of improving user experience and enabling effective trip planning, this study implements the design and development process of a website for travel and itinerary planning. The research looks at several topics, such as user mindset, functionality needs, backend infrastructure and user interface design. Important aspects and design concerns are determined by means of a review of the literature combined with user input. The study also covers the technological frameworks and tools used throughout the development process, emphasizing speed optimization, security, and scalability. The results emphasize how crucial it is to have user-friendly navigation, tailored suggestions, and smooth service integration. The study offers valuable perspectives on the efficient development and operation of trip planning website that meet the changing demands of modern tourists. This project aims to revolutionize the way people plan and experience their travels, making it more convenient, enjoyable, and tailored to individual preferences.

Keywords: travel, itinerary, planner, tourist

1. Introduction

More than ever, effective trip preparation is necessary in this day of cutting-edge technology and growing wanderlust. People want smooth and customized travel experiences, whether they are traveling for work, a single excursion, or a family holiday. As a result, computerized programs called itinerary and travel planners have become increasingly popular. These are meant to make the difficult process of planning travels easier and to make every trip more enjoyable. The age-old activity of exploration and discovery, travel has developed into a complex undertaking in the digital era. Travel and itinerary planners have emerged in response to consumer desire for smooth, customized travel experiences.

This project explores the basic ideas and precepts that guided the creation of this kind of tool, with a strong emphasis on user-centricity, technical innovation, and flexibility in response to the ever-changing nature of travel. Travel has become an essential aspect of our lives in an increasingly connected world because it gives us the opportunity to discover and encounter

[a]shankar_gadhve@yahoo.com, [b]shreyapise09@gmail.com, [c]nabhabhongle27@gmail.com, [d]ayushikolhe18@gmail.com, [e]atharvapalandurkar5@gmail.com, [f]imr4ghav@gmail.com

DOI: 10.1201/9781003567653-37

other cultures, landscapes, and experiences. But even with so many options for places to go and things to do, organizing a trip can often be difficult and time-consuming. This is where our website, which aims to revolutionize travel planning, comes into play as an itinerary planner.

In the digital age, a well-crafted Travel and Itinerary Planner website is not just a tool. It is an immersive experience that caters to the diverse needs and aspirations of modern travelers. This detailed overview explores the key components and considerations in designing and developing such a website, emphasizing user experience, functionality, real-time adaptability, accessibility, data security, and the integration of cutting-edge technologies.

2. Literature Review

In their study author (Dasari et al., 2023) Everybody demands a vacation to relax from their hectic lives, but preparing for these trips takes a lot of energy. A primary cause of this is the dearth of platforms offering customized guidance for holiday arrangements. Users need to look for hotels and restaurants with good reviews on their own and organize a suitable itinerary to visit popular tourist destinations within their budgets. To support them in proposing the route based on their interests, this project will consider the interests of every user. Due to the multifaceted nature of the characteristics this study intends to encompass, a hybrid model has been adopted. Features produced from the gathered data are used to train the created model. Therefore, the model surfaced and can be effectively applied to generate many recommendations for customers. The URLs of various tourist destinations are collected for this hybrid model from websites such as TripAdvisor and Holidays to obtain data about the Point of Passion through Web scraping.

In (Panneerselvam et al., 2022) The "Travel and Tourism Management" project

automates travel and tourist processes, including booking, confirmation, and user data. The travel and tourism management system enables customers to book trips from any location around the world via a centralized customizable website, providing comprehensive information on all trip locations and details. The website's owner can add packages from sure travel agencies and accommodations to the website via an itinerary page. Users can subsequently login in and schedule each project, which is verified by the administrator on their booking administration site. Users may verify the confirmation on the My Bookings page.

In study (Lin et al., 2021) Planning a trip schedule is crucial as more and more individuals are opting to travel to unwind after work due to the advancements in technology and society. The Yangtze River Delta region's twenty-nine picturesque locations were chosen subjects, a trip schedule was established for tour organizing. Three trip scenarios were considered to acquire the route planning: time just, price only, and time and price combined. A simulated annealing approach is used to determine the best routes and costs for each of the three situations once the issue is converted.

In (Lim et al., 2021) This challenging job of suggesting itinerary ideas is exacerbated by the need to account for a few practical limitations such as limited traveling duration, uncertain vehicle traffic, bad weather, tour groups, queuing delays, and congestion. In this survey, they must do a thorough review of the literature on tour itinerary options and propose a fundamental classification for touring-related study.

In (Gokul Krishna et al., 2021) This research paper presents a Budget and Experience-based Travel Planner designed to address the constraints faced by modern travelers, focusing on budget and time limitations while maximizing enjoyment. Through data filtering and machine learning algorithms, the planner recommends destinations tailored to user preferences, leveraging a comprehensive database of over 600 destinations with 77 unique tags. Key findings include optimal travel plans generated through routing and scheduling algorithms, personalized recommendations based on user input, and a focus on user satisfaction. The paper underscores the planner's efficacy in providing efficient and enjoyable travel experiences, considering factors such as distance, budget, and desired experiences, ultimately enhancing the overall travel experience for users.

In survey (Angskun et al., 2020) numerous time-related parameters, involving allowable overall journey duration stated by people, duration related elements at the attraction. Flying, accommodation, and other expenses are aggregated in real time across many airlines and hotels by travel booking websites. It is a terrific method to get a bargain, but almost forty percent of respondents claimed the continuously changing pricing were a major pain point. Apps and websites are not always the greatest. Another 40% of respondents reported dealing with a confused or sluggish travel website.

In (Osmond et al., 2020) Information technology utilization in the lodging sector has the ability to boost revenue for the area while simultaneously improving the standards of amenities offered to travelers and industry stakeholders. Tourism 4.0 refers to the tendency of going digital. Tourism 4.0 is aimed at millennial visitors and has several growths aims. This 4.0 tourism movement is distinguished by changes in visitor behavior, which tends to be independent and unique. Consumer behavior is shifting in a more mobile, personal, and interactive manner because of technological advancements. In this work, we suggest a smart trip planner to let independent travelers arrange their route automatically.

This study (Lopes et al., 2020) suggests a way to help with the organization of travel excursions. Itinerary allows for the declaration of preferences and limitations on a range of topics and provides details about possible areas of interest to study. The main objective is to minimize travel time between locations while maximizing the number of visits within the allotted time. With e-tinerary, you can plan sightseeing excursions, discover new places to visit, and interact with other users, particularly by browsing and using shared travel schedules. Furthermore, travel arrangements can be tailored according to details like preferred locations, available funds, and more. The use of a user-centered design methodology was made. Important criteria for tourism travel planning were first identified, along with the profiles of various target user types and their usage contexts, desires, and motivations. A low-fidelity application prototype was built, and the ideas that resulted were shown to and tested by actual users. This made it possible for us to verify both the conceptual model of the user interface and the recommended functions. Then, using data from external APIs like Wikipedia, Open Weather Map, and Google Maps Platform, the application was developed to offer travel planning information. Its objective was to suggest tours based on travel strategies for optimization while taking the user's preferences and constraints into account.

The paper (Roy et al., 2020) introduces an interactive itinerary planning approach where users provide feedback iteratively to construct personalized itineraries based on their interests and time constraints. Formalizing this process, the study outlines steps where users offer feedback on selected Points-of-Interest (POIs), the system recommends itineraries based on this feedback, and new POIs are suggested for further feedback. Despite the computational complexity of computing itineraries and selecting POIs, the paper presents heuristics and optimizations tailored for scenarios where itinerary scores are proportional to the number of desired POIs. Extensive experiments demonstrate the efficiency and quality of the algorithms proposed, highlighting their effectiveness in generating personalized and satisfactory travel itineraries within reasonable timeframes.

3. Proposed Methodology

The proposed system aims to deliver an innovative travel and itinerary planning website that effectively fulfills the changing demands of modern travelers.

To attain this purpose, a sequential mixed-methods research methodology is used for requirement gathering, with both qualitative and quantitative approaches integrated throughout the developmental and evaluative phases. Initially, the study will include a complete requirement collection and analysis phase. Data will be collected through a variety of means, including online surveys, interviews with travel aficionados, and focus group discussions with regular travelers. A theme-based approach will be used on the obtained data to derive key insights, such as common preferences, pain areas, and vital features for an effective itinerary planner.

The subsequent stage focuses on website development, including both frontend and backend components. Using technologies such as HTML5, CSS3, and JavaScript for frontend development, PHP for backend operations, and XAMPP as the server. The website will be designed to provide a smooth and intuitive user experience across several devices and platforms.

Figure 37.1 is a flow chart diagram of the site navigation begins with the homepage, users can see the features of the website and explore throughout. There are several pages and they are connected to each other through hyperlinks

4. Conclusion

The "Travel and Itinerary" website is a comprehensive resource designed to meet the various demands of tourists from around the globe. With its seamless integration of robust features with user-friendly interfaces, this website enables users to plan, organize, and share their vacation plans with ease. It makes travel planning easier by giving users access to a wealth of destination data, reservations for activities, a variety of lodging options, and teamwork tools.

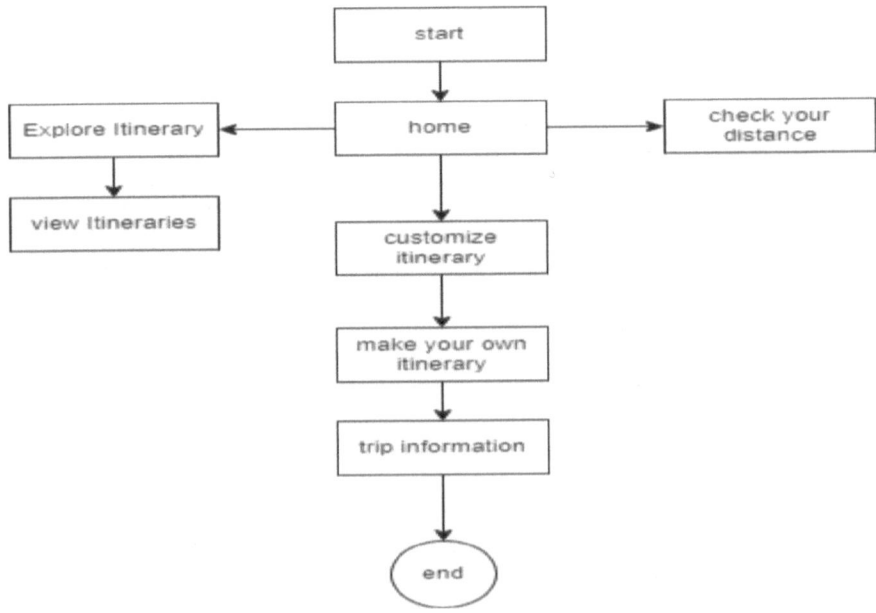

Figure 37.1: Site navigation.

Source: Mudhale et al. (2024).

Travelers of all skill levels can enjoy a great experience on the website because to its user-centric approach. Its user-friendly interface makes it simple for users to move between its many features, which makes creating itineraries, researching destinations, and organizing activities fun and effective.

To sum up, the website is a flexible and essential resource for travelers looking for creativity, dependability, and ease when organizing their trips. It improves the travel experience by utilizing state-of-the-art technology and user-centric design, turning the challenging process of itinerary planning into a smooth and joyful journey.

References

Angskun, J., Korbua, S., and Angskun, T. (2020). Time-related factors influencing on an itinerary planning system.

Dasari, S. B., Vandana, V., and Bhharathee, A. (2023). Smart Travel Planner using Hybrid Model.

Mudhale et al. 2024. Travel Itinerary Planner Using AI, IRJET.

Gokul Krishna, M, Haseeb, Mohammed, Mohammed Siyad, B Zameel, P. A. M., and Raj, S. V. (2021). Budget and experience based travel planner using collaborative filtering.

Lim, K. H., Chan, J., Karunasekera, S., and Leckie, C. (2021). Tour recommendation and trip planning using location-based social media: A survey. Knowledge and Information Systems.

Lin, Y., Teng, S., Fan, Y., et al. (2021). Travel route planning through simulated annealing algorithm and model data.

Lopes, R. B., Silva, E., and Santos, B. S. (2020). E-tinerary: A decision support approach for tourist trip planning.

Osmond, A. B., Supangkat, S. H., and Hidayat, F. (2020). Design and implementation of smart trip planner.

Panneerselvam, K., Kumar, J. V., Rahul, M. R. P., and Kumar, T. (2022). Travel and tourism management system.

Roy, S. B., Das, G., Amer-Yahia, S., and Yu, C. (2020). Interactive itinerary planner.

38 Optimal Power Flow Control Using Constraint Optimization

Manoj Bhaurao Deokate[1,a] and Rajesh S. Surjuse[2,b]

[1]Research Scholar, Electrical Engineering Department, Government College of Engineering, Chandrapur, India
[2]Associate Professor, Electrical Engineering Department, Government College of Engineering, Nagpur, India

Abstract

The aim of a power flow study is to assess the real power, voltages, currents, and reactive power flows in a system under particular load conditions using a steady state analysis. The goal of an OPF (Optimal Power Flow) is to determine a steady state operation point that reduces loss, generation cost, and so on. In the last fifty years, OPF has emerged to be among the most significant and broadly examined nonlinear optimization issues while preserving a system performance that is acceptable in terms of limits on the line flow limitations, real and reactive powers of the generators, the output of different compensating devices, etc. OPF has become complex due to the addition of various devices and the deregulation of the power industry. Economic market management has been placed on top of the conventional power system concept practices. This paper's goal is to present the IEEE-14 Bus system a comprehensive constraints optimization solution to resolve OPF problems.

Keywords: Active power, reactive power, OPF (Optimal Power Flow), consensus, partitioning

1. Introduction

Based on research conducted on optimal power flow over many years, the size of the power network is growing, which causes more power system problems. The aim is to offer the optimise solution on optimal power flow issues and exhaustive method to address the system problems (Eghbalpour, 2019; Nejabatkhah & Li, 2015; Iggland et al., 2015; Teixeira et al., 2016; Guo et al., 2017; Fajal et al., 2016).

2. Literature Review

The literature review studies the issue of increased power network sizes and convergence, as demonstrated by previous studies to improve the total system by employing the constraint optimization technique to solve the issues.

2.1. Limitation of Existing System

The operation is closer to the network's limits as a result of the power system's rising

---
[a]mbdeokate25@gmail.com, [b]surjusemnit@rediffmail.com

DOI: 10.1201/9781003567653-38

complexity as well as system sizes, which raises the iteration's computing costs. The present paper overcomes all issues using constraint optimization.

2.2. Research Background

The objective is to create an optimized mechanism to resolve the issues of complicated energy systems (Eghbalpour, 2019; Nejabatkhah & Li, 2015; Iggland et al., 2015; Teixeira et al., 2016; Guo et al., 2017; Eajal et al., 2016). The most interesting approach to solving the OPF using constraints optimization which involves the limitations to achieve the optimal solution and defined goal.

3. Objectives of the Research

The main objective of the current research is to determine the power flow structure and investigate issues related to equality and inequality constraints and ultimately reduce the generation cost by considering the constraints optimization.

4. Optimal Power Flow

Figure 38.1 shows the IEEE 14-Bus system and also mark the Area A, Area B and Consensus Area. The system has transformers, generators and load.

The problem of dynamic control of the OPF requires precision which requires arrange of equality and inequality constraints. Need to identify the control variable of the system and define the mathematical expression accordingly(Eghbalpour, 2019; Molzahn et al., 2017; Erseghe, 2014; Binetti et al., 2014; Wang et al., 2017; Eajal et al., 2016; Magnússon et al., 2015; Marvasti et al., 2014). Minimization of total generation cost as under.

$$\min_u \sum_{i=1}^{N_G} C_i(P_i) \tag{1}$$

Where, P_i is no. of generators with capacity, C_i is Cost of each generator-$/MWh

Each generator bus has adjustable actual power, voltage magnitude, reactive power outputs, and voltage angles. The objective function comprises generation cost along with constraints including equality constraints for voltage phase and power injection relationship and inequality constraints that define the power generation, as well as line flow limitations (Eghbalpour, 2019; Huebner et al., 2019; Meyer-Huebner et al., 2016; Robbins & Domınguez-Garcia, 2016; Erseghe, 2014; Engelmann et al.,

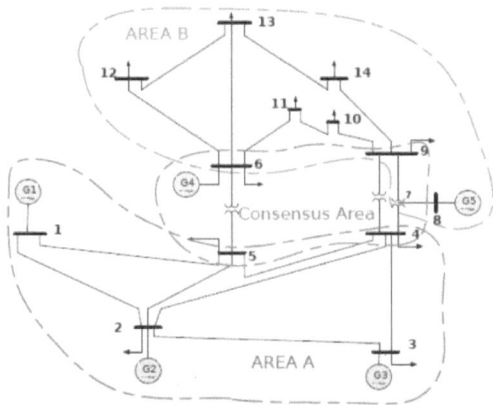

Figure 38.1: IEEE-14 Bus System.

Source: Rabi Shankar Kar et al., 2017.

2019; Iggland et al., 2015; Zimmerman et al., 2009; Guo et al., 2017; Magnússon et al., 2015; Marvasti et al., 2014).

The optimization vector (x) for the conventional problem of OPF includes the *nbx1* vectors representing voltage angles (θ) as well as magnitudes (Vm=V) and the ngx1 generator real vectors and "reactive power injections" Pg & Q_g (Eghbalpour, 2019; Meyer-Huebner et al., 2016; Kim & Baldick, 2000; Erseghe, 2014; Zimmerman et al., 2009).

$$x = \begin{bmatrix} Pg \\ Qg \\ V \\ \theta \end{bmatrix} \qquad (2)$$

5. OPF Constraints

OPF is a nonlinearly constrained optimization. The active and reactive power balances are equality constraints on the ideal power flow (Eghbalpour, 2019; Huebner et al., 2019; Molzahn et al., 2017; Zimmerman et al., 2009; Guo et al., 2017; Marvasti et al., 2014). The limits on the generation of both power with respect to the power source boundary, the angle limit and the voltage magnitude limit are the inequality constraints of the OPF issue.

5.1. Inequality Constraints

The inequality constraint equations show the lower and higher limits of real generator power, reactive power, and bus voltage angles along with magnitudes (Eghbalpour, 2019; Meyer-Huebner et al., 2016; Kim & Baldick, 2000; Molzahn et al., 2017; Erseghe, 2014; Iggland et al., 2015; Zimmerman et al., 2009; Guo et al., 2017; Marvasti et al., 2014).

Subject to

$$\underline{P_{gi}} \leq P_{gi} \leq \overline{P_{gi}} \qquad (3)$$

$$\underline{Q_{gi}} \leq Q_{gi} \leq \overline{Q_{gi}} \qquad (4)$$

$$\underline{V_i} \leq V_i \leq \overline{V_i} \qquad (5)$$

$$\underline{\theta_i} \leq \theta_i \leq \overline{\theta_i} \qquad (6)$$

5.2. Equality Constraints

The equality constraints equations involve real and reactive components are presented as generator generations P_g and Q_g and functions of V, θ. The load demand at each bus denoted as P_d & Q_d is considered to be constant.

$$P_{gi} - P_{di} - P_i(V, \theta) = 0 \qquad (7)$$

$$Q_{gi} - Q_{di} - Q_i(V, \theta) = 0 \qquad (8)$$

6. Decision Variables (P Q V θ)

The variables that are initialized with the default values Zimmerman et al., n.d.; Wiget & Andersson, 2013; Sass et al., 2017; Minyue Ma et al., 2016; Boyd et al., 2011; Zimmerman and Murillo-Sánchez, n.d.; Mudumbai and Dasgupta, 2014 of Reactive Power Q, Voltage magnitude V, Real Power P, as well as voltage angle θ. (i.e., P, Q, V, θ), are identified as Area A and B decision variables, respectively, and are represented by the symbols x_A and x_B. The decision variables for each region will be specified, and it is depending on how the area is divided. Assume that the decision variables for areas A & B are X_A and X_B, correspondingly.

$$x_A = \begin{bmatrix} P_{g1A}, P_{g2A}, P_{g3A}, Q_{g1A}, Q_{g2A}, Q_{g3A}, \\ V_{1A}, V_{9A...}, \theta_{1A}, \theta_{9A} \end{bmatrix} \qquad (9)$$

$$x_B = \begin{bmatrix} P_{g4B}, P_{g5B}, Q_{g4B}, Q_{g5B}, V_{4B}, V_{5B}, \\ V_{6B}, V_{7B}, V_{14B}, \theta_{4B}, \theta_{14B} \end{bmatrix} \qquad (10)$$

Every region determines its buses, phase angles, voltage magnitudes and real & reactive power of the internal generators.

7. Objective of Control Variables

The objective function (Overall Cost) depends on the power generated by the generators in concerned areas subject to inequality constraints of P, Q, V, θ with its lower and upper band or limits and equality constraints subject to real and reactive power generation and demand.

Table 38.1: Case study of IEEE-14 Bus Result

Overall Cost is 8081.53 ($/MWHr)						Convergence Time is 3.32 Second		
Generation Capacity		Actual Generation				Load (MW)		Losses
P MW	Q MVAr	P MW	Q MVAr	P MW	Q MVAr	P MW	Q MVAr	
772.40	148.0	P=268.29 Pg1=194.33 Pg2=36.72 Pg3=28.74 Pg6=0.0 Pg8=8.49	Q=67.63 Qg1=0.0 Qg2=23.69 Qg3=24.13 Qg6=11.55 Qg8=8.27	259	73.50	9.29	39.16	

Source: IEEE-14 Bus System R.S. Kar, Zhixin Miao., et al., ADMM for Nonconvex AC Optimal Power Flow. 2017 North American Power Symposium (NAPS), 16 November 2017, volume 1.

8. Results

The MATPOWER outcomes of the standard bus system are revealed in Table 38.1. The system is converging and case studies of concerned area generators with actual generation and load as well as losses are being compared.

8. Conclusion

Case study concludes the defined objective of paper is being achieved by proper OPF control using constraints optimization, which reduces the overall cost of generation and losses. Whereas the system convergence depends on proper selection of constraints and optimal minima and maxima. Hence, the goal of producing the actual generation by minimizing the cost while taking the economy into account is being achieved.

References

Binetti, G., Davoudi, A., Lewis, F. L. (2014). Distributed consensus-based economic dispatch with transmission losses. IEEE Transactions on Power Systems, 29(4), 1711–1720.

Boyd, S., Parikh, N., Chu, E., Peleato, B., and Eckstein, J. (2011). Distributed optimization and statistical learning via the alternating direction method of multipliers. Foundations and Trends, 1, 1–122.

Eajal, A.A., Abdelwah, M.A., El-Saadany, E. F., and Ponnambal, K. (2016). A unified approach to the power flow analysis of AC/DC hybrid microgrids. IEEE Transactions on Sustainable Energy, 7(3), 1145–1158.

Eghbalpour, H. (2019). Application of distributed optimization technique for large-scale optimal power flow problem. Spring 2019. Memorial University of New foundland, Canada. March 2019.

Engelmann, A., Jiang, Y., Mühlpfordt, T., Houska, B., and Faulwasser, T. (2019). Toward distributed OPF using ALADIN. IEEE Transactions on Power Systems, 4(1), 584–594.

Erseghe, T. (2014). Distributed optimal power flow using ADMM. IEEE Transactions on Power Systems, 29(5), 2370–2380.

Guo, J., Hug, G., and Tonguz, O. (2017). A case for nonconvex distributed optimization in large-scale power systems. IEEE Transactions on Power Systems, 32(5), 3842–3851.

Huebner, N. M., Suriyah, M., and Leibfried, T. (2019). Distributed optimal power flow in hybrid AC–DC grids. IEEE Transactions on Power Systems, 34(4), 2937–2946.

Iggland, E., Wiget, R., Chatzivasileiadis, S., and Anderson, G. (2015). Multi-area DC-OPF for HVAC and HVDC grids. IEEE Transactions on Power Systems, 30(5), 2450–2459.

Kim, B. H., and Baldick, R. (2000). A comparison of distributed optimal power flow algorithms. IEEE Transactions on Power Systems, 15(2), 599–604.

Magnússon, S., Weeraddana, P. C., and Fischione, C. (2015). A distributed approach for the optimal power-flow problem based on ADMM and sequential convex approximations. IEEE Transactions on Control of Network Systems, 2(3), 238–253.

Marvasti, A. K., Fu, Y., Dor Mohammadi, S., and Rais-Rohani, M. (2014). Optimal operation of active distribution grids: A system of systems framework. IEEE Transactions on Smart Grid, 5(3), 1228–1237.

Meyer-Huebner, N., Gielnik, F., Suriyah, M., and Leibfried, T. (2016). Dynamic optimal power flow in AC networks with multi-terminal HVDC and energy storage. IEEE Innovative Smart Grid Technologies—Asia, 1, 300–305.

MinyueMa, LinglingFan, and ZhixinMiao (2016). Consensus ADMM and proximal ADMM for economic dispatch & AC OPF with SOCP relaxation. IEEE Transaction on Power System, 1, 1–6.

Molzahn, D. K. et al. (2017). A survey of distributed optimization and control algorithms for electric power systems. IEEE Transactions on Smart Grid, 8(6), 2941–2962.

Mudumbai, R., and Dasgupta, S. (2014). Distributed control for the smart grid: The case of economic dispatch. In for. Theory Appl. Workshop (ITA). IEEE, 1–6.

Nejabatkhah, F., and Li, Y. W. (2015). Overview of power management strategies of hybrid AC/DC microgrid. IEEE Transactions on Power Electronics, 30(12), 7072–7089.

Robbins, B. A., and Dominguez-Garcia, A. D. (2016). Optimal reactive power dispatch for voltage regulation in unbalanced distribution systems. IEEE Transactions on Power Systems, 31(4), 2903–2913.

Sass, F., Sennewald, T., Marten, A. K., and Westermann, D. (2017). Mixed AC-HVDC benchmark Test system for security constrained OPF calculation. Transmission Distribution IET Generation, 11(2), 447–455.

Teixeira, A., Ghadimi, E., Shames, I., Sandberg, H., and Johansson, M. (2016). The ADMM algorithm for distributed quadratic problems: Parameter selection and constraint preconditioning. IEEE Transactions on Signal Processing, 64(2), 290–305.

Wiget, R., and Andersson, G. (2013). DC optimal power flow including HVDC grids. Proc. Elect. Power and Energy Conf. (EPEC), Halifax, NS, Canada.

YaminWang, LeiWu, and ShouxiangWang (2017). A fully-decentralized consensus-based ADMM approach for DC-OPF with demand response. IEEE Transactions on Smart Grid, 8(5), 2637–2647.

Zimmerman, R. D., and Murillo-Sánchez, C. E., MATPOWER, open-source electric power system simulation and optimization tools for MATLAB. Cornell. http://www.pserc.cornell.edu

Zimmerman, R. D., Murillo-Sanchez, C. E., and Thomas, R. J. (n.d.). MATPOWER: Steady-state operations, planning and analysis, 21 June 2010 tools for power systems research and education.

Zimmerman, R. D., Murillo-Sanchez, C. E., Thomas, R. J. (2009). MATPOWER's Extensible Optimal Power Flow Architecture. 2009 IEEE Power & Energy Society General Meeting. Conference paper 26–30 July 2009 IEEE Xplore.

Rabi Shankar Kar, Zhixin Miao, Miao Zhang and Lingling Fan. (2017). ADMM for Nonconvex AC Optimal Power Flow. 2017 North American Power Symposium (NAPS), 1.

39 Hybrid maker-based indoor positioning system using augmented reality

Shantanu Ingole[a], Sahil Narsale[b], Divyanee Desale[c], Snehal Rade[d], and Savita Adhav[e]

Department on Information Technology, JSPM's Rajarshi Shahu College of Engineering Pune, India

Abstract

Conventional ways of navigation in an indoor environment still face difficulties while navigating complex locations including business buildings, airports, hospitals, and museums. This paper describes an approach that can be used to build a system to create an indoor navigation system using Augmented reality (AR). An augmented reality superimposes the digital data with the real environment data to create effective user navigation. The blend of a hybrid marker-based system with augmented reality is considered. Our research describes the results of different navigational technology. The paper describes the need for the systems and details of small experimentation. The main goal of this paper is to analyse and compute the various methods for indoor wayfinding and find a suitable methodology for implementation purposes.

Keywords: Augmented reality navigation, Indoor Navigation Systems (INS), SLAM methodology, MLKit, superimposition navigation, A* algorithm

1. Introduction

Scientists from all over the world are developing technologies aimed to simplify human life. We have made another attempt to find a solution to one problem. As we visit new places, it becomes really difficult to find a way into that new indoor environment. Considering this problem, this research work provides Augmented Reality Based Indoor Navigation that introduces the concept of a navigation system for smartphones capable of guiding users accurately to their destinations in an unfamiliar indoor environment. Digitally enhanced guides have many advantages over traditional paper-based indoor guides. Global Positioning Systems (GPS) open the door for outdoor navigation through the best routes possible. However, it fails to work indoors due to signal attenuation by infrastructural surfaces and other obstructions. Integrating the proposed system with augmented reality enhances and enriches the user experience by combining real and computer-based scenes and images to deliver a unified but enhanced view of the world. Image comparison techniques are applied to get features

[a]shantanuingole27@gmail.com, [b]sahilnarsale13@gmail.com, [c]divyaneedesale50@gmail.com, [d]snehalrade2309@gmail.com, [e]sadhav_it@jspmrscoe.edu.in

DOI: 10.1201/9781003567653-39

that will be useful in classifying and recognition of images captured by the user. It helps achieve accuracy by relative location detection of the user's current location and comparing to landmark images saved in the database. It will work on devices like smartphones, tablets based on Android which will guide the users accurately to their unknown destination, the services are not restricted to any specific industry the fields of this application are limitless.

The use of real time motion tracking and object detection helps the system to identify the environment where the user is present. further the small experimentation is done to check out the working of proposed system thoroughly it obtain the single feature image considered as the real time data. The

Figure 39.1: Navigation is depicted with the virtual components.

Source: Pavani, n.d.

system adds the digital components to it and return back to the user. Even though augmented reality is a topic having research more than 20 years in timeline, it started trending recently when Google announced their work on Google Glasses or when AR applications, photo filters and games went on a hit. This void in research in this area along with interest in developing user-friendly applications further motivated us to take up this project.

2. Background Study (Literature Survey)

The author Sandra C J et al. were able to efficiently design and find an economical indoor navigation system by using the SLAM methodology of AR core. The main goal is to determine detours that users could take in the event of obstacles. If an obstacle that is not mentioned in the design arises, it expects the system will work efficiently using the NavMesh components. The technique is discovered to provide a way to navigate and guide the people through complex systems. Additionally, the augmented reality features enhance the way to interact with users. The advantage of the application is anyone can simply download and install the mobile application and can take advantage of it (Dileep & Sunny, 2021).

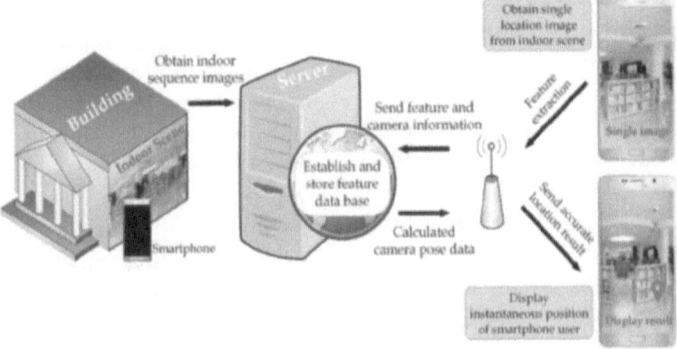

Figure 39.2: Navigational system architecture.

Source: Li et al., 2020.

Girija v and the other authors describe a workable approach to improve navigation; it also provides location-based indoor services to users. For more accurate and reliable indoor positioning, the authors have proposed that the systems include indoor location technologies such BLE, Wi-Fi, LiDAR, and sensor fusion. They describe a way where users can enjoy an immersive and captivating navigation experience with the combination of AR overlays, machine vision, and markers. The use of augmented reality has provided an efficient way to interact with the user, it has also provided great accessibility, and also increasing efficiency in interior environments. The proposed system has not only given users real-time directions but also displays the nearby areas of interest. Further the user can get the local information by utilizing location-based services that are included into these platforms (Ganesh et al., 2020).

The research work by Dimitrina Deliyska et al. offers a methodology for the development of indoor navigation systems. The offered methods include determining the user's position, making the digital map with respect to the location, and selecting the fattest and shortest root among the digital map. Databases, XML, QR codes, and graph theory are the components that are used to develop this methodology. As the research shows the proposed method can be used for various types of construction. The entire system is developed as a mobile based application to meet the needs of the Sofia University of Mining and Geology, and as the research shows the developed system found to be effective (Deliyska et al., 2021).

Dikdhant jopat et al. research on normalized detection parameters of the AR-Core SDK which represent the average distance needed to detect an object as research shows the distance between the camera and object has an impact on the system's object detection accuracy. With the aid of special capabilities like machine learning models and motion testing from many perspectives, the SDK makes the most of any mistake. An appropriate score of 0.5–1 was established for both metrics based on dependable observations in order to accomplish dependable sensing and high-quality visualization. Also, they have developed a QR code-based function that assists users (Jopat et al., 2020).

The author Utkarsha patil proposed a system where the user can go to the destination as quickly as possible shortest path to the destination is found in the small amount of time it is computed by using the A* algorithm. Theuser experience of the application is enhanced by utilizing augmented reality to guide the user towards the destination (Patil et al., 2020).

The authors have proposed a positioning and navigation system that works indoors. The result shows that the kNN classifier has achieved an accuracy of 1.5 m (position predicted as the second closest reference point) and 0.75 m (position predicted as the closest reference point with a reference point spacing of 1.5 m) 66% and 9% of the time, respectively. The kNN classifier shows a better result than the Gaussian naïve Bayes, which has only reliability to reach an accuracy of 0.75m. Also the study found that the presence of Strong signal reception over large distances degrades the performance by producing identical RSSI vectors at successive reference sites (Hussien et al., 2017).

Pradip Kumar G H et al. describes the algorithm that is optimal for indoor positioning the study finds that the among the several methods of A* Path finding is the optimal algorithm; it uses the heuristic function. It gives the best route among the routes (Pradeep Kumar et al., 2020).

Ashraf Saad Shewail et al examine the various indoor navigation technologies. It compares the possible technology for the navigational approach. further the study examines the different technologies in various aspects (Shewail et al., 2023).

The common characteristic of most indoor tracking system shown in the Table 39.1.

The author has concluded that the hybrid based super impositional system is

Table 39.1: Characteristics of most indoor tracking systems

Indoor	Outdoor	With low light intensity	Work in multi-level	Use predefined map	Use hardware requirements	Marker	Markerless	Cloud
✓	✓	NA	✓	NA	✓	✓	NA	NA
✓	NA	✓	NA	NA	✓	NA	✓	NA
✓	NA	NA	✓	✓	NA	✓	NA	✓
✓	NA	NA	✓	NA	NA	✓	NA	NA
✓	NA	✓	✓	✓	✓	NA	✓	NA
✓	NA	✓	✓	✓	NA	✓	NA	NA
✓	NA	NA	NA	NA	NA	NA	✓	NA
✓	✓	NA	✓	NA	✓	✓	NA	NA
✓	✓	NA	✓	NA	✓	✓	NA	NA
✓	NA	NA	✓	✓	NA	NA	NA	NA
✓	NA	NA	✓	✓	NA	NA	NA	NA
✓	NA	✓	NA	NA	✓	✓	✓	NA
✓	NA	NA	✓	NA	✓	NA	✓	✓
✓	NA	NA	✓	✓	NA	NA	NA	✓
✓	NA	NA	✓	✓	NA	NA	NA	✓
✓	NA	NA	✓	NA	NA	NA	yl	✓

Source: Shewail et al., 2023.

the system which is efficient in most of the aspects with the help of markers the tracking becomes easy (Shewail et al., 2023).

The author satya kiranmai tadepalli and other develops the system which is based on marker based navigational approach The described methodology works with the combination of unity , blender and ARCore with the help of AR core it is found that the capabilities like motion tracking, environmental understand, depth understanding, light estimation and user interaction can be possible in enhance manner this all feature collectively called SLAM (simultaneous localization and mapping) the study states that for the required outcome the markers need to be placed at a intended location (Tadepalli et al., 2021).

The author Adam Satan et al examines how a navigational system is to be implemented in a mobile based application. The system is primarily dependent on bluetooth beacons. The experiments examine whether the navigational method works in real life scenarios or not. It was found that in two experiments the position, direction was saved and analysed. The drawback was that the beacons were dedicated and could be used within 5 meters instead of the usual 12 therefore the 2-meter area was uncovered by the system. Total 20 tests were conducted out of which 5 were for source and destination and 15 were for random samples. The beacon system shows drawbacks beyond the bound of 5 meters (Satan & Zsolt, 2018).

Pankaj et al. proposed a system that is prominently having a marker-based approach. It examines the high accuracy it mitigates the drawback of beacon-based technology (Badgujar et al., 2015).

3. Dataset Description

The inputted data of proposed system is dealt with the 2 main categories:

a. augmented data
b. real world data

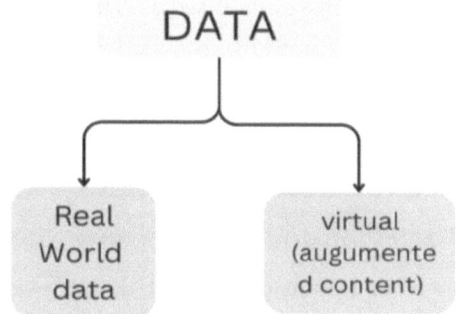

Figure 39.3: Data descriptions.

Source: Author.

Figure 39.4: Virtual and real-world objects.

Source: Author.

a. Augmented data—here It is termed as the virtual data or the virtual components that are required to ads to create an augmented reality in this system we are adding the virtual 3d components arrow, path, node being as virtual data blending it with the real time data as well. The virtual or digital data is the 3d data in the form of text, pictures, 3d assets, or a combination of all of the above. Often the data is generated in the form of 3d elements. The Augmented Reality based system will need to understand the difference between reality and reconstruct it to create its digital twin by combining the real world part.

b. Real world data—The actual data that is associated with the real time environment is considered as the real data. The real time data is captured by the rear camera of the smartphone devices and the features like floor, corner, object are extracted from it.

4. Experimentation Details

4.1. Hardware Requirements

To compute the usability of the proposed systems of superimposition model for the indoor navigation system the experimentation has carried out. It examines the consolidated working of all the components. The hardware of this camera control system proposed consists of three pieces of hardware: an AR marker can be a QR Code or the board with text, and a single RGB camera, and a PC as a processing unit.

ARCore supported smartphone

Camera: Minimum 4Mp resolution

To carry out this experiment we have some minimum hardware setup one is the Arcore supported smartphone to run the application into device and the camera to capture the real time data. The Figures 39.5 and 39.6 depicts the scanning and iteration to capture the frame from the device's camera. This process is iterated multiple time to capture the real time data.

4.2. Software Requirements

Android Studio SDKversion: API 28 and above Cross platform support—this requirement is essential to collaborate all the requirements into android SDK.

4.3. Technology Stack

a. Android Studio (version 3.0 and above)—the android studio is the IDE that provides the various utilities to develop an android based application.

b. Database: An XML database is a data persistence software system that allows data to be specified, and sometimes stored, in XML specified format. To operate on this data the SQL language is used to provide the operation and queried as

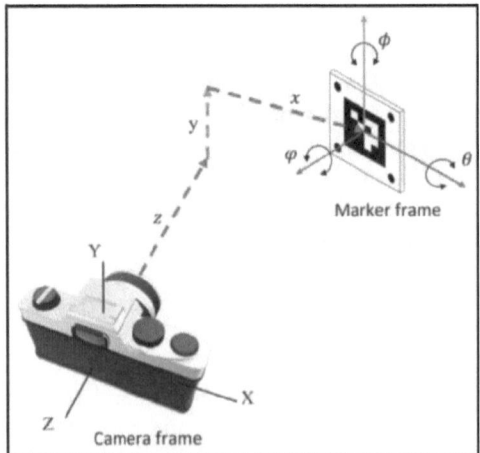

Figure 39.5: Scanning process of Environment to take real time data.

Source: Haraguchi & Miyahara, 2023.

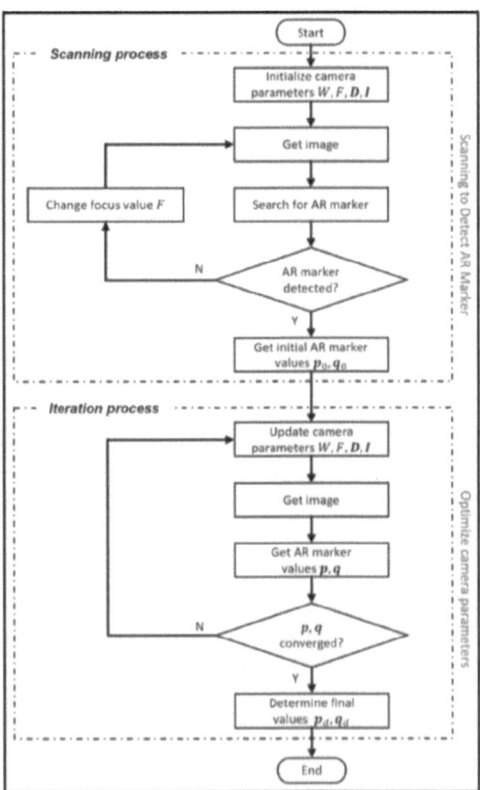

Figure 39.6: Scanning process of Environment to take real time data.

Source: Haraguchi & Miyahara, 2023.

per the requirement. Further the data can be transferred and also exported as per the requirement.

c. Google ARcore SDK—the ARcore SDK allows the operation of the AR components along with the motion tracking of the environment.

d. Google Poly for 3D objects MLkit on Firebase Backend as a Service—this helps in accessing the machine learning utility in android-based applications further it also reduces the processing time.

4.4. Developer System Requirements

a. RAM: 8 GB or above—to develop this system we need a ram above then 8 GB below that will be supported for android studio. Often the system required high performance to develop an android based application.

b. Processor: considering the window it requires the Intel core I5 or above processor and considering the mac the M1 chip or above is required.

A static or head-mounted camera captures the actions of an expert during the execution of a task. Workflow segmentation, modelling, and content creation: The recorded video is then segmented into a set of single atomic actions via an unsupervised

method. The single steps are analysed to estimate required level of accuracy, repetitiveness, etc. (Petersen & Didier, 2015; Petersen & Stricker, 2012; Petersen et al., 2013).

Further to the successful compilation of the experiment. The Figure 39.7 depicts the UI feature and how the nodes are being computed, connected and stored in the system.

The Figure 39.8 illustrates the detection of features particularly the floor with respect to the user's location. When the floor corner or any feature is detected it is represented by cluster of white dots.

5. Proposed system

Based on the research work, we proposed a system which is based on mobile application. The real time environment feed is provided to the system. further the user has to select the starting node and ending node. Further it goes into various steps that are depicted in a figure: a 1. source detection 2. Identification of landmark and marker, 3. Motion tracking, 4. Augmentation, 5. Navigation.

A. Source detection - The user has to scan the nearest landmark present in the form of a marker. It is the very first step which involves the identifying and tracking real-Environment objects, surfaces, or the features within the environment where the

Figure 39.7: The real time implementation of virtual and real components.

Source: Author.

Figure 39.8: The detection of floor.

Source: Author.

user is actually present to overlay the digital content that is virtual content or information onto them. ·

B. The MLKit's kits utility is used for the Identification of landmarks for the tracking. It also helps in the identification of floor, objects, corners texts etc

C. Motion tracking -The motion of a person with respect to the environment is tracked based on the landmarks that are detected.

D. Augmentation- once the real time environment is being set up the real time environment feed from the camera and the virtual feed from the is clubbed to create the augmented environment each and every landmark is already stored in the database in the form of nodes which leads to a creation of graphs.

E. Navigation- the added 3D assets in the system depicts the augmented components which are responsible for navigating the user with the described path. The path is derived from the prestored graph of the system. by computing the current node and the destination node the path among them is set up and given back to the user.

6. System Architecture

The entire system setup is android based application The system comprises of two primary components that work collectively

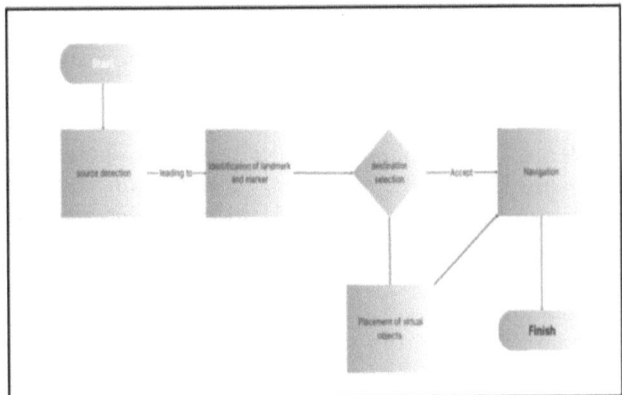

Figure 39.9: Proposed methodology.

Source: Author.

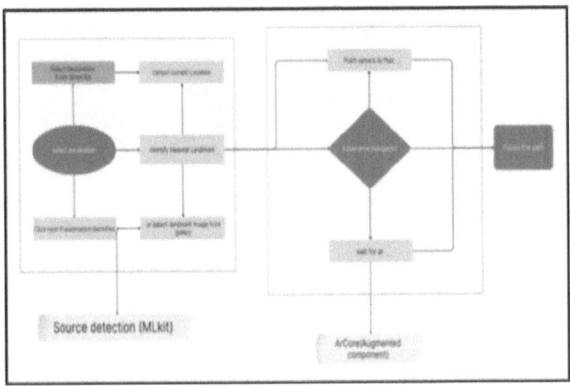

Figure 39.10: System architecture.

Source: Author.

to process real-time data captured data along with the prestored assets

• MLKit

Step 1: The landmark or marker that is images/board to the intended or a required area. These markers are considered as the node and the text associated with the marker is detected and analysed and stored in the database as a form of node by collecting such node the graph is created

Step 2: The initial node or the starting point is scanned by using the nearest marker; this scanning is again done by MLKit.

• ARcore

Step 3: Further the user needs to select the destination as the end point from the preeting map that is stored in the system.

Step 4: As the position chosen by the user the system will navigate to the required destination with the help of an augmented 3D arrow.

Both the components need to work Together. These components and the back-end logic is developed in android studio SDK which comprises several functions that

are associated with the activity (utility in android studio SDK called activity) corresponding with the users action.

7. Result and Discussion

The Accuracy and reliability of augmented reality depends on a variety of factors. The findings are computed based on several factors like Source detection; it basically comprises identification of nearby landmarks mostly boards, logos and objects. We have considered the boards can be detected using text recognition by MLKit. It could also be calculated using OCR. but the time taken by OCR is less compared to the MLkit. Also, it supports more languages than OCR. It needs additional training for a new language.

Below is the result analysis os ocr and ml kit (Grobkopf, n.d.). And the results of various indoor navigation-based technology is computed; the prediction is computed based on the positions predicted as the second closest reference point.

The calibration of three-dimensional (3D) coordinates in augmented reality systems involves the recognition of environmental characteristics and technology that models the 3D space of the device,

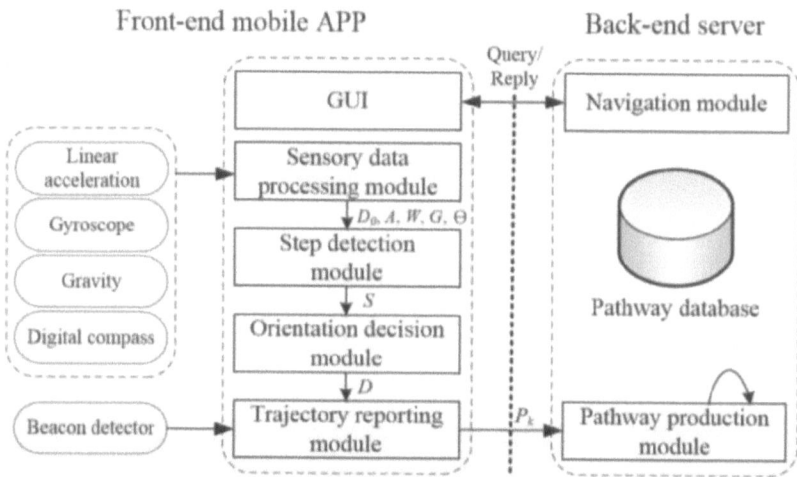

Figure 39.11: Front-end and back-end architecture.

Source: Pan & Li, 2019.

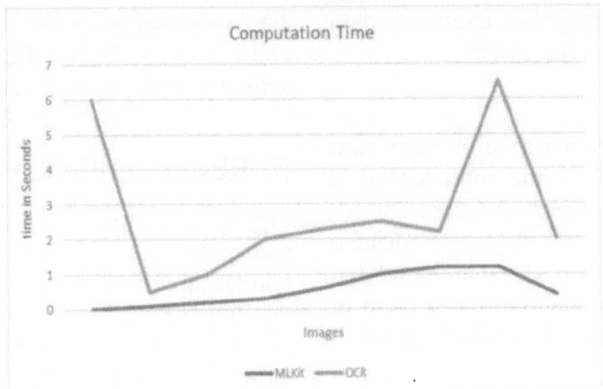

Figure 39.12: Time consumption comparison of OCR and MLKit for text recognition.
Source: Grobkopf, n.d.

Table 39.2: The comparison of accuracy of various methodology of indoor navigation

Indoor Navigation Technology	Accuracy (Small Area)	Accuracy (Wide Area)
Beacon based Indoor navigation	1m	5m–10m
Vision based indoor navigation	0.75m	30m–40m
Marker based indoor navigation	1.0m	-
Superimposition based indoor navigation	0.70m	30m–40m

Source: Author.

Technology	Coverage	Power consumption	Accuracy	Cost
GPS	Outdoor	Very High	6–10 m	High
Wi-Fi	(outdoor/indoor)	High	1–5 m	Low
Bluetooth	Indoor	Low	2-5 m	High
RFID	Indoor	Low	1-2 m	Low
Ultra-wide band (UBW)	(outdoor/indoor)	Low	5-30 cm	High
Infrared	Indoor	Low	1-2 m	Medium
ZigBee	Indoor	Low	3–5 m	Low
Cellular	(outdoor/indoor)	Low	50m-150m	High
Visible light communication (VLC)	Indoor	Low	4-10 cm	Low
NFC	Indoor	Low	4 cm	Low
Frequency modulation (FM)	Indoor	Low	2-4 m	Low
Ultrasound	Indoor	Low	3cm–1m	Medium

Figure 39.13: Comparison of various navigational technology based of some factors.
Source: Shewail et al., 2023.

determining its position and orientation (Brunnstrom et al., 2020).

8. Conclusion

Few computation studies were done and we conclude that the superimposition indoor based navigation shows the accuracy in terms of navigation findings further the system along with the ARcore enhances the UI feature which is easily understandable by any age group.

The key components of any indoor navigation system is source detection and motion tracking which correspond to the environment. The results show that the use of MLKit has enhanced the amount of time consumption it is taking for the recognition of text, source and other landmarks over the OCR.

9. Future Scope

The *Geospatial API (Google, n.d.)* can be combined with GPS and outdoor navigation. Geospatial Creator allows us to access the same 3D map source used by Google Earth, as we know that it consists of the biggest data of the real world through the Map Tiles API. With this powerful API, we can bring to life mobile immersive experiences, customized to your users' location. Further combined with indoor navigation systems.

The voice assistance can also be added to the system it leds to an easier of system for the people who are visually impaired and age group people can able to use it actively without having the much technical knowledge

References

AR core. (2024, 02 26). Retrieved from googles AR core: https://developers.google.com/ar/develop/geospatial

Badgujar, P., Aswani, N., Gurbani, M., & A. S, O. (2015). V.E.S.I.T. indoor NAvigation using Augmented Reality concept. IJSRD, 3, 1776–1778.

Brunnstrom, K., Dima, E., Tahir, Q., Mathias, J., Andersson, M., & Sjostro, M. (2020, 11). Latency impact on Quality of Experience in a virtual reality simulator forremote control of machines. Signal Processing Image Communication, 89, 116005. doi:10.1016/j.image.2020.116005

CJ, S., Sandra, D., C S, S., & Leva, E. (2021). Indoor navigation system using augmented reality. International Reasearch Journal of Engineering and Technology, IRJET, 8, 2408–2412.

deliyska, D., Yaney, N., & Trifonoya, M. (2021). Methods for developing an indoor navigation system. E3S Web of Conferences.

Großkopf, E. (2020, 09). Bitfactory. Retrieved 1 30, 2024, from https://www.bitfactory.io/de/dev-blog/comparing-on-device-ocr-frameworks-apple-vision-and-google-mlkit/

Haraguchi, D., & Miyahara, R. (2023). HIgh Accuracy and wide Range Recongnition og micro AR markers with Dyanamic Camera Parameter Control. Electronics, 12, 4398.

Jopat1, D., Makwana2, K., Dhanmeher3, H., & lemos4, j. (202). Indoor navigation system using augmented Reality. ISOR hournal of Engineering (IOSRJEN).

K. A., W. A., T. G., G. M., & H. H. (2017). Indoor navigation system using fingerprint. International Journal of Engineering Research & Techonology (IJERT), 6, 227–230.

Kumar, P., N, A., R, A., & P, M. (2020). Indoor Navigation USing AR Technology. International Journal of Engineering Applied Sciences and Technology, 4, 356–359. IJEAST.

ming, l., Ruizhi, C., Xuan, L., Bingxuan, G., Weilong, Z., & Ge, G. (2020, 03). A Precise Indoor Visual Positioning Approach Using a Built Image Feature Database and Single User Image from Smartphone Cameras. Remote Sensing, 12, 869.

Nills, P., & didier, S. (2015). Cognitive Augmented Reality. Computer and Graphics, 53, 82–91. doi:10.1016/j.cag.2015.08.009

Nils, P., & Didier, S. (2012). Learning Task Structure from Video Examples for Workflow Tracking and Authoring. IEEE (pp. 237–246). IEEE International Symposium on Mixed and Augmented Reality (ISMAR).

Nils, P., Alain, P., & didier, S. (2013). Real-time Modeling and Tracking Manual Workflows from First-Person Vision. 2013 IEEE International Symposium on Mixed and Augmented Reality, ISMAR 2013.

Pan, M.-S., & Li, K.-Y. (2019). ezNavi: An Easy-to-Operate Indoor Navigation System Based on Pedestrian Dead Reckoning and Crowdsourced User Trajectories. IEEE, 20, 488–501.

Patil1, U., Bicholakar, K., Bhaidkar, A., m, Angadi, M., v, . . . Sawant, S. (2010). Indoor navigfation System Using augumented Reality Intelliegence. International reaserch jouranal of Engineering and technology(IRJET).

Pavani, A. (2018, 1 26). mapbox.com. Retrieved 12 24, 2023, from https://blog.mapbox.com/indoor-navigation-in-ar-with-unity-6078afe9d958

Raj, G., V, G., Saud, G., Mishra, J., & Kuma, R. (2023, 05). AR Based Indoor Navigation System. Journal of Emerging Technologies and Innovative Research, 10, b205-b211.

Satan, A., & Toth, Z. (2018). Development of Blutooth based indoor positioning application. IEEE.

Shewail, A., Abdelaziz Elsayed, N., & Zayed, H. (n.d.). Survey of Indoor Traking Systems Using augumented Reality. IAES, 12, 402–414.

Tadepalli, S., Ega, P., & Inugurthi, P. (n.d.). Indoor Navigation Using Augmented Reality. (pp. 588–592).

40 Design of microcontroller-based smart wristband for safety

Hema Kale[1,a], Kaustubh Joshi[2], Isha Potphode[3], Abhishek Hiwarkar[3], Aniket Kawade[3], Krunal Ujjainkar[3], Tejas Selokar[3], and Tejas Sahare[3]

[1]Department of Electronics and Telecommunication Engineering, St. Vincent Pallotti College of Engineering and Technology, Nagpur, India
[2]Senior Software Consultant, Meritis, Paris, France
[3]Students, Department of Electronics and Telecommunication Engineering, St. Vincent Pallotti College of Engineering and Technology, Nagpur, India

Abstract

The Smart Wristband is a ground-breaking initiative aimed at enhancing personal safety and well-being, especially for vulnerable populations such as Alzheimer's patients, old age persons, women, and children. The Smart Wristband, integrating an ESP8266 microcontroller, represents an advanced answer to monitoring the safety of needy persons. This project work is equipped with advanced sensors and communication modules, it monitors vital signs such as heart rate and oxygen levels in real-time to detect possible health problems. GPS technology precisely tracks the user's location to identify them in an emergency quickly and automatically alerts pre-defined contacts for help. Featuring an intuitive OLED display, it provides easy access to important information, helping users make informed decisions in emergencies and increasing safety and reliability. Through continuous improvement and adaptation to new trends, this project strives to provide a reliable beacon of safety and security to those who need it.

Keywords: GPS, OLED, ESP, MAX, GSM

1. Introduction

The introduction to the "Smart Wristband for Safety with ESP8266 Microcontroller" project heralds a pioneering endeavour aimed at significantly bolstering personal safety and security. Explored wireless control mechanisms in IoT, focusing on communication protocols. Investigated advancements in IoT for improved device management and data transmission. The paper by Gatsis and Pappas (2017) aligns with the communication features of wristbands, emphasizing wireless control and IoT connectivity. The innovative wearable device emerges as a response to the pressing need for comprehensive safety solutions, particularly for vulnerable demographics

[a]hkale@stvincentngp.edu.in

DOI: 10.1201/9781003567653-40

such as Alzheimer's patients, women, and children, Mendoza et al. (2017). Muurling et al. (2021) introduced remote monitoring technologies in Alzheimer's care with patient-centric design and investigated the effectiveness of remote monitoring for improved patient outcomes. Explores remote monitoring in Alzheimer's care, akin to the monitoring capabilities of wristbands. Santos et al. (2016) highlight IoT-based solutions and health environments, similar to the connectivity and data management in wristbands. Ovidiu and Friess (2013), the introduction emphasizes the project's commitment to innovation, ethical responsibility, and inclusivity, as it aims to provide a reliable safety companion tailored to the specific needs and expectations of its users. Additionally, the project draws upon existing research and developments in the field of wearable technology and safety solutions to inform its design and implementation strategies. Vongsingthong and Smanchat (2014), by integrating advanced technology with user-centric design principles, the project endeavours to address the multifaceted challenges faced by these groups in ensuring their well-being. The device, taking the form of a wristband, promises a multifunctional approach to safety, leveraging cutting-edge features such as GPS tracking and communication technologies. Zhou et al. (2020), notably, the incorporation of the ESP8266 microcontroller, renowned for its versatility and efficiency, forms the backbone of the device's functionality.

2. Literature Survey

The literature survey gives a comprehensive exploration of key research studies and technological improvements pertinent to the Smart Wristband for Safety undertaking. It delves into the convergence of wearable era and safety solutions, examining studies on the mixing of IoT in cell fitness environments and the improvement of tracking structures tailor-made for

inclined populations, including Alzheimer"s patients. Additionally, insights from discussions on the evolution of wearable devices and wi-fi managing mechanisms for IoT gadgets provide valuable context for knowledge of the venture"s technological underpinnings. Moreover, moral and privacy concerns are mentioned as pivotal aspects in the layout and implementation of wearable safety gadgets, reflecting the task's dedication to ensuring user belief and confidentiality. Overall, the literature survey underscores the project"s strategic alignment with modern-day developments and research endeavors in the wearable generation, emphasizing its potential to meaningfully address safety and nicely-being issues among prone demographics through revolutionary and user-centric answers.

3. Methodology

3.1. Experimental Design

The experimental design of the "Smart Wristband for Safety with ESP8266 Microcontroller" assignment is depicted through interconnected blocks as shown in Figure 40.1, showcasing the integration of key additives, such as the ESP8266 microcontroller within the shape of the Wemos D1 Mini. At its middle, the ESP8266 manages sensor records processing and communication, while the A9G GSM GPS Module enables GSM and GPS functionalities for correct place monitoring and conversation with emergency services. The OLED Display serves as the intuitive user interface, presenting visible comments and facilitating navigation. Meanwhile, the MAX30100 Sensor ensures continuous monitoring of critical symptoms for real-time health surveillance and emergency detection. A solid energy supply unit ensures uninterrupted operation, even as the conversation interface helps statistics alternate with external devices or offerings. Together, these components form a comprehensive safety solution

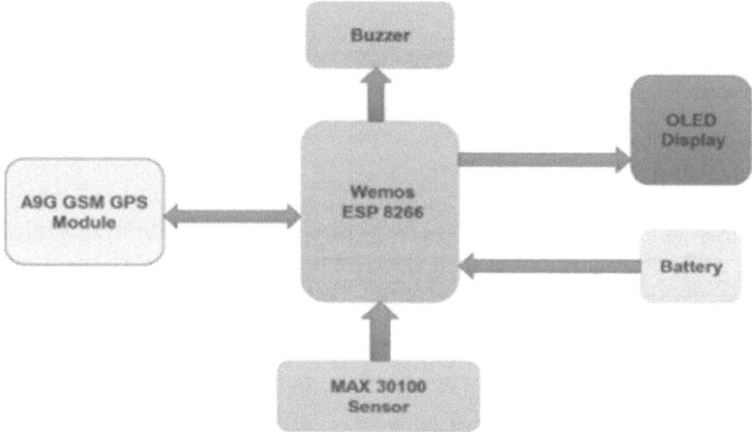

Figure 40.1: Block diagram for proposed Smart Wristband.

Source: Author.

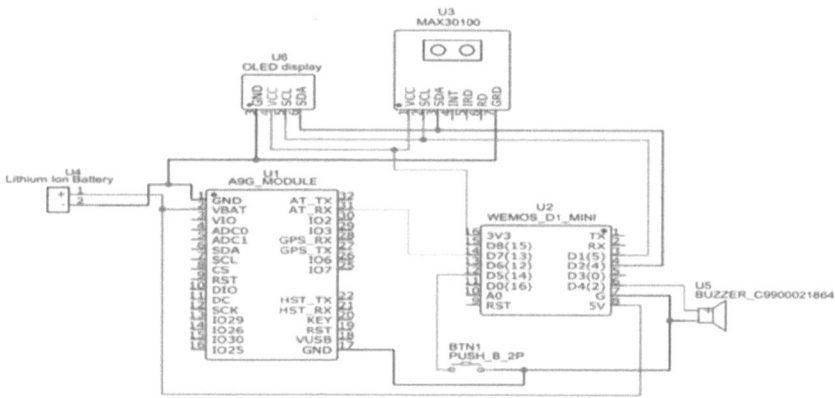

Figure 40.2: Circuit diagram.

Source: Author.

prioritizing nicely-being and security, leveraging advanced era and proactive monitoring abilties.

Figure 40.2 is the circuit diagram for the implementation , connections are Wemos D7 to the RX pin of the A9G module and Wemos D6 to the TX pin. Utilize the 3V pin of the Wemos to power the VCC pins of the OLED display and the MAX30100 sensor. For the push button, connect Wemos D5 to one terminal and ground the other terminal. Attach Wemos D4 to the buzzer and ground the Wemos to the speaker. Connect Wemos D1 to the SCL (Serial Clock) pin and Wemos D2 to the SDA (Serial Data) pin of both the MAX30100 sensor and the OLED display. Link the negative terminal of the battery to the GND pins of the A9G module, OLED display, and MAX30100 sensor. Connect the GND pin of the A9G module to the Wemos ground. Finally, connect the positive terminal of the battery to the VBAT pin of the A9G module and the 5V pin of the Wemos.

4. Circuit Elements

4.1. *Microcontroller ESP8266 (Wemos)*

The Microcontroller ESP8266 (Wemos) serves as the vital processing unit of the clever wristband. Its number one utility lies in dealing with various duties essential for the device's operation. Specifically, the ESP8266 is responsible for handling sensor statistics processing, communique with peripherals, and executing decision-making algorithms. With its built-in Wi-Fi abilities, it manages all networking components, facilitating seamless interaction with other gadgets and offerings. Additionally, the ESP8266's processing energy allows it to execute firmware answerable for data acquisition, transmission, and control good judgment, making sure the green functioning of the wristband's functions.

4.2. *A9G GSM GPS Module*

Integral for communique and place monitoring, imparting GSM and GPS functionalities for voice calls, SMS messages, correct region dedication, and communique with emergency services and predefined contacts, improving user protection and well-being.

4.3. *OLED Display*

Primary person interface showing critical statistics inclusive of health metrics, emergency notifications, and navigation activates, making sure clear visibility and intuitive navigation for knowledgeable decision-making in the course of emergencies, enhancing basic consumer revel in.

4.4. *MAX30100 Sensor*

Critical for continuous tracking of important symptoms like heart rate and oxygen tiers, permitting actual-time fitness surveillance and early detection of capability health problems or emergencies, contributing considerably to improving consumer protection and well-being.

4.5. *Lithium Ion Battery Three.7V 2Ah*

Power source chosen for its high energy density and compatibility, seamlessly included with the ESP8266 microcontroller to make certain stable power deliver for using the tool's algorithms and conversation protocols, emphasizing performance and reliability in powering the wristband.

4.6. *Buzzer*

Enhances safety features by efficaciously alerting customers in emergency situations, aligning with the challenge's multifunctional approach to making sure person safety and nicely-being.

5. Implementation

The implementation of the Smart Wristband for Safety with ESP8266 microcontroller, specifically the Wemos D1 Mini version, entails a scientific approach. It starts with a thorough requirement analysis to apprehend task goals and personal desires. The component choice includes the ESP8266 microcontroller, MAX30100

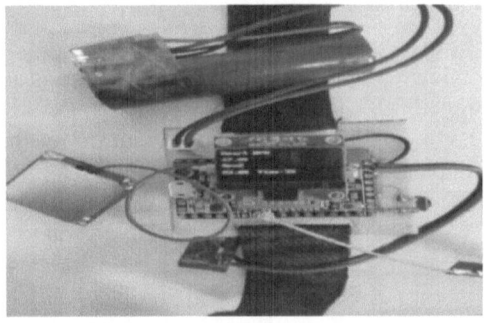

Figure 40.3: Spo2 and heartrate display on OLED.

Source: Author.

sensor for fitness tracking, and A9G GSM GPS Module for place tracking. Hardware design integrates these additives, while software program improvement makes a specialty of growing firmware for information acquisition and verbal exchange protocols. Sensor integration ensures seamless operation and improvement of algorithms for statistics processing. Communication setup establishes protocols for facts transmission, and the consumer interface layout guarantees intuitive interaction through an OLED display. Rigorous testing and validation verify functionality and reliability, with iterative improvement addressing any identified problems. Detailed documentation is compiled for destiny reference, and arrangements are made for deployment to efficaciously convey the smart wristband to users. Additionally, region records may be sent through SMS, even as fitness monitoring records can be visualized on platforms like ThingSpeak for comprehensive tracking and analysis. Through this systematic technique, the task objectives to supply a dependable protection solution assembly user requirements, leveraging the competencies of the ESP8266 microcontroller inside the Wemos D1 Mini for improved functionality and performance.

6. Result

The implementation of the Smart Wristband project brought promising results, which significantly strengthened the personal safety of vulnerable groups of the population. Rigorous testing and validation have proven the device's ability to provide timely assistance in emergency situations, ensure quick response and communication with caregivers and loved ones. Incorporating features such as a timer for user interaction and automatic location sharing via SMS has proven to be a tool to improve device functionality and further ensure user safety. Additionally, the visualization of health monitoring data on platforms like ThingSpeak offers valuable insights for

comprehensive monitoring and analysis, contributing to the overall well-being of users. As the project continues, continuous improvement and collaboration will continue to be critical to effectively addressing evolving security challenges and providing a reliable security companion to those who need it most. Figure 40.4 displays changes in the user's heart rate over time, offering valuable information about cardiovascular health and activity levels. Figure 40.5 displays oxygen saturation levels over time and provides an overview of the wearer's respiratory health and oxygenation status obtained on the thingspeak.

Figure 40.6 shows a location link sent from the wristwatch, which provides precise

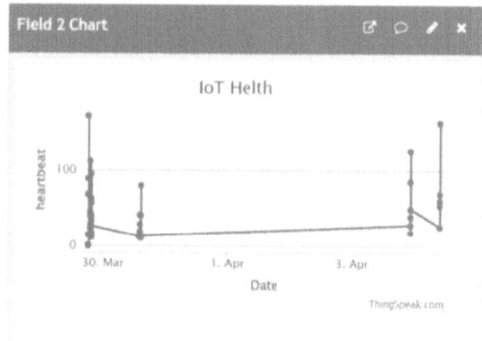

Figure 40.4: The heart rate graph.

Source: Author.

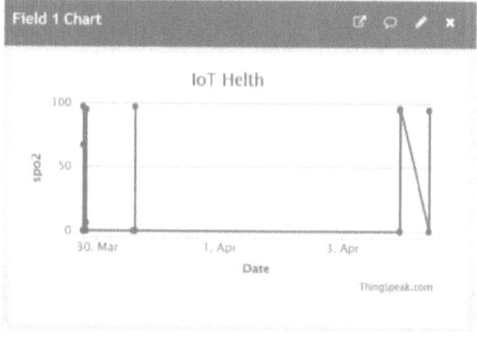

Figure 40.5: The SpO2 chart.

Source: Author.

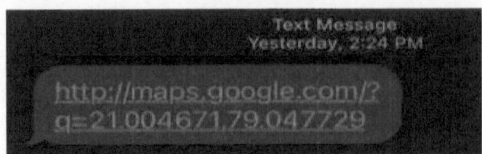

Figure 40.6: The received SMS.

Source: Author.

information about the wearer's location. This link ensures quick and accurate help in an emergency for the patient, women or other persons using the device.

7. Conclusion

The smart wristband is an innovative wearable device that emerged as a response to the pressing need for comprehensive safety solutions, particularly for vulnerable demographics such as Alzheimer's patients, women, and children. The proposed smart wristband is an integration of advanced sensors and communication modules, the device offers a comprehensive safety network ensuring immediate assistance in case of emergency. The inclusion of a timer prompting user interaction within 60 seconds and automatic location sharing via SMS increases the device's effectiveness in providing timely assistance. In cases where the button is not pressed within a set time frame, the project mechanism will automatically send the wearer's location via SMS and initiate auto-answered calls, facilitating communication between caregivers and loved ones for immediate assistance. The project's focus on visualizing health monitoring data on platforms like ThingSpeak significantly contributes to overall user safety and peace of mind. Going forward, the project is committed to further improving its capabilities and creating partnerships

to maximize its impact on safety and security. Looking ahead, the project envisions further developments through the use of IoT and partnerships with healthcare providers, while prioritizing ethical considerations and user privacy.

References

Gatsis, K., and Pappas, G. J. (2017). Wireless control for the IoT. Proceedings of the Second International Conference on Internet-of-Things Design and Implementation, pp. 341–342.

Mendoza, M. B., Bergado, C. A., De Castro, J. L. B., and Siasat, R. G. T. (2017). Tracking system for patients with Alzheimer's disease in a nursing home. TENCON 2017–2017 IEEE Region 10 Conference, Penang, Malaysia, 2017, pp. 2566–2570. doi: 10.1109/TENCON.2017.8228294.

Muurling, M., de Boer, C., Kozak, R. et al. (2021). Remote monitoring technologies in Alzheimer's disease: design of the RADAR-AD study. Alzheimer's Research & Therapy, 13, 89. doi: 10.1186/s13195-021-00825-4

Santos, J. R., Silva, B., Casal, J., Saleem, K., and Denisov, V. (2016). An IoT-based mobile gateway for intelligent personal assistants on mobile health environments. Journal of Network and Computer Applications, 71, 194–204.

Vermesan, O., and Friess, P. (2013). Internet of Things: Converging technologies for smart environments and integrated ecosystems (PDF). River Publishers. ISBN 978-87-92982-96-4.

Vongsingthong, S., and Smanchat, S. (2014). Internet of Things: A review of applications & technologies. Suranaree Journal of Science and Technology, 21(4), 359–374.

Zhou, B. F., Shan, J., Sun, F., and Guo, D. (2020). Survey of the development of wearable devices. 2020 5th International Conference on Advanced Robotics and Mechatronics (ICARM), pp. 198–203.

41 Innovative approaches to water tank cleaning: a design of modern water tank cleaning machine

*Rahul Waghmare [1], Vandita T Shahu[2,a], Prakash Dhopte[2,b],
Safal Shambarkar[2], Pravin Petkar[2], Sanjay Sajjanwar[2],
Anurag Karande[2], Bhushan Deshmukh[2], and Amar Kawale[2]*

[1]Green Solar and Fabrication works, Nagpur, India
[2]Jhulelal Institute of Technology, Department of Mechanical Engineering,
Nagpur, India

Abstract

Safe drinking water is essential for human health and wellbeing. Contaminated water can cause numerous diseases, including diarrhoea, cholera, and typhoid. Therefore, it's important to ensure that water tanks are clean and free of contaminants. Water tank cleaning machines are commonly used to clean water storage tanks. The process is quick, efficient, and ensures that the water stored in the tank remains safe for consumption. The techniques and methods used in our research involved the use of a rotating brush with a knuckle joint arrangement, which allowed for effective cleaning. Additionally, we used a vacuum cleaner to simultaneously remove dirt, further enhancing the cleaning process. The results of our research showed that the combination of the rotating brush and knuckle joint arrangement, along with the use of a vacuum cleaner, led to highly effective cleaning outcomes. This research can benefit people by providing a more efficient and effective method for cleaning with the best product at a lower cost.

Keywords: Water tank, tank cleaning, cleaning technology, innovation in cleaning

1. Introduction

Water storage tanks are essential for meeting our daily water needs, but they are susceptible to various types of contamination, such as algae, bacteria, sediment, and grime. Manual cleaning of these tanks requires significant effort and time and may not be entirely effective, increasing the risk of contamination. As a result, there is a clear need for an innovative and effective water tank cleaning machine that can address the current gaps in this field.

However, despite the apparent need for a more efficient and effective cleaning solution, there are still several research gaps that need to be addressed. For example, some cleaning machines should not be able

[a]v.shahu@jitnagpur.edu.in, [b]p.dhopte@ jitnagpur.edu.in

DOI: 10.1201/9781003567653-41

to clean corners of the tank, including hard-to-reach areas that are often overlooked during manual cleaning. Additionally, the machine's weight and handling need to be optimized to ensure ease of use and prevent accidents during operation.

Therefore, the aim of this research paper is to identify and address the gaps in the water tank cleaning machine and propose innovative solutions to improve the efficiency and effectiveness of tank cleaning. Through a comprehensive review of existing research in this area, we will analyses the current state of the art in water tank cleaning machines, identify the key challenges and limitations, and propose innovative solutions to overcome these challenges. The ultimate goal is to develop a new and effective cleaning solution that can ensure safe and clean water storage for domestic and commercial use.

2. Literature Review

Literature review of the existing water tank cleaning machines is presented where different authors have presented their approaches and mechanisms to clean water tanks.

S.D. Thonge et al. (2017) proposed a reciprocating 4-bar mechanism to clean the wall. The assembly is powered by an electric motor, and the wall brush position is controlled by a rack and pinion mechanism. P. Chaudhary et al. (2019) proposed a mechanism with multiple brushes to cover the entire area of the tank. The assembly is also powered by an electric motor. R. Chaurasiya et al. (2019) proposed a mechanism with a cleaning brush and pump to remove waste. The mechanism has a knuckle joint to control the direction of the brush.

P. Kumar et al. (2018) proposed a robot that uses high-pressure water jets for cleaning. The machine has sensors, a high-pressure pump, and a controller. J. Ghormade et al. (2022) proposed a mechanism with brushes for bottom and side wall cleaning. The brushes are provided with an adjustable arm, and the assembly rotation

happens manually. A. Dwari et al. (2020) proposed a mechanism that uses a long leadscrew to cover the entire height of the tank. Brushes for cleaning the bottom and side walls are attached to the leadscrew, and when the leadscrew rotates, the brushes also rotate and move along the height of the tank. Sawansukha et al. (2021) proposed a mechanism that uses a C-type brush powered by an electric motor. M. Vazarkar et al. (2022) proposed a cleaning mechanism based on the principles of centrifugal force and sedimentation. S.R. Lomte et al. (2022) proposed a mechanism that uses water jets for cleaning the tank. R. Raffik et al. (2018) proposed a mechanism that cleans the tanks using a method similar to that of an irrigation sprinkler but with increased force. The machine uses rotary jet nozzles that are rotated by a servo motor. S. Shirashyad et al. (2021) proposed a mechanism that uses a shaft with multiple brushes to clean the wall. Linkages are used to make the side wall compact. Ramachandran et al. (2021) proposed a mechanism that uses four-bar linkages for the wall brush and a rack and pinion arrangement to control the position of the wall brushes. The power source of a machine is an electric motor. K. John et al. (2018) proposed a mechanism that uses water jets to clean the tank. The position and direction of the water jet are controlled by electric motor links. M.P. Bresil et al. (2022) proposed a mechanism that uses brushes for cleaning the bottom of the tank and a nylon thread to clean the wall. When the electric motor starts, the nylon thread experiences centrifugal force and acts as a brush on the wall of the tank.

2.1. Literature Outcomes

The literature outcomes offer valuable insights into the limitations of current water tank cleaning machines and underscore the need for innovative solutions in this field. What makes this survey unique is its focus on the challenges faced by certain water tank cleaning machines, such as ineffective corner cleaning, unwieldy weight, and difficulty of handling. Although some researchers have employed water jets in their machines, the problem of water wastage in countries such as India where water scarcity is a major issue persists. In addition, while some machines have incorporated automation, the survey found that society does not frequently clean water tanks, thus necessitating the development of semi-automatic machines that offer better outcomes at a reduced cost without depending heavily on automation.

Moreover, the survey emphasizes the importance of semi-automation in water tank cleaning machines, which can provide better results at a reduced cost. Overall, the literature outcomes provide a strong rationale for the development of a more efficient and cost-effective water tank cleaning machine that addresses the limitations of existing machines.

3. Methodology

3.1. Components

Following components were used for constructing the Cleaning machine. The main aim of the design was to result in an economical product which can be used by even an unskilled human resource in least possible cost.

Stainless Steel Pipe: Used as the main structure of the cleaning machine (Figure 41.1). The specifications are given in Table 41.1.

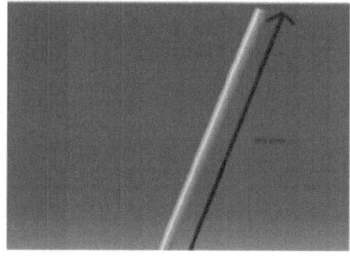

Figure 41.1: Stainless steel pipe.
Source: Author.

Table 41.1: Specification

Material	Stainless Steel
Diameter	2.5 cm
Length	89 cm; with an arrangement, it can extend up to109 cm.

Source: Author.

- Hollow Disc Chamber
  Hollow disc chamber is to make a space for the suction of the dirty water after cleaning of the tank. The suction hole is provided on the top (Figure 41.2) from where the dirty water is collected to be disposed off. And at the bottom of the disc the holes are provided to provide the inlet for the suction (Figure 41.3).

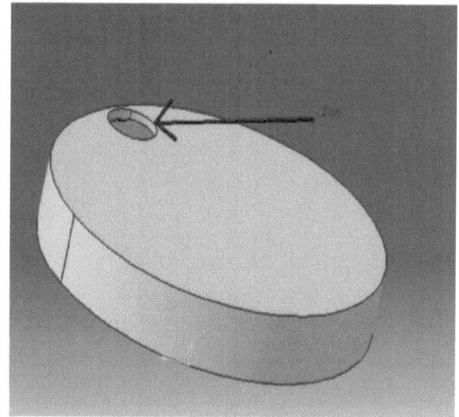

Figure 41.2: Top view of disc.
Source: Author.

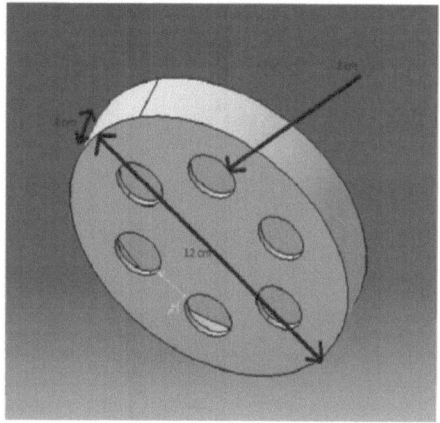

Figure 41.3: Bottom view of disc.
Source: Author.

- Electric Motor : details of the motor used for rotating the brush are given in Table 41.2 (Figure 41.4)

Figure 41.4: Electric motor.

Source: Author.

Table 41.2: Specification

Name	Orange OG555 High-torque DC Motor
Torque	173.6 N-cm
Voltages	12 volts
Current	4.134 amp
Motor Body Length	96cm
Shaft Length	3.5cm
Speed	100 Rpm
Shaft Diameter	6mm

Source: Author.

- Brush: The brush is used to rub and clean the surface and walls of the tank. The specification are given in Table 41.3 (Figure 41.5)
- Dry & Wet Vacuum Cleaner Vacuum cleaner is used to suck and collect the dirt laden water from the tank. The specification are given in Table 41.4.

Figure 41.5: Brush.

Source: Author.

Table 41.3: Specification

Material	Nylon
Diameter of Brush	12cm
No. of Holes	6
Diameter of Holes	2cm
Location of Holes from Center	4cm

Source: Author.

Table 41.4: Specification

Name	Wet & Dry Vacuum Cleaner
Capacity	15L
Power Consumption	1400 watt
Product Dimension	32.5 L × 32.5 W × 45.5 H cm
Items Wight	4200 gm
Suction	18 kPa

Source: Author.

- U Shape Frame (Dimensions are shown in Figure): Frame provide a support structure to hold the brush, motor and suction passage assembly (Figure 41.6)

3.1.1. System Diagram

The complete assembles system is shown in system diagram (Figure 41.7)

3.2. Working

The water tank cleaning machine has been designed to simplify and optimize the

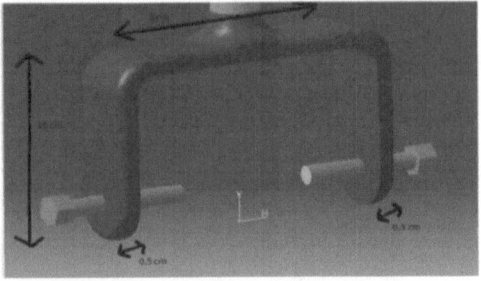

Figure 41.6: U-shape frame.

Source: Author.

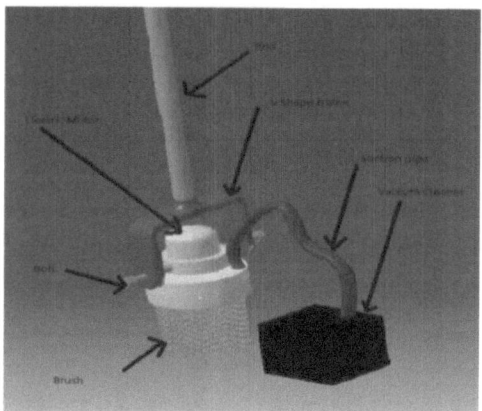

Figure 41.7: System block diagram.

Source: Author.

cleaning process. The process involves a few simple steps to ensure the thorough cleaning of the tank. Initially, the tank water is drained through the drainage valve to ensure that the tank is empty and ready for cleaning.

Next, a cleaning agent or bleaching powder is applied to the walls and bottom surface of the tank to remove any stubborn stains or dirt particles. Once the cleaning agent has been applied, the electric motor is started, and the brush begins to rotate, applying pressure to the surface being cleaned. This effectively removes any remaining dirt or stains on the surface.

If the user wishes to remove sediment and dirty water during the cleaning process, they can turn on the vacuum cleaner. The dirty water will be sucked out through the brush hole and into the fixed hollow disc chamber. The vacuum cleaner is synchronized with the brush movement, and when the brush hole overlaps with the hole on the disc, the dirty water is effectively removed.

The water tank cleaning machine has a unique feature—a revolute joint that allows the user to change the orientation of the entire electric motor to clean any surface of the tank. This makes the cleaning process more efficient and convenient as it is adaptable to various shapes and sizes of the tank.

During the cleaning process, a small amount of water is continuously supplied to facilitate the cleaning process.

Once the cleaning process is complete, the tank can be washed with water, and the dirty water can be removed by the vacuum cleaner. This ensures that the tank is not only clean but also free from any sediment or dirty water, which can compromise the water quality. Overall, the water tank cleaning machine is an efficient and cost-effective way to clean tanks with the added benefit of simultaneously removing sediment and dirty water from the tank, ensuring that the water is clean and safe for consumption.

4. Design and Calculation

- Assumption
  - o No deformation &bending of nylon brush.
  - o Brush covering entire area of plate.
  - o Also, there is negligible increase and decrease in the area.
  - o Neglecting effect of dirt, algae, water etc.
- Torque by Statics

  Assume max permissible load that can apply by person is W. These forces are transmitted to motor shaft to the brush. At the brush side, these forces are distributed uniformly and the local pressure exist at each and every point.

$$\text{Pressure (P)} = \frac{\text{Allowable Load}}{\text{Effective cross−area of brush}} = \frac{W}{\pi R^2}$$

Differential Friction force = coefficient of friction × differential normal force at that point

where, differential normal force (dN) = P × (dA)at that point

We take differential area in form of ring and differential area is variable with respect to r. Hence, normal force is different at each point and therefore friction force is variable with respect to " r" (Figure 41.8).

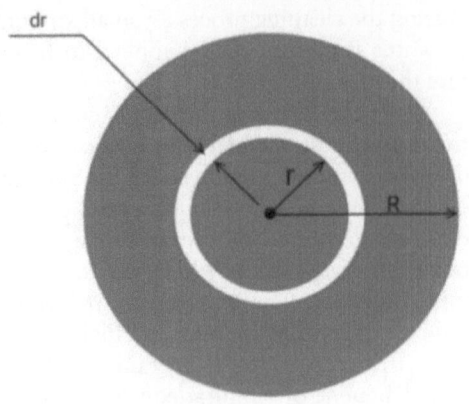

Figure 41.8: Brush area at bottom.

Source: Author.

Differential torque at any point
dT = force × Point of Application of force

$$= dF \times r$$

$$= (\mu \times dN) \times r$$

$$= \mu \times (P \times dA) \times r$$

To Calculate total torque on disc we need to integrate

$$T = \int dT$$

$$= \int \mu \times (P \times dA) \times r$$

$$= \mu P \int r dA$$

$$= \mu P \int r \times (2\pi r dr)$$

$$= 2\pi \mu\, P \int_0^5 r2 dr$$

$$= 2\pi \mu\, P \left[\frac{r^3}{3}\right]_0^R$$

$$= 2\pi \mu P \left[\frac{R^3}{3}\right]$$

$$T = \frac{2\mu WR}{3}$$

- Torque due to viscous effect
  Because there is negligible area of contact and the viscosity of water is very low, the resisting torque developed on the motor is negligible.
- Actual Resisting Torque developed on shaft

Given that: Coefficient of Static friction (μ) = 0.4

(For design calculation assuming friction develop between nylon & steel from engineeringtoolbox.com)

Maximum Permissible Load (W) = 5Kg = 49.05 N

Effective Outer Radius of Brush (R) = 5 cm

$$\text{Resisting Torque (T)} = \frac{2\mu WR}{3}$$

$$= \frac{2 \times 0.4 \times 49.05 \times 5}{3}$$

$$= 65.4 \text{ N-cm}$$

5. Conclusion

In conclusion, the development of an efficient and cost-effective water tank cleaning machine is crucial to ensuring a safe and clean water supply. Existing machines face challenges such as ineffective corner cleaning, heavy weight, water waste, and reliance on automation. Through a market survey, it was found that a semi-automatic machine that reduces overall cost while providing better results can be an effective solution.

Overall, the developed water tank cleaning machine has the potential to address the limitations of existing machines, reduce overall cost, and provide better cleaning results. The research findings can be useful for policymakers, researchers, and stakeholders in the water supply industry to improve water tank cleaning practices and ensure access to safe and clean water.

References

Bresil, M. P., Navaneeth, M., Pranav, C., and Yuvaraj, K. B. (2022). Design and fabrication of cylindrical water tank cleaning machine. International Journal of Scientific Research in Engineering and Management (IJSREM), 06(08). ISSN: 2582-3930.

Chaudhary, P., Darji, B., Patel, H., and Rajput, G.S. (2019). Smart water tank cleaning machine for household applications. International Research Journal of Engineering and Technology, 6(4), 4144–4147. doi: 10.35940/ijrte.D5689.068419.

Chaurasiya, R., Chamoli, P., Ahmed, S., Vashisth, V., and Singh, D. (2019). PVC water tank & slurry cleaning machine. IOP Conference Series: Materials Science and Engineering, 691, 012085. doi: 10.1088/1757-899X/691/1/012085.

Dwari, A., Ingle, G., Belsare, S., Nimkande, S., and Bhanarkar, A. (2020). Design and fabrication of overhead water tank cleaning machine. International Journal for Research in Applied Science and Engineering Technology, 8(4), 148–152. doi: 10.22214/ijraset.2020.4046

Ghormade, J., Patle, K., Shiekh, S., Bhongle, R., Shauh, S., Gupta, A., and Karande, A. (2022). Development & fabrication of water tank cleaning machine. International Journal of Innovative and Research Technology, 9(2), 1–6. doi: 10.15680/IJIRT.2022.0902010

John, K., Ahammed, V. A P, M. P, L. M C, and M. S. M. (2018). Design and fabrication of automatic tank cleaner. International Research Journal of Engineering and Technology (IRJET), 5(4). e-ISSN: 2395-0056, p-ISSN: 2395-0072.

Kumar, P., Deotus, W., Kathir, S., Prasad, T., and Karthikeyan, K. (2018). Autonomous water tank cleaning robot for indian household. International Journal of Research in Advanced Electronics Engineering, 7(3), 5441–5447.

Lomte, S.R., Kumar, S., Bade, K., and Deshpande, S. (2022). Comparative study & conceptual design of water tank cleaning system for residential multi-storey building. International Research Journal of Modernization in Engineering Technology and Science, 4(7), 78–84. doi: 10.46281/irjmets.2022.v04i07.010

Raffik, R., Shameer, S. K, Arun, S., Prabhu Raja, P., and Manoj Kumar, R. (2018). Automatic tank cleaner. International Journal of Mechanical and Production Engineering Research and Development (IJMPERD), 8(special issue 7), 2249–6890. ISSN: 2249-8001

Ramachandran, A., Iyer, A., Iyer, S., Mudaliyar, V., and Ahmed, S. (2021). Design and fabrication of automatic water tank cleaning machine. International Journal of Engineering Research and Applications, 11(5), (Series-V), 37–42. ISSN: 2248-9622.

Sawansukha, A., Humane, A., Bhute, H., Fating, P., Bobhate, P., Raghorte, S., and Arajpure, V.G. (2021). Intelligent based design and development of overhead water tank cleaning machine. Journal of Emerging Technologies and Innovative Research, 8(6), 14–21. doi: 10.30534/jetir.2021.8.6.2197.168

Shirashyad, S., Kumar, S., Ansari, I., Wale, M., and Robinson, P. (2021). Development of automated overhead water tank cleaning machine. Journal of Emerging Technologies and Innovative Research (JETIR), 8(7), 100–106. ISSN: 2349-5162.

Thonge, S.D., Shelke, P.K., Wakte, V.B., Thonge, S.A., and Shinde, R.S. (2017). Automatic water tank cleaning machine. International Research Journal of Engineering and Technology, 4(2), 660–663. doi: 10.2395/ijrtes.2017.4371.

Vazarkar, M., Kar, P., Dhobale, V., Ghawate, P., and Shiralkar, A.D. (2022). Design and development of automatic water tank cleaner. International Journal of Innovative Technology and Exploring Engineering, 11(8), 430–434. doi: 10.35940/ijitee.K1189.0780822.

42 Design and development of automatic shoe cleaning and polishing machine

Anurag Karande[a], Vandita T. Shahu, Sanjay Sajjanwar, and Amar Kawale

Department of Mechanical Engineering, Jhulelal Institute of Technology, Nagpur, India

Abstract

To make the process of polishing shoes easier, an automatic shoe polishing machine was designed and manufactured in this work. The goal of this project is to fully automate the process of polishing shoes without the need for human interaction. An automatic shoe polishing equipment that simplifies and saves time was attempted to be designed and manufactured in this work. The goal of this project is to fully automate the process of polishing shoes without the need for human interaction. Zero human effort is the key objective while constructing an automatic shoe polishing system. The machine is composed of three primary units: the control unit, which manages all tasks in accordance with instructions, the polishing function section, and the transport unit. The goal of our project is to create an automated polishing equipment that is intelligent. Shoes are required for all employees in any firm, regardless of their attire. It takes time to consistently wear clean shoes in today's hectic world. This device aids in lightening the weight of the goods available for purchase. This initiative is innovative and beneficial, so any business can use our product more effectively.

Keywords: Shoe polishing mechanism, DC motor, semiautomatic, fabrication

1. Introduction

In the realm of business, looking professional is greatly dependent on what you wear. A high-quality shoe shine cream is more significant in this context. Hand-polishing a shoe can cause harm to the leather's surface. This will cause the shoe's lifespan to decrease. In addition, polishing takes a lot of time and manpower. On the other hand, the automatic shoe polishing machine currently on the market does not do the recommended polishing and polishing also takes more time.

The purpose of shoes is to protect and cushion the wearer's foot while they perform various tasks. The majority of shoes are made of leather, which requires special care and frequent polishing to maintain its shiny appearance. Shoe polish, a waxy paste or cream that you apply evenly to the shoe with a cloth or brush, is needed for this. The surface is then forcefully rubbed until it is glossy to apply polishing. It also extends the life of the shoe. The manual application of this wax is time-consuming. This machine's job is to carefully hold the shoe, treat it with

[a]a.karande@jitnagpur.edu.in

DOI: 10.1201/9781003567653-42

polishing wax, and then polish it to make it shiny while ensuring.

There are so many machines of this type, but the developed production method makes users use immediately, the desires of users never end. However, further changes can be made to this type. A machine that polishes shoes automatically saves time and work by shining shoes quickly. We created this machine with all of these challenges in mind: professionals want their shoes to look beautiful and survive longer, but they often neglect to take the required actions. Your shoes will look great every day thanks to our machine. Automation has to be given more consideration because of the primary issues with manual machine types. Experts mostly concentrate on more intelligent and straightforward lifestyles. It was decided how innovations would emerge after taking all of these things into account. The purpose, objectives, and results of our innovation are eloquently explained in the parts that follow.

2. Literature Review

Until the middle of the 19th century, shoes were cleaned or polished by hand with a rag, cloth or brush. Back then, people used a wax product to shine or polish their shoes. This wax product was made from natural wax, oil, soda and fat. At that time, there was no concept of a shoelace machine.

In Jan 1997, William A. Beck and River Hills designed a coin-operated shoe shine machine that included two brushes, a motor, and a coin detector. A sensor used to start the motor immediately after a coin is detected. As soon as the coin was placed in the box where the sensor was installed, the motor started and the brushes rotated at a higher speed, then the shoe was oriented along the polishing brushes. (Kanna et al., 2021)

In September 2013 Asst. Professor Sreenivas H T of the Mechanical Engineering Department of VVIT and Shankar Gouda (Sreenivas et al., 2013), who studied at the Mechanical Engineering Department

of EPCET in Bangalore, Karnataka, India, designed a shoe polishing machine. It consisted of four brushes mounted on three motor-driven shafts. Two brushes were mounted on the left and right sides of the shoe, one was used to clean the sole of the shoe, and the fourth was mounted on the pole to polish the back and upper of the shoe. Fourth, one brush-shoe directed along the brush for polishing the back and upper surface of the shoe. In this model, nail polish is applied manually to the surface of the shoes during use. (Campbell, 2006)

In April 2015 Chauhan Vipul M., Swami Harsh R. Daiya Pradip R and Chauhan Vishwas S. designed a shoe polishing machine by Professor. Under the guidance of Rajeshkumar. It consisted of four brushes, with two brushes sliding along the shoe surfaces using a yoke mechanism. This turns the rotary motion into a sliding motion. One motorised brush used to clean or polish the back of the shoe. Fourth, one of the brushes can move up and down because it is still attached by a flexible joint that gives them bidirectional movement. It is used to clean or polish the upper edge of the front part of the shoe. There is a water pump that is used to distribute liquid to the surface of the shoe.

In April 2016, Animesh Kujur, Digvijay Murmu, Ashok Kumar Law, Amardeep Kumar, Kunal and Anup Ojha published a research paper at the International Conference on Engineering and Technology in which they designed a coin-operated shoelace machine. It contained two brushes, one for cleaning and one for polishing the shoe, both mounted on the same motor-driven shaft. As soon as the metal sensor detected the coin, the motor started moving. In this machine, the shoe is oriented along the brush during the polishing operation. (Harver et al., 2004)

In 2016, Amogh, Denver Martis, Gagan, and Chirag worked to design an automated shoe polishing machine. It had two roller brushes, two wax rollers, and a trolley with a handle to safely hold the shoe. The shoe has a hook that fastens to the cart, making it simple to safely fasten shoes of various sizes. Similar studies are found in literature to show the feasibility (Thorat, 2016, Neermarga et al., 2016; Viswanath et al., 2020; Ramesh et al., 2019; Kujuret al., 2016; Srihami et al., 2021; Aaron et al., 1995).

The addition of a polishing function to the shoe cleaning machine is regarded as an innovative advancement. To ensure their own safety and the cleanliness of the workplace, all professionals are required to wear clean shoes before entering. A shoe polishing tool was integrated with the bottom cleaning station that earlier researchers had built, taking these tasks into account.

3. Problem Statement

Involving people is the largest obstacle in the design and production of an automatic shoelace maker. When polishing shoes by hand, human involvement is vital. In addition to shoe cleaners, there are those who help align the shoe with the brush. The procedure will require longer time to finish.

4. Objective

A novel gadget that could perform the necessary functions even more efficiently and economically was attempted to be created, keeping in mind the features of the shoe washing machine that was previously stated. The following are the main objectives of this work.

- Reduce the demand for human labour.
- Enhanced qualities of high-quality polishing
- Cutting down on shoe-related time;
- Including sole polishing as an extra service;
- Offering a reasonably priced solution that considers ergonomics, cuts down on time, and gives the customer a high-quality polished product.

5. Project Description

5.1. Flow Chart

The block diagram shows the process flow of the shoe cleaning mechanism (Figure 42.1).

6. Working

In order to obtain the required shoe polishing effect—which is comparable to manual polishing or produces exceptional results—this project offers a shoe polishing equipment that solves the issues with conventional procedures.

In essence, this model uses a shoe washing method. The circular motion in this instance is changed to a sliding motion. Our project now has the most controllable course thanks to this phase (Figure 42.2). Here, power is transmitted by a single 12 volt motor. This motor has a shoe cleaning system attached to it. For residential use, there is a single slide that is utilised to slide the adversary. Aluminium angles in the shape of a L are used to fasten this slide plate to the shoe cleaning mechanism. These corners have a screw holding a scoop brush to the rear of them. The brushes are rotated by a second 12V DC motor located

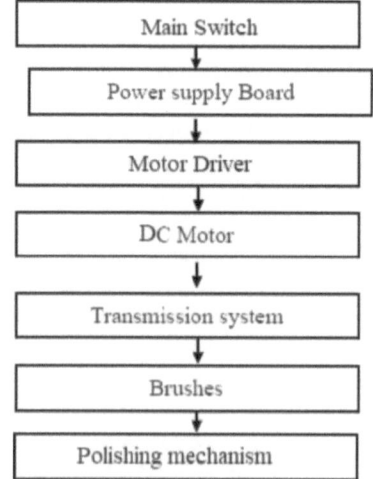

Figure 42.1: Block diagram of shoe polishing machine mechanism.

Source: Author.

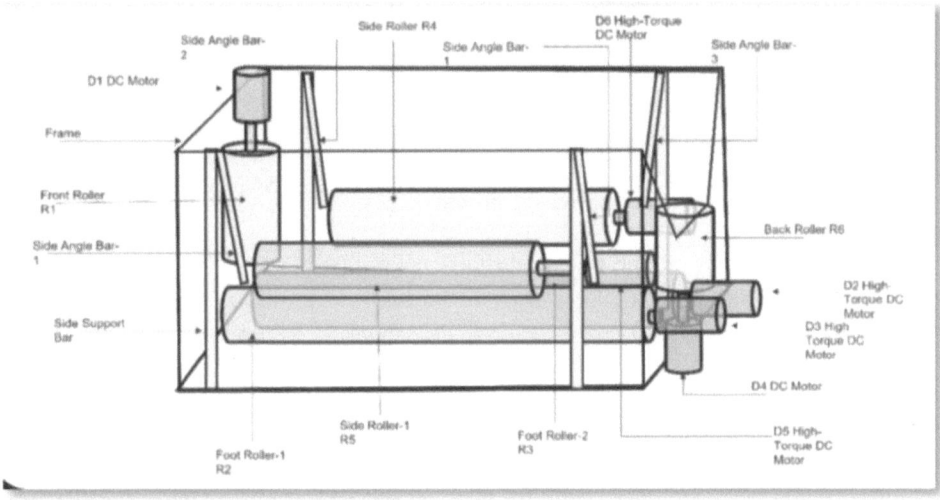

Figure 42.2: Schematic diagram of shoe polishing machine mechanism.

Source: Author.

at the front and rear. Figure 42.1 displays the block diagram of the shoe polishing machine's mechanism. The belt transmission, shaft, and sole cleaning brush shaft transform the motor power into the sole cleaning brush shaft. A motor connects the various components of the shoe polish brush. The uppermost layer has a device for washing and polishing boots. This features four nylon-made brushes. In this case, four brushes were employed, and they were separated into classes.

1. Left hand brush
2. Right hand brush
3. Front side brush
4. Back side brush

Figure 42.3 shows the actual model developed for the innovative shoe polishing machine.

Components
- Main Power supply
- High torque DC Motor 100 RPM
- Brushes
- Shoe polishing mechanism
- Frame
- Adapter
- LED Light
- Switch & Others

Figure 42.3: Actual model.

Source: Author.

7. Conclusion

We came to the following conclusions in this work. A few adjustments can be made to this work to make it better. The following findings guided the design and production of a DC semi-automatic intelligent shoe polishing machine:

The shoe's sole and the revolving brushes that are linked to it are both effectively cleaned by the brushes' quick rotation. When the user adds wax polish to the shoe, the shaft aids in shoe polishing.

Since the machine is DC powered, it would be wiser to use it commercially. In this way, machine maintenance can be effectively managed.

It also performs a polishing function to give shoes a shiny appearance.

Master switch control integrated. It is easy to update.

Shoe polish is an easy and practical solution to keep your shoes looking like new for a longer period of time.

References

Aaron, D. D., Walter, J. M., and Charles, E.W., (1995). Machine design, theory and practice. Macmillan Publishing Co. Inc.

Campbell, I. E. (2006). Foot-support for shoe polishing machines. International Machine Corporation.

Harver, J. C., Slater, S. F., and Maclarchora, D. L. (2004). Response and proactive market orientation and new – Produce access. Journal of Product Innovation Management, 21(5), 334–347.

Kanna, L. S., Lavnis, A. K., and Lagdive, H. B., (2021). Design and development of automatic shoe polishing machine. International Advanced Research Journal in Engineering and Technology, 8(4), 238–250.

Kujur, A., Kumur, A., Murmu, D., Kumar, A., and Ojha, A. (2016). Design of a coin operated shoe polishing machine. International Research Journal of Engineering and Technology, 3(4), 1644–1648.

Neermarga, A. A., Chirag, V. R., and Martis, D. P. (2016). Design and fabrication of automatic shoe polishing machine. National Conference on Advances in Mechanical Engineering Science (NCAMES), pp. 249–252.

Rajeev, N. (2009). Design and Development of an automatic shoe polishing device used in offices. PT.

Ramesh, P., Anish, M., Sunday, J. B., and Raj, A. D. (2019). Design and fabrication of semi-automatic sole cleaner. International Research Journal of Multidisplinary Technovation (IRJMT), 1(4), 9–16.

Sreenivas, H. T., Gouda, S., and Shankar, G. (2013). Design of shoe sole cleaning with polishing machine. International Journal of Innovative Research in Science, Engineering and Technology, 2(9), 5022–5029.

Srihami, D., Kumar, B. R., and Yuvaraj, K. (2021). Development of Indian coin based automatic shoe polishing machine using Raspberry pi with open cv. International Journal of Advanced Research in Electrical, Electronics and Instrumentation Engineering, 1(3).

Thorat, S.. (2016). Design and Fabrication of automatic shoe polishing machine. Mechanical Project Report. learnwork.com.

Viswanath, V. S., Kumar, B. P., Raju, O. K., Venkateswahi, P., and Rajesh, D. (2020). Design and Fabrication of smart automatic shoe polishing machine. Junikhyat UGC Group 1listed Journal, 10(5), 535–545.

43 Design and development of portable briquette machine

Bhushan Deshmukh[a], Prakash Dhopte[b], Safal Shambharkar, and Pravin Petkar

Department of mechanical Enginering, Jhulelal Institute of Technology, Nagpur, India

Abstract

The portable biomass briquette machine is designed with considerations for portability and ease of use. It incorporates a compact and lightweight design, allowing for easy transportation to different locations. The machine utilizes a robust mechanism for crushing and shredding various organic waste materials, including agricultural residues and municipal solid waste. These materials undergo a drying process to achieve optimal moisture content, followed by compaction under high pressure to form uniform fuel briquettes. The briquetting process ensures the complete combustion of biomass, resulting in reduced ash production and lower CO2 emissions compared to traditional fuel sources. Extensive testing and optimization are conducted to evaluate the quality, energy content, and combustion characteristics of the produced briquettes. Economic viability is also assessed, considering factors such as material costs, machine operation, maintenance, and potential market demand for the briquettes. The briquette machine have the potential to positively impact the environment, public health, and resource conservation. By transforming unwanted organic waste into valuable energy sources, this portable biomass briquette machine contributes to the cleaning of surroundings, minimization of landfills, and conservation of precious natural resources. The project aligns with the principles of sustainability and offers a greener alternative for energy generation, paving the way for a cleaner and more sustainable future.

Keywords: briquette machine, biomass, valuable energy

1. Introduction

The development of a biomass briquette machine that aims to address environmental challenges like the accumulation of organic waste and municipal solid waste, and the unsustainable reliance on finite fossil fuel resources. By transforming this unwanted waste into clean-burning fuel briquettes, we aim to clean our surroundings, minimize landfills, and contribute to a greener and more sustainable future. To address these challenges, we focuses on the utilization of biomass waste as a renewable energy source. Biomass briquettes, derived from organic waste and municipal waste, provide a clean-burning alternative to traditional fuels. The objective is to design and develop a portable biomass briquette machine that can efficiently convert the identified waste materials into high-quality fuel briquettes. The machine will be

[a]b.deshmukh@jitnagpur.edu.in, [b]p.dhopte@jitnagpur.edu.in

DOI: 10.1201/9781003567653-43

compact, lightweight, and user-friendly, allowing for easy transportation and operation in various locations. By optimizing the briquetting process, we aim to produce briquettes with excellent combustion characteristics, ensuring maximum energy efficiency. Additionally, the we will evaluate the economic viability and environmental benefits of utilizing biomass briquettes as an alternative fuel source. By reducing the amount of waste sent to landfills, we contribute to minimizing the strain on the environment and conserving valuable land resources.

2. Literature Review

2.1. *Ownership Concentration and Stock Return*

Design and Development of a Compact Screw-Press Biomass Briquetting Machine for Productivity Improvement was studied by Wessapan, Teerapot, et al. (2010). This study focuses on the design and development of a compact screw-press biomass briquetting machine. The researchers highlight the importance of portable machines in rural areas with limited access to electricity. The machine's design incorporates locally available materials, aiming to improve productivity and ensure ease of operation. The study concludes that the developed machine has the potential to enhance biomass briquette production and contribute to sustainable energy solutions.

Design and Fabrication of a Low Cost Briquetting Machine and Estimation of Calorific Values of Biomass Briquettes (Palanikumar et al., 2016). This research paper presents the design and fabrication of a low-cost briquetting machine suitable for small-scale biomass briquette production. The authors discuss the machine's components, including the power transmission system, hopper, and screw extruder. They also provide an estimation of the calorific values of biomass briquettes produced using different feedstock materials.

Development of an Improved Portable Biomass Briquetting Machine (Omoniyi and Ojo, 2023). This research paper presents an improved portable biomass briquetting machine designed to address challenges faced by existing models. The authors discuss the importance of briquette production in waste management and sustainable energy production. They highlight the machine's features, including its compactness, simplicity, and use of hydraulic components for increased efficiency. The study concludes that the improved machine shows promise for small-scale biomass briquette production.

Development and Performance Evaluation of a Hand-Operated Biomass Briquetting Machine (Rafeeque et al., 2012). This study focuses on the development and evaluation of a hand-operated biomass briquetting machine suitable for rural communities. The authors discuss the importance of biomass briquettes as an alternative fuel source and outline the machine's design, fabrication, and testing process. The study includes performance evaluation parameters such as briquette density, moisture content, and calorific value. The results indicate that the hand-operated machine can effectively produce biomass briquettes for household use.

Design and Development of a Compact Biomass Briquette Machine for Rural Areas(Amarnath et al., 2019). This study focuses on the design and development of a compact biomass briquette machine suitable for rural areas. The authors discuss the machine's design considerations, fabrication process, and performance evaluation. They emphasize the importance of portable machines in rural settings and highlight the potential of biomass briquettes as a sustainable fuel source. The study concludes that the developed machine can contribute to rural development and environmental conservation.

Performance Evaluation of a Hand Operated Biomass Briquetting Machine. (Rafeeque et al., 2012). This study focuses on the performance evaluation of a hand-operated biomass briquetting machine.

The authors discuss the machine's design, construction, and operation. They evaluate various parameters such as density, moisture content, compressive strength, and energy consumption of the produced briquettes. The study concludes that the hand-operated machine is suitable for small-scale briquette production and can contribute to sustainable energy solutions.

Asamoah et Al. presented an experimental evaluation of a manual briquetting machine for different biomass types (Asamoah et al., 2016). The authors discuss the machine's operation, briquette production process, and the effects of various parameters on briquette quality. They evaluate the performance of the machine using different biomass feedstocks and assess briquette characteristics such as density, durability, and combustion properties. The study provides valuable insights into manual briquetting machines' potential and limitations.

Development of a Compact Screw-Press Biomass Briquetting Machine for Productivity Improvement. (Agidi et al., 2006). This research paper highlights the development of a compact conical screw-press biomass briquetting machine with a focus on productivity improvement. The authors discuss the design considerations, fabrication process, and performance evaluation of the machine. They assess parameters such as briquette density, moisture content, and energy consumption. The study demonstrates the potential of the compact machine in enhancing biomass briquette production

2.2. Details of Components Used in Briquette Machine

The goal is to create a mobile biomass briquette machine that can effectively transform the indicated waste materials into high-quality fuel briquettes. The device will be portable, lightweight, and user-friendly for convenient use and transit in a variety of settings (Figure 43.1). We seek to make briquettes with good combustion characteristics, assuring optimum energy efficiency, by streamlining the briquetting process.

In the context of a portable biomass briquetting machine, the hopper is an essential component that plays a crucial role in the feeding and storage of biomass materials. The hopper is a critical part of a portable biomass briquetting plant that is essential to the feeding and storage of biomass feedstock. The hopper is a key element that is necessary to the feeding and storage of biomass materials in the context of a portable biomass briquetting plant. The hopper is a critical part of a portable biomass briquetting machine that is essential to the feeding and storing of biomass materials (Figure 43.2).

2.2.1. Electrical Motor

In a portable biomass briquetting machine, the motor is a key component responsible for providing the necessary power to drive various mechanical parts and processes involved in the briquetting operation (Figure 43.3).

Figure 43.1: Briquette machine.
Source: Author.

Figure 43.2: Hopper of the Briquetting machine.
Source: Author.

Figure 43.3: Electric motor.

Source: Author.

2.2.2. *Screw Jack*

In a portable biomass briquetting machine, a screw jack is often utilized as a mechanical component to facilitate the compression process. The working of a screw jack in a portable biomass briquetting machine allows for continuous and controlled compression of the biomass material, resulting in the formation of solid briquettes (Figure 43.4). The specific design and configuration of the screw jack may vary depending on the machine model and the desired briquette specifications. Manufacturers or suppliers of such machines can provide detailed information on the screw jack's specifications and its working in their specific models.

2.2.3. *Outer Surface of Delivering Cone (Nozzle)*

In the context of a portable biomass briquetting machine, the outer surface of the

delivering cone nozzle is an important component involved in the discharge of the briquettes from the machine. The working of the outer surface of the delivering cone nozzle ensures the smooth and efficient discharge of the formed briquettes from the portable biomass briquetting machine. It helps to guide and control the flow of briquettes, reducing friction and facilitating their collection or further processing (Figure 43.5). The specific design and features of the delivering cone nozzle may vary depending on the machine model and the requirements of the briquetting process.

3. Designing a Portable Biomass Briquetting

Designing a portable biomass briquetting machine involves several considerations to ensure its functionality, efficiency, and ease

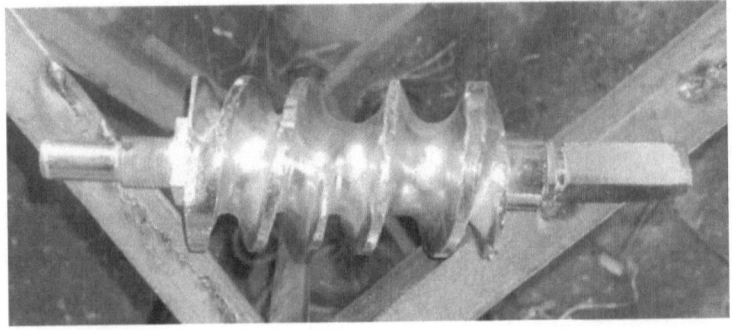

Figure 43.4: Screw Jack.

Source: Author.

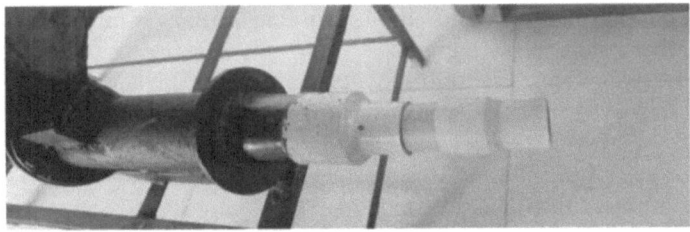

Figure 43.5: An image of nozzle.

Source: Author.

of use. Here are some key aspects to consider in the design process:

Machine Structure: The machine should have a compact and lightweight structure to facilitate easy transportation and on-site operation. It should be sturdy and robust to withstand the forces generated during the briquetting process. The design should include appropriate reinforcements and supports to ensure stability and minimize vibration.

Power Source: Consider the availability and accessibility of power sources at the intended operating locations. Design the machine to be compatible with different power options, such as electricity, diesel engines, or hydraulic systems. Ensure that the power source is efficient and provides sufficient energy to drive the machine components effectively.

Feeding Mechanism: Design a suitable feeding mechanism to ensure a continuous and uniform supply of biomass material into the briquetting chamber. The feeding system should be efficient, minimizing material spillage and blockages. Consider the possibility of incorporating a hopper or conveyor system for easier material loading.

Briquetting Chamber: Design the briquetting chamber to accommodate the biomass material and facilitate the compaction process. Optimize the chamber dimensions and shape to ensure optimal compaction and uniform distribution of pressure. Include appropriate guides and surfaces to ensure proper alignment and formation of briquettes.

Compaction System: Select a suitable compaction mechanism, such as a piston, screw, or hydraulic press, based on the desired briquette size and production capacity. Design the compaction system to exert adequate pressure for efficient densification of the biomass material. Ensure that the system is capable of applying consistent pressure throughout the briquetting process.

Heating or Drying Mechanism (if applicable): If the design includes a preheating or drying mechanism, incorporate an efficient heat source or drying system to reduce the moisture content of the biomass material. Consider using methods such as hot air, infrared heating, or waste heat recovery to optimize energy utilization.

Control System: Include a control system to monitor and regulate the key parameters of the briquetting process, such as pressure, temperature, and production rate. Implement appropriate sensors, switches, and control mechanisms to ensure safe and efficient operation. Consider incorporating automation features to simplify operation and improve process control.

Safety Features: Design safety features to protect operators and prevent accidents or equipment malfunctions. Include emergency stop buttons, safety guards, and interlock mechanisms to ensure safe operation. Consider ergonomic design principles to minimize operator fatigue and enhance user-friendliness.

Materials and Components: Select materials and components that are durable, corrosion-resistant, and suitable for the intended operating conditions. Consider factors such as cost, availability, and environmental impact when choosing materials. Ensure that all components are properly sized and fitted to prevent leakage, wear, or failure.

Maintenance and Accessibility: Design the machine for easy maintenance and access to critical components, such as the compaction system, heating elements (if applicable), and control systems. Include features such as removable panels, access doors, and quick-release mechanisms for efficient servicing and repairs.

Testing of the Machine: For the purpose of this study, jatropha husk was used for the testing of the machine. The jatropha husk sample was collected from a jatropha seed oil extraction plant. Cassava starch was prepared with cassava bought from a local market and used as a binding agent mainly to overcome the major problem of material compaction and post compaction recovery, which represents enormous waste in energy input

3.1. Biomass-binder Mixture

Jatropha husk sample in three different particle sizes (original particle size, particle size less than or equal to 6mm and particle size less than or equal to 2mm) was mixed with an already prepared cassava starch in proportions of 100:15, 100:25, 100:35 and 100:45 by weight respectively. The starch and the biomass sample was well mixed

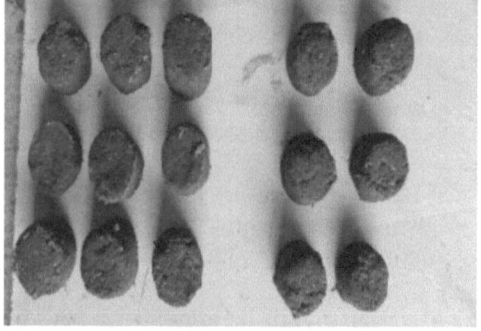

Figure 43.6: Sample of Briquettes.

Source: Author.

without forming a mixture with high moisture content because the formation of a mixture with higher moisture content due to excess addition of water reduces both the durability and density of the briquette.

4. Conclusion

The biomass briquette machine has successfully addressed the challenges of organic waste management, energy generation, and environmental sustainability. The machine has demonstrated the feasibility and effectiveness of converting biomass waste into high-quality briquettes, providing a renewable and cleaner alternative to traditional fossil fuels. Through the utilization of biomass waste materials such as agricultural residues and municipal solid waste, the project has contributed to waste reduction, minimized landfill usage, and promoted a cleaner environment. The briquettes produced have exhibited excellent quality, with desirable attributes such as shape, density, and durability, making them suitable for various heating applications. Furthermore, the biomass briquette machine holds economic benefits, as it utilizes locally available waste materials and offers opportunities for income generation and cost savings. It demonstrates the potential for sustainable development, where waste can be transformed into a valuable resource. While the machine has showcased positive outcomes, there are areas for further improvement and exploration. Continual research and development can focus on optimizing the machine design, enhancing the briquette production process, and expanding the range of feedstock materials. Collaboration with local communities, industries, and policymakers is crucial for scaling up the project's impact and promoting its adoption on a broader scale.

References

Amarnath, C., and Guha, A. (2019). Briquette compacting machine: A design for rural applications. In: Badodkar, D., Dwarakanath,

T. (Eds), Machines, Mechanism and Robotics. Lecture Notes in Mechanical Engineering. Springer, Singapore. https://doi.org/10.1007/978-981-10-8597-0_64

Asamoah, B., Nikiema, J., Gebrezgabher, S., Odonkor, E., and Njenga, M. (2016). A review on production, marketing and use of fuel briquettes. Colombo, Sri Lanka: International Water Management Institute (IWMI). CGIAR Research Program on Water, Land and Ecosystems (WLE). 51. (Resource Recovery and Reuse Series 7). doi:10.5337/2017.200

Abakr, Y. A., and Abasaeed, A. E. (2006). Experimental evaluation of a conical-screw briquetting machine for the briquetting of carbonized cotton stalks in Sudan. *Journal of Engineering Science and Technology*, 1(2), 212–220.

Rafeeque, M., Murugan, C., and Subrahmanian, N. (2012). Development and evaluation of hand operated low pressure biomass briquetting machine. Journal of Agricultural Engineering, 49(2), 46–49.

Omoniyi, T. E., and Ojo, O. (2023). Design and fabrication of an improved low-cost biomass briquetting system for rural communities. *Journal of the Ghana Institution of Engineering*, 23(4), 89–98.

Palanikumar, C., and Chinnaswamy, P. (2016). Design and Fabrication of low cost briquetting machine and estimation of calorific values of biomass briquettes. 5(7). doi:10.15680/IJIRSET.2016.0507055.

Wessapan, T., Somsuk, N., and Borirak, T. (2010, October). Design and development of a compact screw-press biomass briquetting machine for productivity improvement and cost reduction. In *Proceedings of the First TSME International Conference on Mechanical Engineering, Ubon Ratchathani* (pp. 20–22).

44 Exploring the role of mathematics in artificial intelligence and machine learning: an in-depth examination

Smita Chandar Tolani

Department of Applied Physics, St. Vincent Pallotti College of Engineering and Technology, Nagpur, India

Abstract

Artificial Intelligence is the future of technology that has permeated our daily existence in unprecedented and unforeseen manners. The world is slowly shifting towards functional gadgets, virtual reality, self-driving cars, robots, autobots and automated tools. Artificial Intelligence is just mathematics. This article explores the use of mathematical concepts like probability, reasoning, mathematical programming and rules for understanding AI problems.

Keywords: Artificial intelligence, mathematics and AI, mathematical reasoning

1. Introduction

Artificial Intelligence (AI) concerns with building smart and automated machines which are capable of performing varied tasks that typically require human intelligence. The possibility of intelligent machines who can copy human behavior was conceived with the help of mathematical reasoning and logics. Artificial Intelligence can be classified broadly as problems of Search and Representation which involves Mathematical Logics, Rules, Frames, nets and logic in interconnected tools and models (Garrido, 2010). The concepts of Linear Algebra, Probability, Mathematical Programming, Calculus, Logic Regressions, Vectors, Matrix and statistics find major use in Data Science, Machine Learning and AI advances. The solutions to virtual problems can be solved by mathematical concepts which enhance our ability to reason. Any change in the components can be dealt with effectively due to rules, structures and principles which are well defined and universally true. Mathematics and Statistics are very important subjects, without which understanding AI would be like having a body without soul. The understanding of Mathematical concepts is minimally required for concepts of Machine Learning and AI. A higher knowledge of factorization is needed for procedures in AI which is a bit tedious.

smitatolani@gmail.com

DOI: 10.1201/9781003567653-44

2. Mathematics in Artificial Intelligence and Machine Learning

The fundamental knowledge of Mathematical Modeling is the basis of design and automation of intelligent systems of artificial intelligence (Yu, n.d.). The four main streams of mathematics that are essential in the field of AI are **Probability, Linear algebra, Calculus and Statistics.** To get acquainted with Machine Learning, AI and Data Sciences one has to have a basic understanding of the above four branches of Mathematics. It involves writing and using complex algorithms to solve real world issues. The artificial systems require reasoning which is beyond just crushing numbers which is challenging. It needs the ability to draw inference, follow the order, knowledge of axioms, cognitive skills and understanding the question thoroughly (O'Neill, 2019). Broadly the areas of Mathematics which are useful for complex AI algorithms are depicted in Figure 44.1.

2.1. Probability

Probability theory is the heart of artificial intelligence which permits us to create statements and rational into uncertainty and gives the probability distribution to measure the uncertainty. In AI and machine learning, probability is used to predict the occurrence of future events. It is calculated as,

$$P(Event) = \text{Favorable Outcomes} / \text{Total Number of Possible Outcomes}$$

For predicting and estimating the outcomes and parameters such as Likelihood of Maximum Estimation, Bayesian Networks, A Posterior Maximum, Probabilistic Models or Graphical strategies are extensively used. We can do analysis of random phenomenon using probability distribution. Probability theory can be utilized in all the fields of AI including uncertainty analysis, reasoning, learning, problem solving and representation. Figure 44.2 depicts the different models applied with probability theory used in AI.

Machine Learning makes use of statistics which can be classified into Descriptive Statistics and Inferential Statistics. Descriptive Statistics mainly concerns with datasets which are small and it summarizes the target population. Inferential Statistics works with large datasets and is used to predict future outcomes.

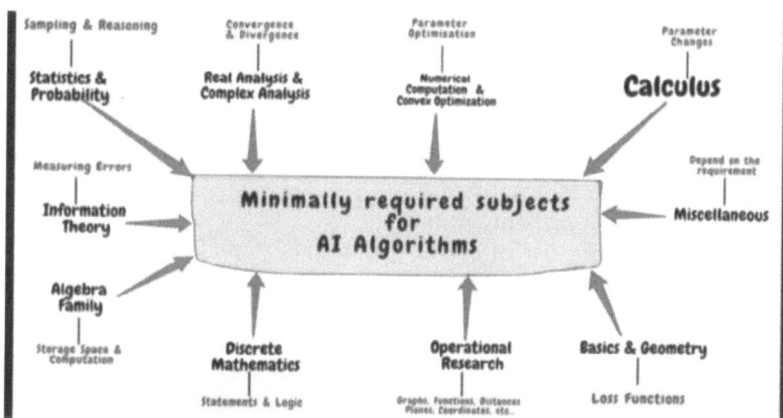

Figure 44.1: Areas of mathematics used in AI.

Source: AI and Mathematics, Shafi, 2020.

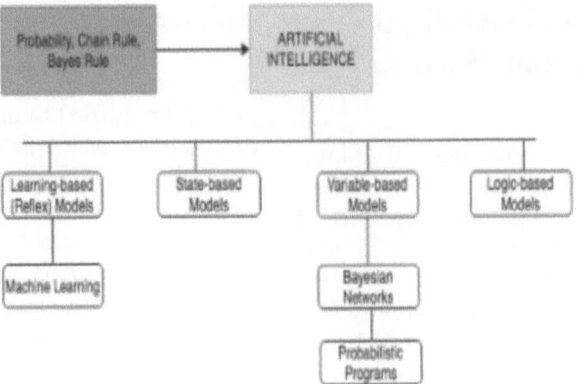

Figure 44.2: Different models applied with probability theory used in AI.

Source: Shafi, 2020.

2.2. *Linear Algebra*

Linear Algebra is a computational field which deals manipulation of complex data structures, development of machine learning models and operations performed on Matrices. It is a platform which enhances decision making and gives a better understanding of algorithms used. Linear Algebra uses concepts of Vectors, Matrices, Eigen values and Eigen functions, Tensors Game Theory, Graph Theory, and Scalars; to generate new ideas, models and data. Types of matrices used are Inverse, Adjoint and Identity. Some of the advanced concepts like Hilbert Spaces, Unitary, Eigen Value decomposition, Hermitian can be utilized in AI and machine learning.

2.3. *Calculus and Statistics*

Calculus determines how fast or slow the quantities change. It is used for optimizing the performances of models in AI and machine learning. It includes use of derivatives, integration, functions and limits. It can help in developing fast solutions. The different calculus used in models are Multivariate calculus , Advanced logistic regressions, Limits, Differential calculus Error minimization and optimization via gradient descent, Integral calculus. It is used to compute probabilities on data to optimize the possible better outcomes. The outcomes of data mining and machine learning algorithms are better understood by applying fundamental statistical theory. The important concepts in Calculus include:

Vectors or Matrices: Operators like Laplacian, Jacobian and Gradient.

Derivatives: Partial Derivatives and hyperbolic derivatives.

Algorithms: Gradient Algorithms

Machine Learning makes use of statistics which can be classified into Descriptive Statistics and Inferential Statistics. Descriptive Statistics mainly concerns with datasets which are small and it summarizes the target population. Inferential Statistics works with large datasets and is used to predict future outcomes. Basic Statistics like Mean, median, mode, variance, covariance, and so on are easily mastered and very important.

2.4. *Optimization Theory*

Artificial intelligence is mostly used to make optimal decisions in difficult contexts. Hence, Optimization theory is the used in problem solving and problem optimization. The fundamental objective of artificial intelligence is essentially to enhance decision-making by achieving optimal outcomes in intricate environments and interactions involving multiple entities. The resolution of nearly all artificial intelligence issues fundamentally revolves around tackling optimization problems, underscoring the critical importance of optimization theory in artificial intelligence. At the core of optimization theory research lies the challenge of ascertaining the existence of the maximum or minimum value of a specified objective function and identifying the input that yields said maximum or minimum value. Analogously, envisioning an objective function as a peak, the optimization process entails pinpointing the summit's location and devising a strategy to ascend to that peak.

3. Conclusion

"Artificial Intelligence" essentially means the models which make a machine to solve problems, solutions of these problems have

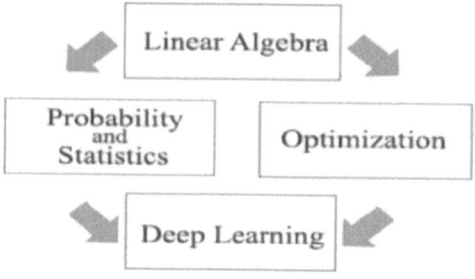

Figure 44.3: Linear algebra, probability and statistics, optimization, and deep learning. Courtesy of Jonathan Harmon.

Source: https://ocw.mit.edu/courses/18-065-matrix-methods-in-data-analysis-signal-processing-and-machine-learning-spring-2018/

not been encoded, but they have been constructed originally by the machine. The primary goal of AI is to develop a model that is acceptable for human knowledge (Garrido, 2010). Its subject is, therefore, "pure form". If you get acquainted with mathematical concepts like calculus, linear algebra, statistics, optimization theory and probability, nothing should get in the way of you getting into Artificial Intelligence. Without the mathematical ability to reason, draw conclusions and predict possible outcomes it is not possible to mimic the human intelligence. Mathematical models are an interconnected and integral part of machine learning and models of Artificial Intelligence and its algorithms.

References

Garrido, A. (2010). Mathematics and Artificial Intelligence, two branches of the same tree. Procedia Social and Behavioral Sciences, 2, 1133–1136.

Koopialipoor, M., and Noorbakhsh, A. (2020). Applications of artificial intelligence techniques in optimizing drilling. Emerging trends in mechatronics. doi: 10.5772/intechopen.85398

O'Neill, S. (2019). Mathematical reasoning challenges artificial intelligence. Engineering, 5, 817–818.

Shafi. (2020). AI & Mathematics.

Shafi. (2020). The Power of Probability in AI.

Stoyan Yu, G., Eschenko, V. G., and Vinarsky, V. Y. (1992). Mathematical methods of geometric design in artificial intelligence system. Institute for Problems in Machinery, Ukrainian Academy of Sciences, Kharkov, Ukraine.

45 Smart water dispenser for animals: automated filling, monitoring, water quality assessment

S. B. Pokle, Tushar Bhardwaj, Aniket Goldar, Ashwin Pande, and Sanskruti Wagh

Department of Electronics and Communication, Shri Ramdeobaba College of Engineering, Shri Ramdeobaba College of Engineering, Nagpur, India

Abstract

A critical issue that arises from the difficulties in controlling animal drinking water, especially in agricultural contexts, is making sure that cattle consistently have access to clean, safe water. In response, this research paper presents a revolutionary approach that will transform animal water management systems by utilising contemporary technologies. The system keeps an eye on a number of water quality measures using a variety of sensors. These sensors cooperate with a central control unit to guarantee that animals always have access to clean water. After receiving data from the sensors, the control unit adjusts the water system. The real-time monitoring capabilities allow for early detection of potential issues like water shortages or contamination. Our system's ability to prevent water shortages while maintaining animal welfare through constant water level monitoring is essential to its effectiveness. Sensors discreetly evaluate water quality, assuring the correct nutrient content and pH balance required for animal health. This research papers proposes a method for balancing performance with cutting-edge technology to meet the diverse demands of animal husbandry and agriculture.

Keywords: Water level monitoring, remote control, water quality assessment, cloud Integration

1. Introduction

Water is essential for all living things and plays an important role in maintaining hydration, supporting digestion and maintaining overall health. It is a vital element of survival in the wild, and its importance in domestic environments extends to supporting animal health and productivity. Providing clean water has become a central part of the responsibility of animal care, with implications not only for the health of animals but also for the overall idea of production. Clean water is essential to ensure animal health. The effects of dirty water are far-reaching and include health problems, reduced productivity, and potential damage to the environment. Water pH is important because high pH can cause problems for animals. From mouth and throat allergies to

DOI: 10.1201/9781003567653-45

eye irritation, animals also refuse to drink water with an unbalanced pH. In addition, pH values of the air can cause materials from various sources to dissolve, producing toxic substances and undesirable tastes. For example, cows do not like the metallic taste of high pH water. High pH water is often rich in nutrients, which causes stress in animals, especially cattle (Srivastava et al., n.d.).

Maintaining a good pH value is important to ensure animal health, especially in the context of the cattle industry. In addition to determining pH, it is also important to focus on the total dissolved solids (TDS) level in the water. While the recommended TDS values should be in proper range to ensures that essential nutrients are not in excess. The difference between TDS tolerance levels of different species must also be recognised. Excessive levels of TDS in water can cause adverse effects such as decreased palatability, which can lead to reluctance to drinking, that can lead to dehydration and health problems. The presence of too much TDS increases the risk of food contamination, impairs the absorption of nutrients and causes corrosion of infrastructure. By integrating our project, we present a thorough approach to reducing the various issues related to water management in animal care. Our system maintains continuous monitoring of water quality parameters, such as pH levels and total dissolved solids (TDS), which are essential for preserving animal health. By delivering clean and precisely balanced water, our system promotes optimal hydration and digestion, thereby enhancing overall well-being and productivity.

2. Literature Survey

Several existing projects in the market are designed to provide a continuous supply of water for animals. These dispensers are often gravity-fed or pressure-based systems, ensuring a steady flow of water. However, these conventional dispensers lack the advanced monitoring and quality assessment features that this research paper offers.

A livestock monitoring system utilizing IoT technology for dairy monitoring lacks in-depth discussion on sensor accuracy, data security, and scalability, while neglecting sensor calibration, maintenance, and connectivity reliability issues crucial for real-world deployment. These omissions underscore the necessity for a comprehensive approach to address practical challenges in implementing IoT-based livestock management systems (Igbinoba & Okhaifoh, 2019).

A smart dispenser system is designed to provide fresh water for companion animals remotely, utilizing IoT and ICT technology. However, the research paper lacks specific details on functionalities, scalability, and limitations, while neglecting crucial considerations like power consumption, sensor accuracy, and user accessibility for practical implementation (Yendri et al., 2019).

A study on employing a cost-effective Wireless Sensor Network (WSN) for dairy cow health monitoring identifies technical challenges, notably data transportation from cow-mounted sensors. Despite emphasizing potential benefits, the paper lacks specifics on addressing challenges and assessing overall efficacy. Enhancing data transmission reliability, sensor accuracy, and practical considerations would augment the paper's comprehensiveness (Von Borstel Luna et al., 2016).

A powered automatic system for animal farm management utilizes Arduino UNO, sensors, real-time clock, and GSM module to minimize labour and promote social distancing. While initial tests show promising efficiency, further details on scalability, robustness, and long-term performance are needed, along with considerations for sensor accuracy, power supply reliability, and system maintenance for a comprehensive evaluation (Manoj et al., 2022).

Our study offers a revolutionary response to the problems mentioned in the literature review. Our project excels over current systems in terms of efficacy and reliability by including robust monitoring capabilities and taking into consideration useful factors like sensor accuracy and scalability. Moreover, by blending technological innovation with a compassionate approach, our project offers a sustainable solution that contributes significantly to improving animal welfare and promoting agricultural sustainability.

3. Proposed Design

The tank will get filled from the water coming from the reservoir. The tank will have an ultrasonic sensor attached to it at its top as shown in Figure 45.1. It will measure the distance of the water level from the top of the tank and when its level goes down the threshold value, the tank gets automatically filled with water by fetching water from the reservoir. The tank will also consist of PH sensor and tds sensor to measure the PH value and the amount of organic and inorganic substances dissolved in water respectively. If the water has the desired range of PH value and tds value, the water will get poured into the container from where animals will drink the water. The container also consists of an ultrasonic sensor, which will monitor the water level of the contains. To do the remote monitoring of the system, it will be accessible by the Blynk app. Due to this, the owner can have a close look over the system functioning and also will have the data about the water usage.

4. Arduino UNO & ESP 8266

The Arduino Uno, with 14 digital and 6 analogue pins, simplifies sensor integration and enables real-time water management. Operating at 16 MHz, it interfaces with the ESP8266 Wi-Fi module to connect to Blynk Cloud for remote monitoring and control, facilitating alerts based on water quality and container levels in our project (Figure 45.2). The ESP8266 and Arduino UNO together provide a complete solution for managing animal water. The Arduino UNO's extensive GPIO pins and analog inputs facilitate smooth sensor interface, and its powerful processor guarantees effective data processing and control algorithms. Task distribution between the two controllers also improves redundancy and reliability of the system (Louis, 2018; Singh Parihar, 2019).

5. PH Sensor & TDS Sensor

In our project, both the pH sensor and TDS sensor will be integrated into the tank. The pH sensor measures acidity or alkalinity, crucial for ensuring water quality, while the TDS sensor quantifies dissolved solids, essential for assessing water purity according to

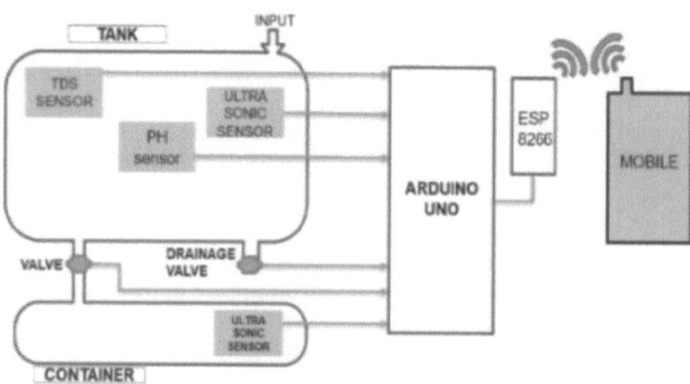

Figure 45.1: Block diagram.

Source: Author.

Table 45.1. These sensors, commonly used in various fields like chemistry and hydroponics, play a vital role in maintaining optimal water conditions for animals.

6. Ultrasonic Sensor

Ultrasonic sensors emit sound waves above 20 kHz, measuring distances by timing reflections. They are vital for water level monitoring in our project, providing reliable performance suitable for obstacle detection and proximity sensing in robotics. Despite their limited range, these non-contact sensors offer versatile functionality essential for our application.

7. Workflow

The system's operation initiates with the Arduino Uno gathering data from all attached sensors as shown in Figure 45.4. If the pH or TDS values exceed predetermined thresholds, indicating suboptimal water quality, the water motor activates to drain the tank. Subsequently, the tank undergoes refilling with fresh water. Once replenished, the Arduino reevaluates sensor readings. Upon confirming that pH and TDS values are within the desired range, the Arduino Uno triggers another motor to dispense water into the container below. Moreover, the Arduino Uno establishes connectivity with the ESP8266, which is equipped with Wi-Fi capabilities. This integration enables remote operation and monitoring of the system via the Blynk app installed on a smartphone. Through the Blynk app, users can efficiently oversee system functionalities and receive real-time updates on sensor readings, ensuring seamless management

Table 45.1: Range for suitable drinking water

Sr. No.	Animal	PH Range	TDS Range (ppm)
1	Cattle	6.5–8.5	500–5000
2	Goats	6.0–8.0	500–4000
3	Dogs	6.5–7.5	100–500
4	Cats	6.0–7.0	50–250
5	Other Animals	varies	Varies

Source: Based on Data from the Wikipedia.

Figure 45.2: Picture of sensors module connected to Arduino UNO.

Source: Author.

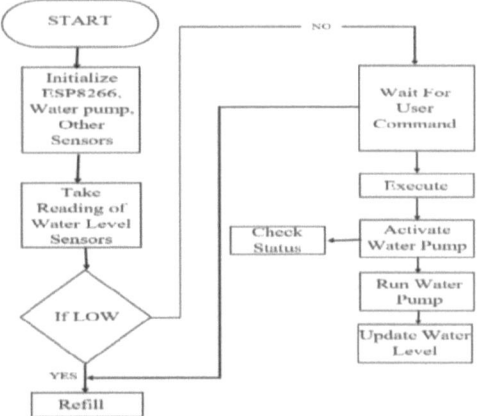

Figure 45.3: Flow chart.

Source: Author.

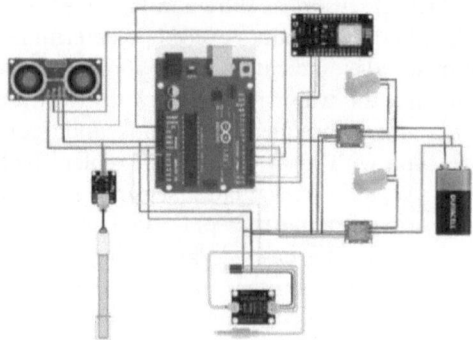

Figure 45.4: Circuit diagram.

Source: Author.

and optimization of the water dispensing process (refer Figure 45.3).

8. Results and Conclusion

In order to identify low water levels in a timely manner and activate the automatic dispensing mechanism, the system makes use of ultrasonic sensors. Furthermore, PH and TDS sensors discern between conditions that are appropriate for animals and those that are not by precisely evaluating the quality of the water (Figure 45.5 & 45.6). The water supply is steady and dependable since automated dispensing starts when the water levels are low and the quality of the water is up to par. Real-time monitoring is made possible by integration with Blynk Cloud,

which also provides email and pop-up alerts to inform stakeholders of important water quality and level evaluations. Further ensuring prompt alerts in the event of water scarcity in the container are push messages from the Blynk application (as Figure 45.7).

Output	Serial Monitor ×
Message (Enter to send message to 'Arduino Uno' on 'COM5')	

```
Distance: 8 | TDS Value: 25 | pH Value: 7.95
Distance: 9 | TDS Value: 25 | pH Value: 7.95
Distance: 9 | TDS Value: 25 | pH Value: 7.95
Distance: 9 | TDS Value: 25 | pH Value: 7.95
Distance: 9 | TDS Value: 25 | pH Value: 7.95
```

Figure 45.5: Output of clean water.

Source: Author.

Output	Serial Monitor ×
Message (Enter to send message to 'Arduino Uno' on 'COM5')	

```
Distance: 3 | TDS Value: 154 | pH Value: 7.95
Distance: 3 | TDS Value: 154 | pH Value: 7.95
Distance: 3 | TDS Value: 155 | pH Value: 7.95
Distance: 3 | TDS Value: 154 | pH Value: 7.95
Distance: 3 | TDS Value: 155 | pH Value: 7.95
```

Figure 45.6: Output of contaminated water.

Source: Author.

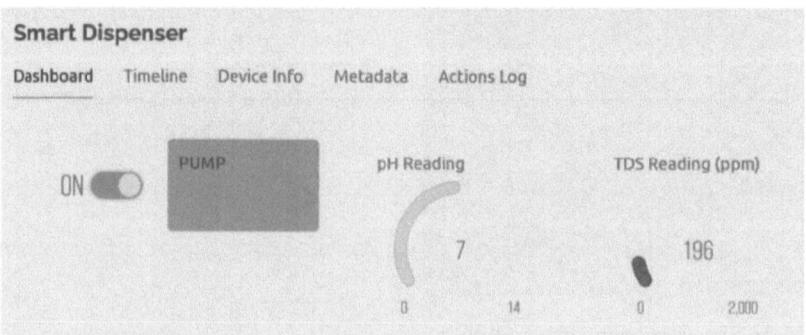

Figure 45.7: BLYNK app output.

Source: Author.

9. Future Scope

In the future, adding a water filtering system to improve the water quality before giving it to the animals could increase the system's functionality. By giving the animals clean water, incorporating such a system into the current setup would guarantee their general health and well-being. The addition of a GSM module would allow for real-time communication and monitoring, which would improve the system's efficacy and efficiency in meeting the water needs of animals.

References

Eugenio, A. E., and Malabago, N. K. (2015). Water dispenser: A unique innovative design that matters. https://eujournal.org/index.php/esj/article/view/6327

Igbinoba, C.K., and Okhaifoh, J.E. (2019). Automatic indoor water dispensing machine. *Federal University of Petroleum Resources Effurun, Effurun, Nigeria.* doi: 10.13140/RG.2.2.12875.52006

Louis, L. (2018). Working principle of arduino and using it as a tool for study and research. International Journal of Control, Automation, Communication and Systems (IJCACS). doi: 10.5121/ijcacs.2016.1203

Manoj, M., Dhilip Kumar, V. Arif Muhammad, Bulai, Elena-Raluca, B., Petru, B., and Geman, O. (2022). State of the art techniques for water quality monitoring systems for fish ponds using IoT and underwater. Sensors. doi: 10.3390/s22062088

Natasha, S., Mohd Daud, S., Dziyauddin, R., Adam, M., and Azizan, A. (2020). A review

on computer vision technology for monitoring poultry farm – Application, hardware, and software. IEEE Access, 1. doi: 10.1109/ACCESS.2020.3047818.

Rajani Devi, M., Jyothi, V., and Naga Jyothi, D. (2022). IoT and cloud-based automated pet care system. *Department of Electronics and Communication Engineering.* doi: 10.1109/ICECA55336.2022.10009347

Singh Parihar, Y. (2019). Internet of Things and Nodemcu: A review of use of Nodemcu ESP8266 in IoT products, National Informatics Centre, India. 337656615.

Srivastava, A., Dwivedi, S., Bhardwaj, S. and Joshi, H. C. Study of automatic water dispenser. *Institute of Technology and Sciences, Haldwani, Nainital, U.K.*

Taneja, K., and Bhatia, S. (2017)Automatic irrigation system using Arduino UNO. Thapar University, Patiala. doi: 10.1109/ICCONS.2017.8250693

Uddin, M. T., Chowdhury, T. A., and Rahmanm R. M. (2021). Design and implementation of automatic smart irrigation system. *North South University.* doi: 10.1109/UEMCON53757.2021.9666518

Von Borstel Luna, S. F. D., de la Rosa Aguilar, E., Naranjo, J. S., and Jagüey, J. G. (2016). Robotic system for automation of water quality monitoring and feeding in aquaculture, *Engineering Group, Centro de Investigaciones Biológicas del Noroeste, La Paz, Mexico.* doi: 10.1109/TSMC.2016.2635649

Yendri, D., Rizza, H., Rahmadya, B., and Derisma. (2019). Hygienic and energy saving of water dispenser machine. IOP Conf. Ser.: Mater. Sci. Eng, 846, 012039. doi: 10.1088/1757899X/846/1/012039

46 Classification of anomaly detection attacks in IoT devices using machine learning

Shilpa Keshri[1,a], Sunil Wanjari[1,b], Kapil Gupta[1,c], and Ankita Shukla[2,d]

[1]Computer Engineering, St. Vincent Pallotti College of Engineering and Technology, Nagpur, India
[2]Department of Computer Science, Northeastern University, Boston, USA

Abstract

Concerns about the security of Internet of Things (IoT) devices are growing, and the abstract emphasizes how vulnerable they are to anomaly attacks. The project suggests a way to find strange things in the Internet of Things (IoT) using Machine Learning (ML) methods like Support Vector Machine (SVM) and Random Forest (RF), along with votes and stacked models. Experiments using the NSL-KDD dataset show that RF and stacked models can get high accuracy rates with few false positives. RF significantly beats current material, which shows how useful it could be. Some ensemble methods, like Voting Classifier (RF + AB) and Stacking Classifier (RF + MLP with LightGBM), are very good at detecting and preventing anomalies because they have high accuracy, memory, and precision. In addition, the project includes user testing through a front end built on the Flask framework and user identification, which makes IoT anomaly detection more useful in real life.

Keywords: *IOT* devices, Support Vector Machine (SVM) and Random Forest (RF)

1. Introduction

With the Internet of Things (IoT), more than just standard gadgets can connect to the internet. This makes it easier for users, businesses, and other groups to send data to each other. IoT devices, which range from toasters to freezers, are divided into three groups: consumer, business, and industrial. They are also very different in size and function. Margaret Lee says that by 2025, there will be 64 billion Internet of Things (IoT) gadgets online (Lee, 2022). Anomaly identification is an important part of IoT because it finds trends that don't behave the way they should, like outliers, exceptions, and irregularities (Haji & Ameen, 2021). This kind of analysis helps find technology problems or changes in how people act. Even though it's important, not much

[a]shilpakeshri12@gmail.com, [b]swanjari@stvincentngp.edu.in, [c]kaps04gupta@gmail.com, [d]shukla.ank@northeastern.edu

DOI: 10.1201/9781003567653-46

study has been done on Machine Learning (ML) methods to finding problems with IoT (Haji & Ameen, 2021; Nassif et al., 2021; Das et al., 2021; Gavrilova, 2021; Benqdara & Ngadi, 2013; Hasan et al., 2019; Mathworks., n.d.). There are a lot of security issues because IoT devices can be attacked, like what happened with Western Digital's My Book Live, where hackers deleted data because of holes in the system (Firedome, 2021). Also, the number of IoT devices that were hacked doubled in Japan in 2018, showing how vulnerable they are (Zmudzinski, 2019). Xu et al. stress that anomaly analysis is very important in many areas, such as data mining and machine learning, to find trends that don't match what is expected (Xu et al., 2019). So, strong tools for finding strange behaviour are needed to lessen the damage that could come from hacks or intrusions.

2. Literature Review

The Internet of Things (IoT) has grown very quickly, letting billions of smart gadgets use sensors to gather data for many uses. But because there are so many IoT devices, there are also more security risks (Haji & Ameen, 2021; Nassif et al., 2021; Das et al., 2021; Gavrilova, 2021; Benqdara & Ngadi, 2013; Hasan et al., 2019; Mathworks., n.d.). Machine Learning (ML) has become smart node devices, SVM models like C-SVM and OC-SVM are used, which results in high classification accuracy rates (Ioannou & Vassiliou, 2021). ML has made a lot of progress in anomaly an important tool for dealing with these problems because it opens up new areas for study and security technologies (Lee, 2022; Haji & Ameen, 2021; Firedome, 2021). Machine learning methods, like Support Vector Machines (SVM), are very important for finding risks and strange things in IoT networks (Ioannou & Vassiliou, 2021; Nassif et al., 2021). For intrusion detection systems (IDS) to keep an eye on strange actions in identification, which is a basic job for finding problems in data

(Haji & Ameen, 2021; Nassif et al., 2021; Das et al., 2021; Gavrilova, 2021; Benqdara & Ngadi, 2013; Hasan et al., 2019; Mathworks., n.d.). A Systematic Literature Review (SLR) was done in (Nassif et al., 2021) that looked at 290 research papers from 2000 to 2020 and showed different machine learning models and datasets that were used to find anomalies. Because they work so well, unsupervised anomaly detection methods are becoming more popular in study (Nassif et al., 2021). Using machine learning in real-life situations, like healthcare, can also change the way anomalies are found, giving faster answers and important information about bodily data (Das et al., 2021). Even though a lot of study has been done on finding anomalies, not much has been done on analysing trained and unstructured methods used on physiological datasets (Das et al., 2021). Also, detecting intrusions is still an important part of cybersecurity study, and getting rid of false alerts is still a problem (Benqdara & Ngadi, 2013). Anomaly detection is an important part of intrusion detection systems because it finds changes from regular behaviour that could mean an attack or a problem (Benqdara & Ngadi, 2013). Even though there are still problems, new study points the way to better ways to handle finding anomalies using both controlled and unstructured methods.

3. Methodology

3.1. Proposed Work

Support Vector Machine (SVM) and Random Forest (RF) are two machine learning techniques that have been suggested for finding strange activity in Internet of Things (IoT) devices. Other methods include stacking and voting classifiers. The NSL-KDD dataset (Mathworks., n.d.) was used to test these algorithms and show that they are good at both recognition and feature selection. To rate how well a model works, we use evaluation measures like accuracy, recall, precision, and f1-score. Notably, the vote predictor is 100% accurate, while stacking is only 99% accurate. This shows how well they work to improve prediction skills. Also, to make it easier to use in real life, a front-end interface made with the Flask framework that is easy for anyone to use makes sure that entry is safe through user registration.

3.2. System Architecture

A plan for the study process is shown in Figure 46.1. The suggested way to use the two algorithms in the Weka tool program to check how they compare to other ways they have been used in the past that are most relevant to this study paper. The names of these two methods are Support Vector Machine and Random Forest. Support Vector Machine, or SVM, is a strong Supervised Learning method that can be used for both regression and classification problems. That being said, it is mostly used in (Haji & Ameen, 2021; Nassif et al., 2021; Das et al., 2021; Gavrilova, 2021; Benqdara & Ngadi, 2013; Hasan et al., 2019; Mathworks., n.d.) Machine Learning for Classification problems. On the other hand, the random forest algorithm is a Machine Learning method that is easy to use and flexible. It uses group

learning to deal with problems like regression and classification.

3.3. Dataset Collection

The main goal of the project is to look at the NSL-KDD dataset (Mathworks., n.d.), which is a common set of anomalies used to compare intrusion detection systems. The NSL-KDD dataset comes from the KDD cup99 dataset (Tavallaee et al., 2009) and has 42 training intrusion strikes and 41 characteristics. Notably, 21 characteristics are about the link itself, and 19 are about the nature of the relationship within the same host (Tavallaee et al., 2009). To rate how well a model works, we use evaluation measures like precision, recall, F-measure, accuracy, false positive rate, and true positive rate. The statistical study of the cup99 dataset showed problems that make intruder detection less accurate (Tavallaee et al., 2009).

3.4. Data Processing

Data handling is the process of turning unstructured data into knowledge that businesses can use. In general, data scientists handle data, which means they gather

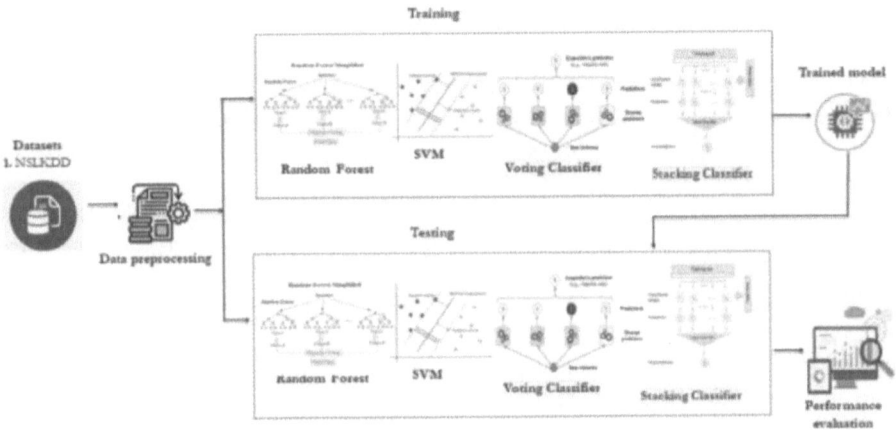

Figure 46.1: Proposed architecture.

Source: Author.

Table 46.1: NSL KDD dataset

duration	protocol_type	service	flag	src_bytes	dst_bytes	land	wrong_fragment	urgent	hot	...	dst_host_same_srv_rate	dst_host_diff_srv_rate	dst_host
0	0	tcp	ftp_data	SF	491	0	0	0	0	0	...	0.17	0.03
1	0	udp	other	SF	146	0	0	0	0	0	...	0.00	0.60
2	0	tcp	private	S0	0	0	0	0	0	0	...	0.10	0.05
3	0	tcp	http	SF	232	8153	0	0	0	0	...	1.00	0.00
4	0	tcp	http	SF	199	420	0	0	0	0	...	1.00	0.00

5 rows × 43 columns

Source: Author.

it, organize it, clean it, check it, analyse it, and turn it into forms that can be read, like graphs or papers. There are three ways to handle data: by hand, mechanically, or electronically. The goal is to make knowledge more useful and decision-making easier. This helps companies run better and make smart strategy decisions more quickly. This is made possible in large part by automated data handling tools, like computer programs. It can help turn big data and other types of data into useful information for decision-making and quality control.

3.5. Feature Selection

A very important part of feature engineering is choosing which features are the most useful to feed into Machine Learning methods (Haji & Ameen, 2021; Nassif et al., 2021; Das et al., 2021; Gavrilova, 2021; Benqdara & Ngadi, 2013; Hasan et al., 2019; Mathworks., n.d.). By getting rid of traits that aren't needed or aren't important, this process aims to improve model performance and make computations simpler. Feature selection improves the performance of forecasting models by gradually decreasing the size of datasets. It lets models focus on the most important traits, which makes predictions more accurate and easier to understand. This proactive method makes sure that models are taught on the most useful

traits, which makes Machine Learning systems work better and faster.

3.6. Algorithms

A well-known guided learning method called Random Forest can be used to solve both Classification and Regression issues. It uses ensemble learning to make predictions more accurate by mixing several decision trees on different parts of the information. Random Forest is tested with 10 and 20 folds using k-FOLD cross-validation, going through training and testing groups over and over to get an idea of how well it does in generalization. Random Forest was chosen because it can handle large datasets. Its group nature helps it find different patterns, which makes it good for IoT apps and lowers the risk of overfitting.

The Support Vector Machine (SVM) is a well-known guided learning method that is mostly used for sorting jobs. Its goal is to find the best decision border, also known as a hyperplane, to divide n-dimensional space into groups. SVM uses K-Fold Cross-Validation with 10 and 20 folds to split the data into groups that can be used for training and testing over and over again. SVM was chosen because it is good at dealing with large amounts of data. It works well in IoT settings, especially when looking for strange behaviour, where there are many possible choices.

A common ensemble modelling method called "stacking" combines weak learners with meta-learners at the same time. The goal is to make better predictions about the future by figuring out the best way to combine the predictions from different models. Stacking improves the spotting of anomalies in IoT data by using the results of sub-models and a meta-classifier. It does this by recording a wider range of patterns, which helps with generalizing about how anomalies change and adapt in IoT settings.

A voting classifier takes results from several models and predicts an output based on the most likely class. It then uses a majority vote method to choose the final output. The Voting Classifier takes the best parts of different models and uses them together to make predictions. This improves overall performance and makes sure that decisions are fair, which makes the system more effective in IoT settings.

4. Experimental Results

Precision: Precision is the percentage of correctly classified cases or samples compared to those that were correctly classified as hits. So, here is the method to figure out the precision. Figure 46.2 shows the precision comparison graph.

$$Precision = True\ positives/\ (True\ positives + False\ positives) = TP/\ (TP + FP)$$

Recall: In machine learning, recall is a parameter that shows how well a model can find all the important cases of a certain class. It shows how well a model captures cases of a certain class. It is calculated by dividing the number of correctly predicted positive observations by the total number of real positives. Figure 46.3 shows the recall comparison graph.

$$Recall = \frac{TP}{TP + FN}$$

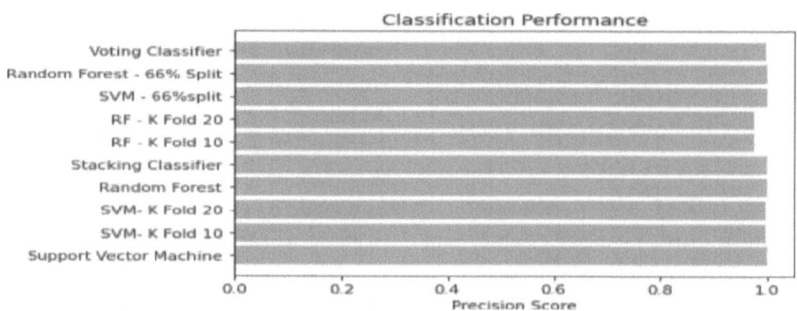

Figure 46.2: Precision comparison graph.

Source: Author.

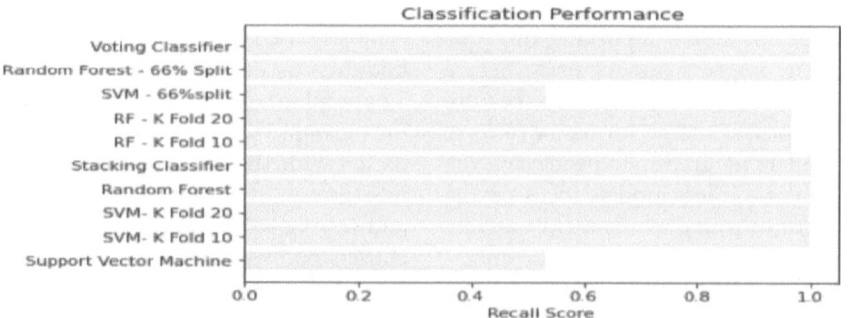

Figure 46.3: Recall comparison graph.

Source: Author.

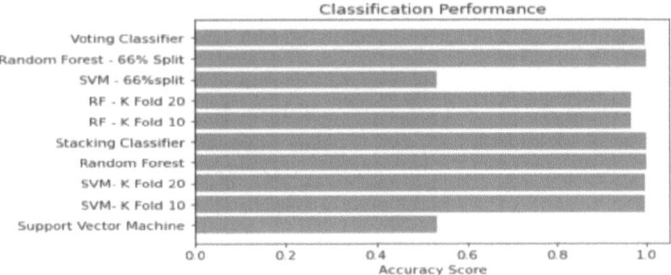

Figure 46.4: Accuracy graph.

Source: Author.

Accuracy: Accuracy is the percentage of right guesses in a classification job. It shows how accurate a model's forecasts are generally. Figure 46.4 shows the accuracy comparison graph.

$$Accuracy = \frac{TP + TN}{TP + FP + TN + FN}$$

F1 Score: The harmonic mean of accuracy and recall is F1. This fair measure accounts for erroneous positives and negatives, therefore it may be used with unbalanced datasets. Figure 46.5 shows the F1 score comparison graph. Figure 46.6 shows the Performance evaluation and Figure 46.7 shows User input interface.

$$F1 \text{ Score} = 2 * \frac{Recall \times Precision}{Recall + Precision} * 100$$

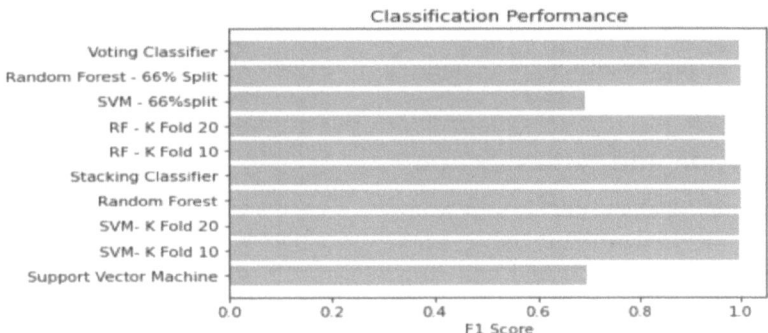

Figure 46.5: F1 score.

Source: Author.

ML Model	Accuracy	Precision	Recall	F1 - score
Support Vector Machine	0.534	0.999	0.534	0.696
SVM- K Fold 10	0.998	0.998	0.998	0.998
SVM- K Fold 20	0.998	0.998	0.998	0.998
Random Forest	1.000	1.000	1.000	1.000
Stacking Classifier	1.000	1.000	1.000	1.000
RF - K Fold 10	0.966	0.975	0.966	0.970
RF - K Fold 20	0.966	0.975	0.966	0.970
SVM - 66%split	0.533	0.999	0.533	0.694
Random Forest - 66% Split	1.000	1.000	1.000	1.000
Voting Classifier	0.998	0.998	0.998	0.998

Figure 46.6: Performance evaluation.

Source: Author.

Service	Same_srv_rate
20	1
Flag	Diff_srv_rate
9	0
Src-Bytes	Dst_host_srv_count
491	25
Dst-Bytes	Dst_host_same_srv_rate
0	0.17
Count	Dst_host_diff_srv_rate
2	0.03
Serror_rate	Dst_host_serror_rate
0	0
Srv_serror_rate	Dst_host_srv_serror_rate
0	0

Predict

Figure 46.7: User input.

Source: Author.

Result: There is an No Attack Detected, it is Normal!

NIDS

Figure 46.8: Predict result for given input.

Source: Author.

5. Conclusion and Future Scope

Support Vector Machine (SVM) and Random Forest (RF) algorithms have been used to show that they can find and stop anomaly attacks in Internet of Things (IoT) devices (Firedome, 2021). The results are better than what has been written before, with high accuracy and consistently low false positive rates, which are important for accurately classifying anomalies (Lee, 2022; Haji & Ameen, 2021). Using the NSL-KDD dataset (Mathworks., n.d.), the project carefully tests the performance of the algorithm, showing how reliable it is in real-world IoT settings. Ensemble methods, such as Voting Classifier and Stacking Classifier, are very accurate, which proves that they work. This project makes IoT security a lot better by tackling the important problem of anomaly threats and making devices more resilient.

In the future, it might be possible to improve the ability to find anomalies by using more advanced Machine Learning methods, such as deep learning models (Ioannou & Vassiliou, 2021; Hasan et al., 2019). Real-time monitoring in IoT devices is very important, and to fight changing cyber dangers, we need algorithms that can handle data in real time (Ioannou & Vassiliou, 2021; Hasan et al., 2019). Adaptive models are needed to deal with new gadgets and strange situations, and they need to be improved all the time. In later versions, the focus may be on better security features like encryption and methods for responding to strange events in order to deal with more advanced threats.

References

Benqdara, S., and Ngadi, M. A. (2013). Machine learning techniques for anomaly detection: An overview. International Journal of Computer Applications. 79(2).

Crunch, T. (2017). The evolution of machine learning. TechCrunch. https://techcrunch.com/2017/08/08/the-evolution-of-machine-learning/ (16 January 2023).

Das, C., Rasool, A., Dubey, A., and Khare, N. (2021). Analyzing the performance of anomaly detection algorithms. International Journal of Advanced Computer Science and Applications, 12(6), 1–7.

Firedome (2021). Top cyber attacks on IoT devices in 2021. https://firedome.io/blog/top-cyber-attacks-on-iot-devices-in-2021/

Gavrilova, Y. (2021). Anomaly detection in machine learning. Software Development Company. https://serokell.io/blog/anomaly-detection-inmachine-learning.

Haji, S. H., and Ameen, S. Y. (2021). Attack and anomaly detection in IoT networks using machine learning techniques: A review. Asian Journal of Research in Computer Science, 9(2), 30–46. https://doi.org/10.9734/ajrcos/2021/v9i230218.

Hasan, M., Islam, M., Md, M., Zarif, I., and Hashem, M. M. A. (2019). Attack and anomaly detection in IoT sensors in IoT sites using [2, 7, 8, 9, 10, 11, 12] machine learning approaches. Internet of Things, 7, 100059.

Ioannou, C., and V. Vassiliou (2021). Network attack classification in IoT using support vector machines. https://www.mdpi.com/2224-2708/10/3/58/pdf

Lee, M.(2022). Anomaly detection: Glimpse into the future of IoT data. The New Stack. https://thenewstack.io/anomaly-detection-glimpse-into-thefuture-of-iot-data/

Mathworks. (n.d.). Machine Learning. Www.mathworks.com. https://www.mathworks.com/discovery/machinelearning.html#:~:text=Machine%20learning%20uses%20two%20types

Nassif, B., Abu Talib, A., Nasir, M., and Dakalbab, F. (2021). Machine learning for anomaly detection: A systematic review. IEEE Access, 9, 78658–78700.

Xu, X., Liu, H., and Yao, M. (2019). Recent progress of anomaly detection. Complexity, 1–11. https://doi.org/10.1155/2019/2686378. 2019.

Zmudzinski, A.(2019). Japan: Hacked IoT devices and cryptocurrency networks doubled in 2018. Cointelegraph. https://cointelegraph.com/news/japan-hacked-iot-devices-andcryptocurrency-networks-doubled-in-2018

47 Driver drowsiness detection using Convolution Neural Network (CNN)

Surbhi Girde[1,a], Reema Roychaudhary[1,b], and Aashay Wanjari[2,c]

[1]Computer Engineering, St. Vincent Pallotti College of Engineering and Technology, Nagpur, India
[2]Master of Engineering (Materials Science and Engineering), University of Toronto, Toronto, Canada

Abstract

Imagine a world where technology is able to prevent accidents caused by drowsy driving. Drowsy Driver Detection system use advanced radiation to monitor driver's behavior, like eye movements and steering patterns, when sign of fatigue is detected, the system will have the capability to signal the driver to stop or even automatically apply breaks to prevent accidents. With this innovative technology, we can make our roads safer and save lives. There has been a substantial amount of research and innovation focused on drowsy driver detection. Many car manufacturers and tech companies have been working on implementing this technology into vehicles. Some existing systems use facial recognition. Steering pattern analysis, and even brainwave monitoring to detect signs of drowsiness. These systems aim to provide timely alerts to drivers to prevent accidents caused by fatigue. The paper elucidates the architecture and functioning of CNNs, detailing their adaption to driver drowsiness detection tasks, various CNN architectures, including AlexNet, VGG, and ResNet are examined along with preprocessing techniques for optimizing model performance. Additionally, dataset and performance metrics commonly used in evaluating CNN models are discussed.

Keywords: Drowsiness detection, eye detection, face detection, yawning detection, alert system

1. Introduction

Driver Drowsiness stands as a pervasive on roadways worldwide, casting a shadow of peril over commuters and pedestrians alike. Its detrimental impact on road safety cannot be overstated, because it leads to a large number of car accidents, resulting in injuries, deaths untold grief. Recognising and mitigating drowsiness among drivers is an imperative task in the ongoing battle to safeguard lives and prevent tragic mishaps (Satish et al., 2020). In the past computer vision algorithm were mainly used in old systems to identify driver drowsiness. Historically, changes in weather, environmental factors, and extended driving were primary reasons for driver fatigue detection in the

[a]surbhigirde@gmail.com; sgirde@stvincentngp.edu.in, [b]rroychaudhary@stvincentngp.edu.in, [c]aashay.wanjari@mail.utoronto.ca

DOI: 10.1201/9781003567653-47

past (Abba, 2020). Every year, the National Highway Traffic Safety Administration (NHTSA) and World Health Organization (WHO) have reported that approximately 1.35 million people die due to vehicle crashes across the world. Generally, road accidents mostly occur due to in ad-equate way of driving (Biswal et al., 2021). In the driver Fatigue detection system, researchers typically leverage various characteristics by analyzing behaviors, physiological data, and information collected from the vehicle to achieve successful detection (Shaik Mhd., 2023). This review paper embarks on a comprehensive journey to unravel the intricacies of leveraging CNNs in the noble pursuit of driver drowsiness detection. It commences by shedding light on the ominous perils associated with drowsy driving, underscoring the urgency of implementing robust detection mechanisms. Furthermore, the paper meticulously surveys the landscape of existing methodologies and technologies employed in the detection of driver drowsiness, elucidating their shortcomings and paving the way for the advent of CNN-based solutions (Jabbar et al., 2020).

Delving deeper into the realm of artificial intelligence, the review unravels the fundamental principles underpinning CNNs, elucidating their architectural intricacies and operational mechanisms. Subsequently, it navigates through the labyrinth of adaptation and optimization techniques tailored to harness the full potential of CNNs for driver drowsiness detection tasks (Karthikeyan, 2021). In particular, the pre-processing steps essential for refining input data and enhancing model performance are meticulously examined.

By synthesizing a rich tapestry of existing literature and cutting-edge research findings, this review aims to provide deep insights into how effective CNN-based methods are in detecting driver drowsiness. Moreover, act as a compass guiding scholars and professionals towards uncharted territories, charting a course for future advancements in this burgeoning field (Albadawi et

al., 2022). As the contours of the road to progress unfold, it becomes increasingly evident that the integration of CNNs into driver drowsiness detection systems holds the promise of a safer, more secure future for all road users, where the specter of drowsy driving fades into obscurity, and the incidence of tragic accidents precipitated by fatigue is consigned to the annals of history (Stancin et al., 2023).

2. Literature Review

As per the research byK. Satish et al. (2020) the main goal has to lower the number of road accidents by using drowsiness detection and monitoring systems that rely on real-time data. Used the Haar algorithm and OpenCV libraries to detect facial features and measure eye closure. Used of maximum variance thresholding (Satish et al., 2020) method to identify and extract facial features accurately and trained a non-straight SVM determine the status of the eye. The system suggested in the paper combines facial and eye recognition with Arduino-based hardware and pressure sensor integration. Qaisar Abbas (2020) stated that how to use of intelligent algorithms, hybrid system, multi-cam approaches, and transfer learning techniques for driver drowsiness recognition (Abba, 2020). Anil Kumar Biswal et al. (2021) reported that for drowsy driver detection using the IoT-based smart alert system and analysis, face detection and analysis, in addition to vehicle-related, psychological and behavioral measurement executed utilizing predictive algorithms (Biswal et al., 2021). Md. Ebrahim Shaik (2023) in this the provided sources discuss various methods to detect driver drowsiness, including behavioral, physiological, vehicle-based, and subjective measures. Hybrid approaches that combine multiple measures are also mentioned (Shaik Mhd., 2023). Jabbar et al. (2020) according to this paper introduces a novel algorithm that addresses the need for lightweight and accurate driver drowsiness and identifies developing blueprints for embedded systems and Android devices (Jabbar et al., 2020). Elena Magán et al. (2022) Conducting further research and experimentation with real-world datasets, like UTA-RLDD, can provide more accurate and reliable results for drowsiness detection systems(Magán et al., 2020). (H. Varun Chand et al., 2021) in this provides a foundation for the proposed model in this paper, which combines Convolution Neural Networks (CNN) and emotion analysis for effective driver drowsiness detection (Aman Albadawi et al., 2022; Karthikeyan, 2021). This paper offers an in-depth review of recent developments in drowsiness detection systems, including their features, algorithms, datasets, challenges, and future directions (Albadawi et al., 2022). The paper highlights the urgent need for improved road security due to the increasing number of car accident fatalities worldwide. Advances in vehicle technology now allow for better detection and analysis of road conditions to help prevent accidents and protect passengers. One significant area of recent research is addressing driver drowsiness, a leading cause of accidents and fatalities. This study introduces a system that employs Convolutional Neural Networks to predict driver drowsiness by analyzing facial sequences, achieving an accuracy rate of approximately 92%. The ultimate objective is to create a real-time driver monitoring system to reduce road accidents by detecting and managing driver drowsiness effectively (Kusuma Sri et al., 2022).

3. Methodology and Model Specifications

Figure 47.1 as per the proposed methodology driver drowsiness detection system has been showcased. Initially, the real-time video is captured using a webcam. The camera will be placed in front of the driver to capture the frontal face image. They extract frames from the video to acquire 2-D images. The face will be detected from the frames using the face detection method. Once the face is detected, they mark facial landmarks such as the position of the eye, nose, and mouth are marked in Figures 47.3 and 47.4 (Stancin et al., 2023).

After identifying the facial landmarks, the location of the eyes and mouth are assessed. By utilizing these extracted features and machine learning techniques, a determination is made regarding the drowsy driver. MTCNN is applied to classify the eyes, which assesses the driver's drowsiness by monitoring eye blinking. To enhance the system, a feature extraction method is

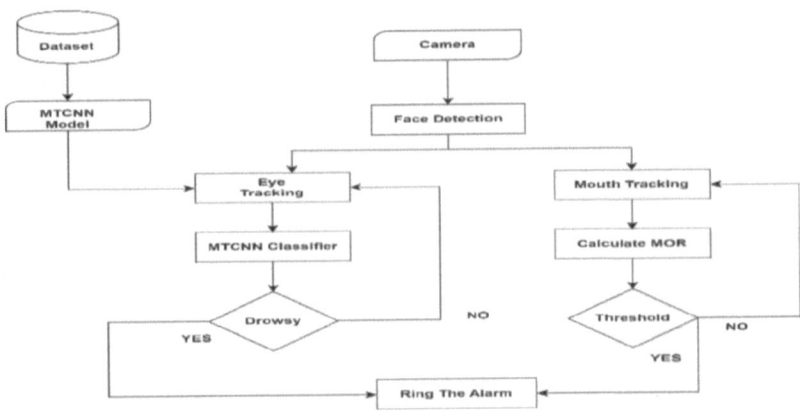

Figure 47.1: Proposed methodology.

Source: Satish et al., 2020.

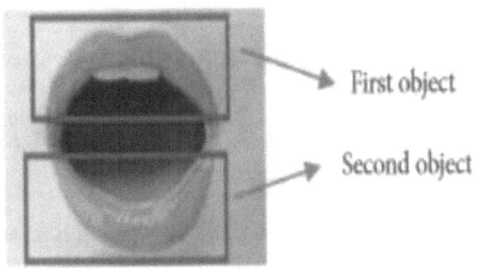

Figure 47.2: Yawning detection.

Source: Wasin AlKishri et al. Enhanced Image Processing and Fuzzy Logic Approach forOptimizing Driver Drowsiness Detection Hindawi Applied Computational Intelligence and Soʏ Computing.

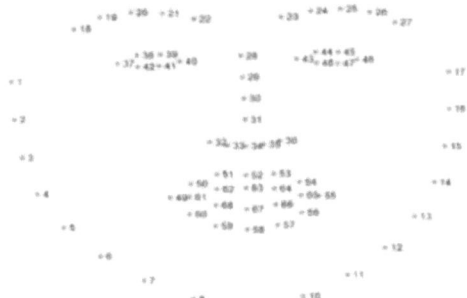

Figure 47.3: Facial landmarks coordinates.

Source: A Survey on Driver Drowsiness Detection Techniques,International Journal of Latest Technology in Engineering, Management & Applied Science (IJLTEMAS).

 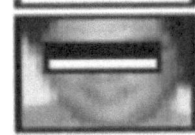 

Figure 47.4: Eye detection.

Source: M. Kusuma Sri and T. Annamani,Driver Drowsiness Detection System Using Convolutional Neural Networks,International Journal of Advanced Research in Science, Communication and Technology (IJARSCT)April 2022.

employed to compute the motion opening ratio, aiding in determining if the driver is drowsy. If drowsiness is detected, an alarm will be sent to alert the driver. Figure 47.2 In the MTCNN algorithm yawning detection, facial landmarks such as the mouth and jaw are used as input. The MTCNN learns to recognize patterns and features associated with yawning. By analyzing the changes in these facial landmarks over time, the algorithm can determine if a person is yawning or not. This can be a valuable component is drowsy driver detection systems, as yawning is often a sign of fatigue or drowsiness. Several techniques exist for identifying facial landmarks, but most approaches involve pinpointing specific regions like eyebrows, eyes, nose, mouth and jaw using series of regression trees. This recognition method is included in the dlib framework.

This technology in cars help identify driver tiredness and notifies them with an alarm sound. Figure 47.4 (AlKishri et al., 2022) shows how the system utilizing OpenCV to retrieve the face, the region of interest (ROI), which is the eyes, from a series of images captured by the webcam. In our project, we applied the Eye Aspect Ratio (EAR) algorithm to our model. The model gives key points for both eyes and then, using the EAR algorithm, it determines if the eyes are open or closed.

4. Conclusion

Using MTCNN with eye tracking and mouth tracking can provide an effective way to detect driver drowsiness. This technique trains the MTCNN model with data from the eyes and mouth to assess the driver's condition. It offers high accuracy and consistency. As per the previous research the using of CNN model the accuracy was 83.3%. So using of MTCNN model we will achieved approximate 98% accuracy.

References

Abba, Q. (2020). A real-time driver drowsiness detection using hybrid Features and transfer learning. College of Computer and Information Sciences & Muhammad, A.I. Ibn Saud Islamic University (IMSIU), Riyadh, Saudi Arabia. https://www.researchgate.net/publication/339020569HybridFatigue.

Albadawi, Y., Takruri, M., and Awad, M. (2022). A review of recent developments in driver drowsiness detection systems. Department of Computer Science and Engineering, American University of Ras Al Khaimah, Ras AlKhaimah.

AlKishri, W., Abualkishik, A., and Al-Bahri, M. (2022). Enhanced image processing and fuzzy logic approach for optimizing driver drowsiness detection. Faculty of Computer Studies & Department of Information Technology and Computing (1596). Qurum, Madinat Sultan Qaboos-P.O.B. Arab Open University: Oman. Oman, p. 130, p. C. Faculty of Computing, I.T. & Sohar University Sohar.

Biswal, A. K., Singh, D., Pattanayak, B. K., Samanta, D., and Yang, M. (2021). IoT-based smart alert system for drowsy driver detection. Wireless Communications and Mobile Computing, 2021, 1–13. doi: 10.1155/2021/6627217

Jabbar, R., Shinoy, M., Karrueche, M., Al-Khalifa, K., Krichen, M., Barkaoui, K., and Lab, C. (2020). Computer Science Department, Conservatoire National des Arts et Meteirs. France.

Karthikeyan, H. V. (2021). School of Information technology and engineering, Vellore Institute of Technology, Vellore, India.

Magán, E., and Sesmero, M. P., Alonso-Weber, J. M., and Sanchis, A. (2020). Computer Science Engineering Department, masm@inf.uc3m.s.

Satish, K., Lalitesh, A., Bhargavi, K., Prem, M. S., and Anjali, T. (2020). Driver drowsiness detection T. International Conference on Communication and Signal Processing. doi:10.1109/ICCSP48568.2020.9182237

Shaik Mhd., E. (2023). A systematic review on detection and prediction of driver drowsiness. Departement of Civil Engineering & Bangabandhu Sheikh Mujibur Rahman Science & Technology Univ. Gopalganj, Dhaka, Bangladesh.

Stancin, I., Cifrek, M., and Jovic, A. (2023). A review of EEG signal features and their application in driver drowsiness detection systems. Faculty of Electrical Engineering and Computing, University of Zagreb, Unska, Vol. 3. Zagreb, p. 10000.

48 Experimental investigation of performance, combustion and emission characteristics of a CI engine fuelled by blend of plastic oil biodiesel with diesel

Sushant Satputaley[a], Shreyash Shende, Madhav Dhage, Suraj Bhakate, Ritesh Ambildhuke, and Sam Jason

Associate Professor, Department of Mechanical Engineering, St. Vincent Pallotti College of Engineering and Technology, Nagpur, India

Abstract

The study presents an exploratory investigation into the working of compression ignition (CI) engine and the effect on the pollutants coming out of the engine fueled by blends of plastic oil biodiesel B20 (20% plastic oil biodiesel and 80% diesel) with conventional diesel. In the quest for economical energy sources and reduction of environmental impact, biodiesel derived from waste plastic oil has emerged as a promising alternative. In this research, plastic oil biodiesel with diesel (B20) and diesel fuel were tested in a single-cylinder CI engine. Performance parameters including combustion, efficiency, and emissions was studied and matched to those of pure diesel operation. Peak combustion pressure was deliberate to understand the combustion behavior of the fuel blends. Similarly, brake thermal efficiency (BTHE) and brake specific fuel consumption (BSFC) were also studied. Additionally, emissions of carbon monoxide (CO), hydrocarbons (HC), and carbon dioxide (CO_2) were measured to evaluate the environmental impact of the fuel blends. The obtained results illustrate that the B20 performs alike to the diesel fuel, with some improvements observed in BTHE and reduced pollutants of CO, CO_2, and HC. The discoveries from this study enhance our comprehension of the viability and advantages of incorporating plastic oil biodiesel-diesel blends into CI engines, offering a promising avenue towards environmentally sustainable fuel alternatives.

Keywords: Biodiesel, combustion, economical, emission, performance, plastic oil

1. Introduction

Lately, it is really important to find new kinds of fuel for car engines because we are worried about running out of energy, making the environment dirty, and changes in the weather. Out of all the options, biodiesel made from renewable sources is getting a lot of consideration as a possible additional for regular diesel. Biodiesel is good because it causes less air pollution, it can break down naturally, and it works with regular

[a]ssatputaley@stvincentngp.edu.in

DOI: 10.1201/9781003567653-48

diesel engines and fuel systems. Additionally, using waste materials like plastic oil to make biodiesel helps deal with waste and make energy in a way that lasts a long time (Kumar et al., 2019).

Plastic trash, especially kinds that can't be recycled, is a big problem for the environment because it doesn't break down easily in landfills or in the ocean. Turning plastic waste into biodiesel helps the environment by reducing plastic pollution, and it also creates a renewable energy source to reduce the need for fossil fuels. Biodiesel made from plastic waste can be a good way to turn trash into energy. It's made by heating up the plastic or using a special process to change it into fuel (Sukjit et al., 2020).

In diesel engines, how well biodiesel blends perform, burn, and produce emissions is really important in deciding if they can be used as other kinds of fuel. Mixing biodiesel with regular diesel can change how well the engine works. It can affect how much fuel the engine uses, how hot the exhaust gas gets, and how efficiently the engine runs. Also, how biodiesel blends burn in the engine can affect how well the engine works and how efficient it is. This includes how they ignite, how long they burn, and how much pressure they create in the engine (Awang et al., 2021).

Additionally, the way biodiesel mixes release gases like nitrogen oxides, carbon monoxide, hydrocarbons, and carbon dioxide is worrying because it can affect the quality of the air we breathe and our health. It's important to know how biodiesel blends release emissions, so we can see how they affect the environment and if they follow emission rules (Basha et al., 2009).

The objective of this research work is to examine the working characteristics and pollutants traits of a CI engine when fueled by combinations of plastic oil biodiesel and conventional diesel fuel. By conducting experimental tests using a B20 blend, a comprehensive understanding of the result of biodiesel blending on engine operation can be accomplished. The results of this research are anticipated to enhance current understandings of employing plastic oil biodiesel diesel blends as eco-friendly fuel options for CI engines. These insights could hold significance for environmental sustainability efforts and strategies related to waste management (Miyuranga et al., 2023).

2. Biodiesel

In recent times, due to the non-renewable nature of petroleum-derived fuels and their adverse environmental impacts, it has become imperative to explore alternative energy sources. In recent decades, biodiesel has risen as a highly promising energy source, boasting non-toxic, biodegradable qualities that align with environmental preservation goals and can significantly bolster societal sustainable development efforts. (Yormesor et al., 2023).

Biodiesel is derived from animal fats and vegetable oils, categorised as long-chain fatty acid esters. It can be manufactured through various methods using different raw materials, but the most common process is transesterification with alcohol and alkaline catalysts. Biodiesel offers numerous advantages, such as non-toxicity, biodegradability, and sustainability. However, its most significant advantage over conventional diesel fuel is that it produces substantially reduced emissions of carbon dioxide and unburned hydrocarbons (Qi et al., 2010).

2.1. Potential of Waste Plastic (WPO) as Biodiesel

Escalating concerns over environmental impacts have driven the pursuit of optimised alternative energy sources. Biodiesel sourced from plastic oil has surfaced as a compelling alternative to traditional diesel fuel. Biodiesel, a renewable, biodegradable, and non-toxic fuel, stands in stark contrast to waste plastic, which is a major contributor to global plastic pollution. However, in recent years, researchers have developed innovative methods to mitigate this problem by recycling waste plastic into useful resources, including biofuels. The study explores the prospective of WPO as a source for biodiesel production. It discusses various techniques for extracting and purifying WPO and evaluates its suitability for the transesterification procedure required for the biodiesel synthesis. Additionally, the paper examines the environmental advantages and challenges associated with using WPO as a biodiesel feedstock and explores strategies to enhance the efficiency and sustainability of biodiesel derived from this source (Sukjit et al., 2020).

2.2. *Production of Biodiesel*

Biodiesel, a renewable elective to fossil fills, is created through a process called transesterification. This handle includes chemically changing over fats or oils into biodiesel by responding them with a liquor, ordinarily methanol or ethanol, within the nearness of a catalyst. The taking after steps layout the generation handle of biodiesel (Sorate et al., 2013).

1. Feedstock Determination:

   The primary step in biodiesel generation is selecting reasonable feedstocks. These feedstocks can be inferred from different sources such as vegetable oils (e.g. oil from soybean, rapeseed, palm etc.), creature fats, or reused cooking oil. The choice of feedstock depends on components such as accessibility, taken a toll, supportability, and territorial inclinations.

2. Pretreatment:

   Sometime recently transesterification, the feedstock experiences pretreatment to expel debasements and move forward its quality. This may include sifting the oil to evacuate strong particles, evacuating water substance, and neutralising free greasy acids (FFAs) on the off chance that display, as tall levels of FFAs can meddled with the transesterification response.

3. Transesterification Response:

   The transesterification reaction is the central process in biodiesel production. In this stage, the feedstock oil is added with an alcohol either methanol or ethanol and a catalyst, usually potassium hydroxide (KOH) or sodium hydroxide (NaOH). The reaction of alcohol with the triglycerides present in the oil to form biodiesel (methyl or ethyl esters), while glycerol is produced as a byproduct (Patel et al., 2013).

   The transesterification process, as depicted in Figure 48.1, is employed for biodiesel preparation. A mixture comprising 100 ml of waste plastic oil and 25 ml of methanol is stirred and heated with 0.8 g of potassium hydroxide as a catalyst for one and a half hours at a temperature range of fifty to fifty five degree celsius, as shown in Figure 48.2. The resulting

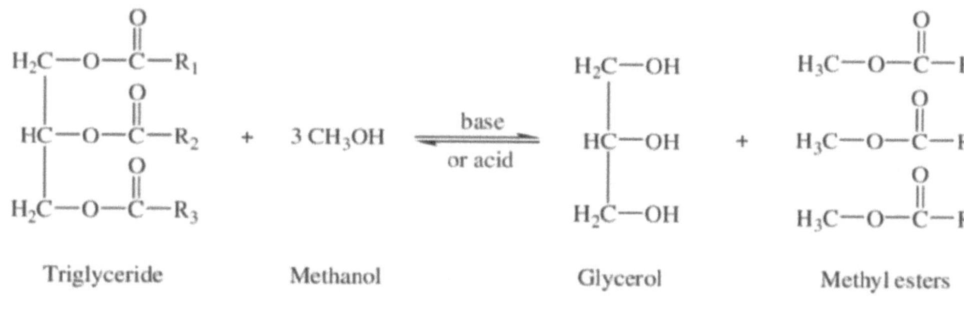

Figure 48.1: Transesterification chemical reaction.

Source: Dulawat (2020). Study on biodiesel production and characterization for used cooking oil. International Research Journal of Pure & Applied Chemistry, 21(24), 76–86.

mixture of biodiesel and glycerol is then allowed to settle for 8 to 10 hours to facilitate the separation of the biodiesel and glycerol layers, as illustrated in Figure 48.3. The upper layer is biodiesel which is subsequently separated from the raw glycerin lower layer through gravity settlement. The obtained biodiesel undergoes water washing process with warm water. The pH of the water washed biodiesel is measured and the process is considered complete when the pH reaches seven, as shown in Figure 48.4.

4. Division and Refinement:

After the transesterification response, the blend is permitted to settle, driving to the arrangement of two particular layers: biodiesel at the beat and glycerol at the foot. The glycerol layer is at that point isolated from the biodiesel layer utilising strategies such as gravity settling or centrifugation. The biodiesel is advance filtered to evacuate any remaining debasements, such as abundance liquor and catalyst buildups, through forms like washing with water and drying.

5. Quality Control:

Quality control measures are fundamental to guarantee the biodiesel meets industry guidelines and details. Parameters such as thickness, thickness, corrosive esteem, dampness substance, and ester content are analysed to evaluate the quality of the biodiesel. Alterations may be made to the generation prepare as required to attain the specified biodiesel quality.

6. Byproduct Utilisation:

The glycerol byproduct gotten from the transesterification prepare can be assist handled to create profitable substances such as glycerin, which has different mechanical applications. Moreover, endeavours are underway to investigate elective employments for glycerol to maximise its financial esteem and minimise squander era.

7. Result of Biodiesel Preparation

The transesterification process was carried out to create biodiesel from WPO, achieving a yield of 70–75 percent. A comprehensive study was conducted to analyse the physiochemical properties of both the WPO and WPOB. These properties were compared to the values established by the American Society of Testing Materials (ASTM). The findings from this comparative analysis are discussed in the following sections.

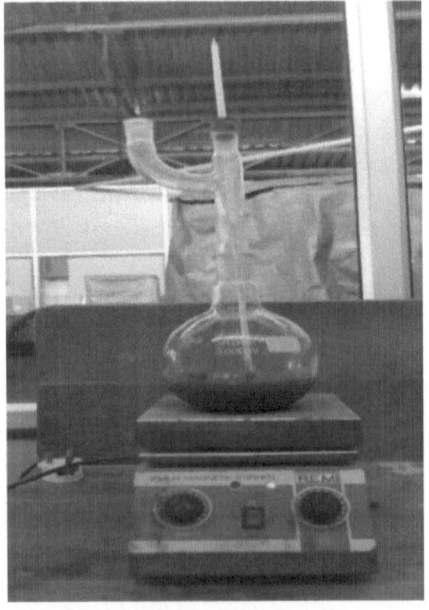

Figure 48.2: Stirring and heating of plastic oil and methanol with KOH on magnetic heater and stirrer.

Source: Author.

Figure 48.3: Separation of biodiesel and glycerol.

Source: Author.

Figure 48.4: Separation of Biodiesel and Water after water-wash.

Source: Author.

2.3. *Physicochemical Characteristics of WPO and WPOB*

Table 48.1 presents a comparison of characteristics of WPO, the WPOB, the ASTM Biodiesel Standard, and the ASTM diesel standards. By examining this comparative data, it is evident that the biodiesel produced from waste plastic oil exhibits specific physical and chemical properties that align with the ASTM standards for biodiesel. The analysis reveals that the waste plastic oil biodiesel possesses qualities and characteristics that conform to the established norms outlined in the ASTM guidelines for biodiesel fuel.

Based on above study it can be stated that biodiesel from waste plastic oil are in the range of ASTM biodiesel standards and comparable to diesel fuel.

3. Experimental Investigation

An experimental investigation was accompanied on a Kirloskar TV1 3.5 kW single-cylinder water cooled, multi-fuel engine to measure the combustion, performance, and emissions characteristics of the biodiesel blend. The engine has eddy current, water cooled dynamometer with an ENGINESOFT, engine performance analysis software. The variable parameter considered was the load. For combustion analysis, the in-cylinder pressure concerning the crank angle was examined. Performance analysis focused on BTHE BDFC and BSEC concerning brake power. The exhaust analysis parameters included CO, CO_2, and HC concerning brake power. The fuels utilised for testing were conventional diesel and B20.

4. Result and Discussion

4.1. Combustion Analysis

In combustion analysis the study of cylinder pressure obtained during combustion is recorded and analysed to see the effect of

Table 48.1: Evaluation of physicochemical characteristics of WPO and WPOB with diesel and ASTM standards

Properties	WPO	WPOB	Diesel	ASTM Biodiesel Standards	ASTM Diesel Standards
Density (kg/m³)	—	0.87	0.830	0.862–0.9	0.85–0.89
Kinematic Viscosity 40°c (cst)	18.1	5.6	3.1	1.9–6.0	1.9–4.1
Calorific value (MJ/kg)	40	43.5	44	=>37.5	42–45
Fire point(°c)	150	98	65	170	62–82
Flash point(°c)	<40	<20	68	100–170	60–80
Cetane number	—	—	54	46–60	40–55

Source: Author.

different fuel on the combustion process. It has been detected that the B20 blend exhibited a greater peak pressure related to diesel fuel within the crank angle range of 361°–365°, reaching a pressure of 43.5 bars. This 5.58% increase in peak pressure for the B20, related to diesel, can be credited to the blend's slightly higher torque and power output, resulting in improved efficiency.

4.2. Performance Analysis

4.2.1. Break Thermal Efficiency (BTHE vs B.P)

THE represents the proportion of brake power output generated by a heat engine relative to the thermal energy input derived from fuel. It was observed that the BTHE for B20 blends of WPO was greater than that of diesel during engine operation. This improvement can be credited to the increased oxygen content in the B20 blend, which gives in enhanced combustion related to pure diesel. Under full load conditions, as illustrated in Figure 48.6, the B20 blend exhibited a BTHE that was 3.4% greater than that of pure diesel. This increase is attributed to the higher oxygen content in the B20 blend.

4.2.2. Brake Specific Fuel Consumption (BSFC Vs B.P)

BSFC is crucial for evaluating engine performance and fuel efficiency. It was

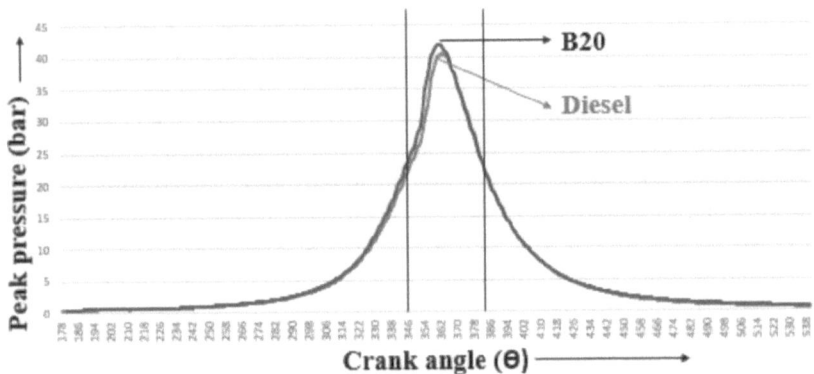

Figure 48.5: In-cylinder pressure Vs crank angle (Θ).

Source: Author.

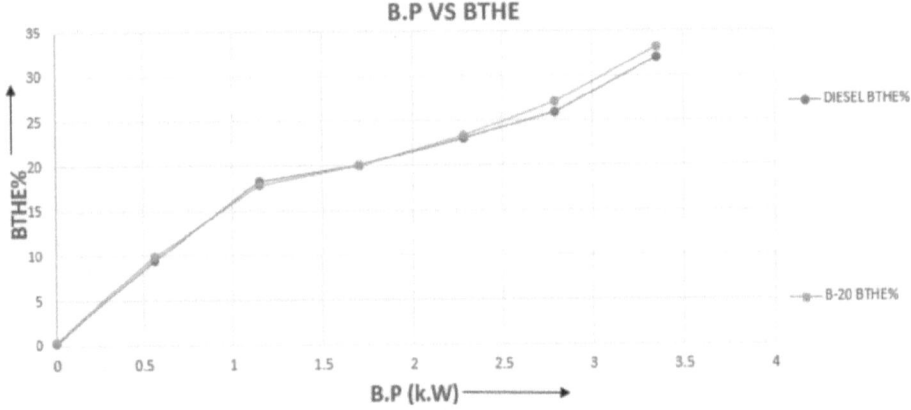

Figure 48.6: Break thermal efficiency Vs B.P.

Source: Author.

observed that as the brake load and blend ratio increased, the BSFC decreased. Under full load conditions, as illustrated in Figure 48.7, the BSFC of the B20 blend was 4.11% lower than that of pure diesel. This decrease in BSFC for the B20 blend is due to the additional oxygen in the fuel blend, which enhanced combustion efficiency.

4.2.3. Brake Specific Energy Consumption (BSEC)

Brake specific energy consumption (BSEC) is the ratio of the energy released by burning fuel over the course of an hour to the actual brake power delivered to the wheels.

It serves as a measure of how efficiently the energy produced from the fuel is transferred to the wheels. As the brake power increases in the figure, the BSEC decreases at lower loads. It was observed that the BSEC of the B20 blend was 4.88% lower than that of pure diesel, as shown in Figure 48.8. This trend follows a similar pattern as the relationship between BSFC and BP, where lower values are desirable for improved efficiency.

4.3. Emission Analysis

4.3.1. CO Emissions Vs B.P

Carbon monoxide (CO) is formed when there is an insufficient amount of oxygen

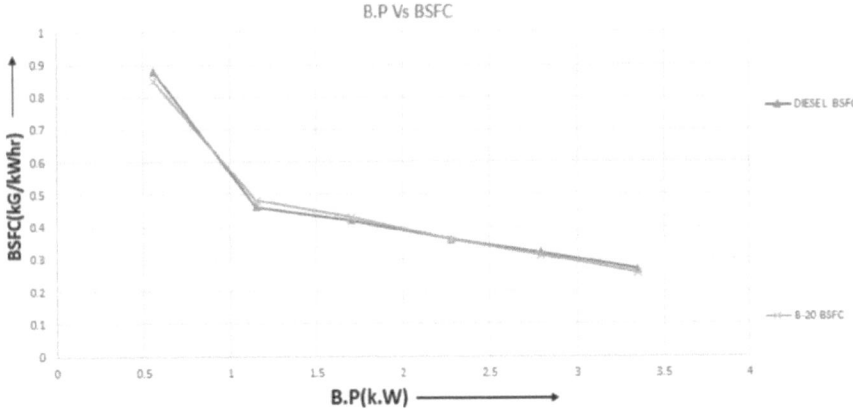

Figure 48.7: BSFC Vs B.P.

Source: Author.

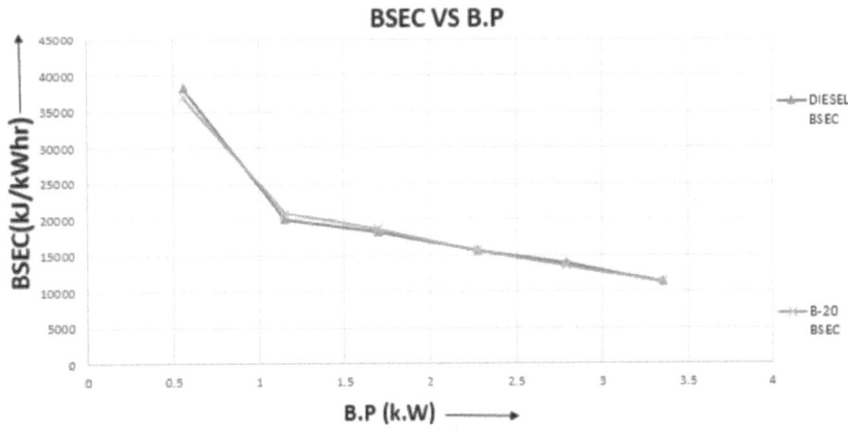

Figure 48.8: BSEC Vs B.P.

Source: Author.

to convert all carbon molecules into carbon dioxide (CO_2). The CO emissions decrease with the B20 blend. This trend can be credited to the occurrence of oxygen molecules in the blends, which facilitates the whole combustion of CO formed within the cylinder. Under full load conditions, as shown in Figure 48.9, the CO emission from the B20 blend was 9.04% lesser than that of pure diesel due to the improved burning process facilitated by the biodiesel blend.

4.3.2. CO_2 Emissions Vs B.P

Carbon dioxide (CO_2) not only contributes to the greenhouse effect in the atmosphere but also increases the acidity of oceans by approximately 30%, adversely affecting a wide variety of marine organisms. During fuel combustion, the oxygen present in the air is converted into CO_2. It was observed that carbon dioxide emissions increased with higher brake loads on the diesel engine, due to greater fuel combustion. At full load conditions, the CO_2 emissions from the B20 blend (20% waste plastic oil biodiesel and 80% diesel) were 5.178% higher than those from pure diesel, as shown in Figure 48.10. This increase can be attributed to the more complete conversion of carbon monoxide (CO) to CO_2, facilitated by the improved combustion process with the biodiesel blend.

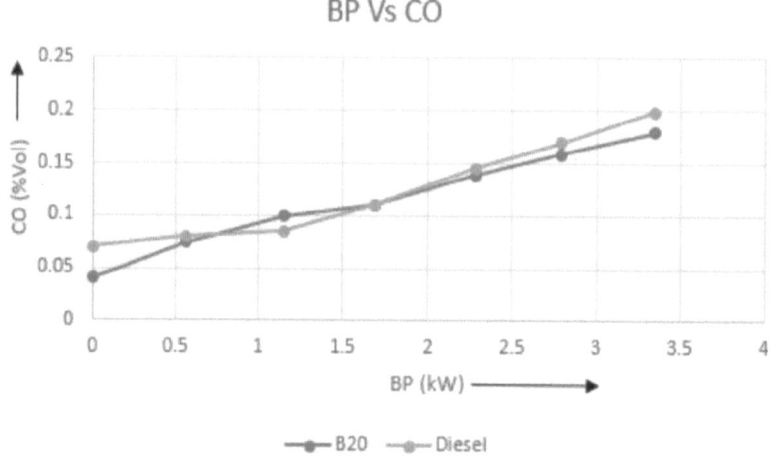

Figure 48.9: CO emission Vs B.P.

Source: Author.

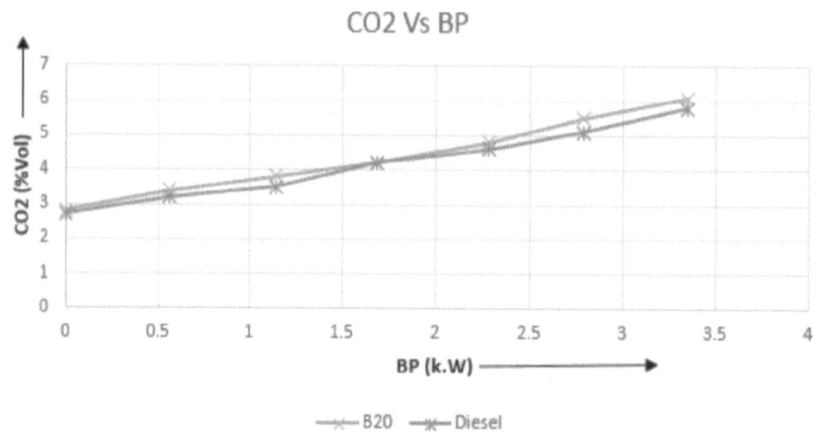

Figure 48.10: CO_2 emission Vs B.P.

Source: Author.

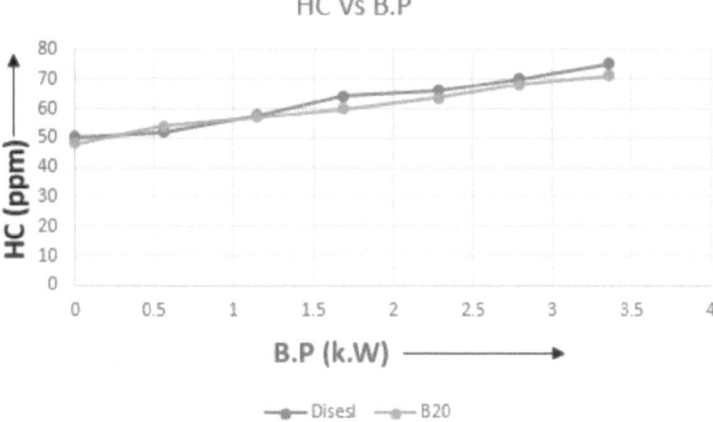

Figure 48.11: HC vs BP.

Source: Author.

4.3.3. HC Emissions

HC, consisting of unburned or partially burned fuel, significantly contribute to urban smog and are toxic. High hydrocarbon emissions typically indicate poor fuel ignition. It was observed that hydrocarbon emissions increased with higher brake loads on the diesel engine due to added fuel being supplied. However, at full load conditions, the HC emissions from the B20 blend (20% waste plastic oil biodiesel and 80% diesel) were 5.33% lower than those from pure diesel, as shown in Figure 48.11. This reduction in HC emissions can be credited to the efficient burning process facilitated by the biodiesel blend.

5. Conclusion

After analysing the combustion, performance, and emission parameters, it can be concluded that the B20 blend is a suitable fuel for CI engines. Experimental investigation shows that the in-cylinder peak pressure with the B20 blend is slightly greater than with diesel fuel due to improved combustion, leading to increased efficiency. The BTHE of the B20 is 3.4% higher than that of pure diesel, attributable to its higher oxygen content. The BSFC of the B20 is 4.11% lower than pure diesel, also due to

the additional oxygen content. Similarly, the BSEC of the B20 blend is 4.88% lesser than that of pure diesel.

Regarding emissions, the CO emission from the B20 blend is 9.04% lower than pure diesel due to the proper combustion process. However, the CO_2 emission of the B20 blend is 5.178% higher than pure diesel, which can be credited to the more complete burning of CO into CO_2. Furthermore, the HC emission of the B20 blend is 5.33% lower than pure diesel, also resulting from the improved combustion process facilitated by the biodiesel blend.

References

Awang, M.S.N., et al. (2021). Effect of addition of plastic pyrolytic oil and waste cooking oil biodiesel in palm oil biodiesel-commercial diesel blends on diesel engine performance, emission, and lubricity. Energy & Environment, 1–29.

Basha, S. A., et al. (2009). A review on biodiesel production, combustion, emissions and performance. Renewable and Sustainable Energy Reviews, 13(6), 1628–1634.

Dulawat (2020). Study on biodiesel production and characterization for used cooking oil. International Research Journal of Pure & Applied Chemistry, 21(24), 76–86.

Kumar T. D., et al. (2019). Performance, Combustion And Emission Characteristics of

B20 fish biodiesel blended with waste plastic oil on a diesel engine. International Conference on Thermo-fluids and Energy Systems.

Miyuranga, et al. (2023). Biodiesel production through the transesterification of waste cooking oil over typical heterogeneous base or acid catalysts. Catalysts, 13(3), 546.

Patel, T., et al. (2013). Performance and emission analysis of CI engine using biodiesel. 23rd National Conference on I.C. Engine and Combustion, SVNIT, Surat, India, 13–16 December 2013.

Qi, D. H., et al. (2010). Performance and combustion characteristics of biodiesel–diesel–methanol blend fuelled engine. Applied Energy, 87(5), 1679–1686.

Sorate, K. A., et al. (2013). Combustion characteristics of CI engine fuelled with high free fatty acid oil biodiesel. 23rd National Conference on I.C. Engine and Combustion, SVNIT, Surat, India, 13–16 December 2013.

Sukjit, E., et al. (2020). Evaluation of waste plastic oil-biodiesel blends as alternative fuels for diesel engines. Energies, 13(11), 2823.

Yormesor, al. (S., et 2023). Biodiesel with fuel additive: An analysis of engine performance, combustion and emission characteristics. Renewable Energy Technologies, 75-XX.

49 Design of tuned damped dynamic vibration absorber for optimum vibration mitigation

Amit R. Bhende

Mechanical Engineering Department, St. Vincent Pallotti College of Engineering and Technology, Nagpur, India

Abstract

In undamped tuned dynamic vibration absorber (DVA), the vibration response of primary system can be reduced at tuned natural frequency of secondary system but introduces two new resonant frequencies, one is less and other is greater than tuned frequency. The tuned frequency is always the operating frequency of the machine or the natural frequency of the machine. It means machine must pass through lower resonant frequency once to reach up to operating frequency during startup and stopping. When the speed of the machine matches with the lower resonant frequency, resonance occurs and machine vibrates with large amplitudes. Similarly, if the machine operates at different frequencies, then machine may vibrate heavily at these resonant frequencies. The problem of introduction of two resonant frequencies in case of undamped vibration absorber can be reduced by using damped vibration absorber. This paper presents the optimization of design considerations of damped DVA, tuning frequency which would make the vibration absorber most efficient. This paper also presents the optimization of damping in vibration absorber and its amplitude response. The optimization is based on flattening the fixed-points on amplitude response curve of damped DVA. The study also helps the designers to determine the narrow frequency band in which the absorber works effectively.

Keywords: Tuned vibration absorber, natural frequency, damping ratio, stiffness, frequency response

1. Introduction

The presence of vibration in machine in any industry would often leads to reduction in component life, performance of machine, quality of product, production time; production cost etc. Long exposure of vibration to human body would also lead to joint pains, discomfort and fatigue (Roseiro et al., 2016).

Hence vibration is an undesirable phenomenon especially when it increases beyond the acceptable limits. Theoretical vibration analysis of machine at the design stage itself may eliminate vibrations by introducing properly designed vibration absorbers or isolators. But that would lead to increase in total weight of machine, high manufacturing costs and high manufacturing time. Some machines vibrate

abhende@stvincentngp.edu.in

DOI: 10.1201/9781003567653-49

less when they are new but over a period of its use, component starts wear and tear which changes the whole dynamics of the machine. The machine starts vibrating beyond acceptable limits. In such cases, mathematical analysis of machine vibration becomes empirical to decrease the vibration response of the machine. This paper presents the design considerations of DVA, tuning frequency which would make the vibration absorber most efficient. This paper also presents the optimization of damping in vibration absorber and its amplitude response. The function of vibration absorber is to absorb the vibrations and restrict it to transfer from one body to another. A DVA is attached to a single-degree of freedom system (SDOF). When DVA is mounted on another SDOF system then the complete assembly becomes a two-degree of freedom system. DVA has many engineering applications. Vibration absorber can work in very narrow band of frequency range because they are tuned for certain frequency range hence they are also called tuned vibration absorbers in some situations. In constant speed applications, vibrations of primary system can be reduced by selecting the natural frequency of the absorber system close or equal to the external excitation frequency. A mathematical model of damped DVA is presented in the research (Lee et al., 2001). A DVA is attached to cutting tool to control the vibration in turning operation. The experimental investigations are carried out on steel and aluminum bar using different damping ratios and frequency ratios. The experimental result shows the effectiveness of damped DVA in suppressing the vibrations during turning operations. The study presented two key findings at the end of the experimental investigation. Firstly, keep the frequency ratio very close to unity. And secondly, keep the damping ratio as high as possible to make the absorber more effective. Similarly, many researchers have tried to optimize the performance of damped & undamped DVA using linear as well as non-linear behavior of the system. H_∞ & H_2 optimization methods are some of the oldest

methods of optimization of damped linear DVA system. H_∞ is minimization problem of maximum amplitude response of the primary system by using fixed-point theory. Whereas H_2 is minimization of total vibration energy of primary system subjected to various excitation frequencies. Optimum vibration energy is achieved by minimizing the area under frequency response curve (Noori and Farshidianfar, 2013). Sobol and Statnikov optimization algorithm is developed to optimize the frequency response of damped and undamped DVA. Optimization of both linear and non-linear systems is examined. Here, multi-objective optimization problem is solved by using Pareto optimization method. Total six objective functions are examined as multi-objective optimization problem (Jordanov and Cheshankovet, 1988). Uncertainty and sensitivity analysis along with multi-objective genetic optimization concept is applied for the optimization of DVA to determine the optimum frequency band. Uncertainty and sensitivity analysis was based on a probabilistic framework using the Monte Carlo simulation (Martins et al., 2020). Optimization of DVA is proposed by negative stiffness method at the fixed points on frequency-amplitude graph. Optimization is obtained in the form of optimum damping ratio, stiffness ratio and tuning ratio. The optimization problem is solved based on the fourth order Runge-Kutta method (Shen et al., 2016). Optimum value of spring constant and damping factor are determined by minimizing the maximum amplitude response of the primary mass. The concept of optimization is based on the fact that the increase in the mass ratio results in decrease in response amplitude of primary system (Younes, 2017).

2. Mathematical Model of Undamped DVA

Figure 49.1 shows DVA for undamped system. In 2 DOF system, there are two natural frequencies and two mode shapes. Here, both the masses will vibrate with same

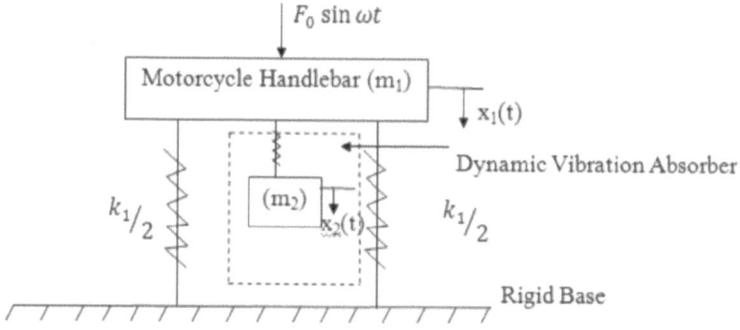

Figure 49.1: Undamped dynamic vibration absorber.

Source: Author.

frequency (ω) equal to the forcing frequency because the steady state response will be governed by external excitation force ($F_0 \sin \omega t$). So the steady-state response of masses (m_1 & m_2) are given by

$$X_1 = \frac{F_0(k_2 - m_2\omega^2)}{m_1 m_2 \omega^4 - [m_1 k_2 + m_2(k_1 + k_2)]\omega^2 + k_1 k_2} \quad (1)$$

$$X_2 = \frac{F_0 k_2}{m_1 m_2 \omega^4 - [m_1 k_2 + m_2(k_1 + k_2)]\omega^2 + k_1 k_2} \quad (2)$$

The natural frequency of the primary mass is ($\omega_1 = \sqrt{k_1/m_1}$) which is given by the square root of the ratio of stiffness of primary system (k_1) to mass of primary system (m_1). Similarly, the natural frequency of the absorber mass is ($\omega_2 = \sqrt{k_2/m_2}$) which is given by the ratio of stiffness of primary system (k_1) to mass of primary system (m_1). Mass ratio ($\mu = m_2/m_1$) is ratio of absorber mass to primary mass. Static deflection (X_{st}) of primary mass is the ratio of excitation force (F_0) to stiffness of primary system (k_1). Now by substituting these above terms in Eq. (1) and (2), the response equations of primary mass and absorber mass can be rewritten as

$$\frac{X_1}{X_{st}} = \frac{\left[1 - \frac{\omega^2}{\omega_2^2}\right]}{\frac{\omega^4}{\omega_1^2 \omega_2^2} - \left[\frac{\omega^2}{\omega_2^2} + (1+)\frac{\omega^2}{\omega_1^2}\right] + 1} \quad (3)$$

$$\frac{X_2}{X_{st}} = \frac{1}{\frac{\omega^4}{\omega_1^2 \omega_2^2} - \left[\frac{\omega^2}{\omega_2^2} + (1+)\frac{\omega^2}{\omega_1^2}\right] + 1} \quad (4)$$

From Eq. (3), it is very clear that the response amplitude of the primary mass can bring to zero by

$$X_1 = 0 \quad \text{when} \quad \frac{\omega^2}{\omega_2^2} = 1 \quad \text{or } \omega = \omega_2$$

It means that when the natural frequency of the absorber system is equal to the excitation frequency, then the response of the primary system becomes zero. So, if the stiffness and mass of the absorber system are selected such that the value of $\sqrt{k_2/m_2}$ is equal to the excitation frequency (ω) then the primary system will have zero amplitude vibration. It means that the vibrations will be mitigated from the primary system. Whereas, the absorber mass will have some finite response. This type of absorber is effective only when the operating frequency or forcing frequency is constant. But when either of them is varying then such tuned absorber becomes ineffective. The obvious reason for the use of DVA is to control the vibrations of primary system. It is also obvious that the response of primary system is extremely high when excitation frequency (ω) matches with the natural frequency of primary system (ω_1). This condition is called resonance condition. So to design an effective vibration absorber then one should design for the natural frequency of primary system. In another words, natural frequency of the absorber system (ω_2) is to be matched with the natural frequency of the primary system (ω_1). So the name tuned mass absorber is given to such system because the natural frequency of the absorber is tuned to the natural frequency of the primary system. Tuned DVA is particularly effective when the excitation frequency varies in certain range. It becomes even more empirical when the resonance condition lies within the excitation frequency range. So for tuned DVA, put $\omega_1 = \omega_2$ in Eq. (3) and (4). The response amplitudes primary system and absorber system for tuned condition are given in Eq. (5) and (6) respectively.

$$\frac{X_1}{X_{st}} = \frac{\left[1 - \frac{\omega^2}{\omega_2^2}\right]}{\frac{\omega^4}{\omega_2^4} - (2+)\frac{\omega^2}{\omega_2^2} + 1} \tag{5}$$

$$\frac{X_2}{X_{st}} = \frac{1}{\frac{\omega^4}{\omega_2^4} - (2+)\frac{\omega^2}{\omega_2^2} + 1} \tag{6}$$

As discussed earlier, dynamic absorber system is a two DOF system and vibrates with two resonant frequencies. The expression for these resonant frequencies can be obtained by equating the denominator part of Eq. (5) and (6) to zero. The two resonant frequencies are given in Eq. (7).

$$\left(\frac{\omega}{\omega_2}\right)^2 = \left(1 + \frac{\mu}{2}\right) \pm \sqrt{\left(1 + \frac{\mu}{2}\right)} \tag{7}$$

Where μ is mass ratio of absorber mass to primary mass. From Eq. (7), it is clear that the resonant frequencies of the absorber system are the functions of mass ratio only. So by properly selecting the mass of absorber system, resonant condition of primary system can be avoided. Another advantage of selecting proper mass ratio is to spread the natural frequencies of absorber system beyond the operating external excitation frequencies and primary system can be safe and protected.

In tuned undamped DVA, the amplitude of primary system is reduced at tuned frequency but introduces two resonant frequencies, one is less than and other is greater than tuned frequency. Tuned frequency is always the operating frequency of machine. It means, machine must pass through lower resonant frequency to reach upto operating frequency during start-up and stopping. Similarly, if the machine operates at deferent frequencies, then machine may vibrate heavily at resonant frequencies. When the speed of the machine matches with the lower resonant frequency then the machine vibrates with large amplitudes. Similarly, if the machine operates at different frequencies, then machine may vibrate heavily at resonant frequencies. The problem of introduction of two resonant frequencies in case of undamped vibration absorber can be reduced by using damped vibration absorber.

3. Mathematical Model of Damped DVA

In damped DVA system, machine is supported by isolators with stiffness (k_1) and an auxillary system is attached to primary machine which is damped vibration absorber. The steady state solution of damped DVA is presented here. The amplitude of vibration of machine and absorber system is given by Eq.(8) and (9) respectively.

$$X_1 = \frac{F_0(k_2 - m_2\omega^2 + ic_2\omega)}{[(k_1 - m_1\omega^2)(k_2 - m_2\omega^2) - m_2k_2\omega^2] + i\omega c_2(k_1 - m_1\omega^2 - m_2\omega^2)} \tag{8}$$

$$X_2 = \frac{X_1(k_2 - ic_2\omega)}{(k_2 - m_2\omega^2 + ic_2\omega)} \tag{9}$$

Eq. (8) and (9) can be rewritten using following terms. The modified equations are shown in Eq. (10) and (11).

$$mass\ ratio = \frac{absorber\ mass}{primary\ mass} = \mu = \frac{m_2}{m_1}$$

$$Static\ deflection\ of\ the\ system = \delta_{st} = \frac{F_0}{k_1}$$

$$Square\ of\ natural\ frequency\ of\ absorber = \omega_a^2 = \frac{k_2}{m_2}$$

$$Square\ of\ natural\ frequency\ of\ absorber = \omega_n^2 = \frac{k_1}{m_1}$$

$$Ratio\ of\ natural\ frequencies = f = \frac{\omega_a}{\omega_n}$$

$$Forced\ frequency\ ratio = g = \frac{\omega}{\omega_n}$$

$$Critical\ damping\ constant = c_c = 2m_2\omega_n$$

$$Damping\ ratio = = \frac{c_2}{c_c}$$

$$\frac{X_1}{\delta_{st}} = \left[\frac{(2g)^2 + (g^2 - f^2)^2}{(2g)^2(g^2 - 1 + g^2)^2 + \{f^2g^2 - (g^2 - 1)(g^2 - f^2)\}^2} \right]^{\frac{1}{2}} \quad (10)$$

$$\frac{X_2}{\delta_{st}} = \left[\frac{(2g)^2 + f^4}{(2g)^2(g^2 - 1 + g^2)^2 + \{f^2g^2 - (g^2 - 1)(g^2 - f^2)\}^2} \right]^{\frac{1}{2}} \quad (11)$$

If the damping ration is zero ($\xi = 0$), then the damped vibration absorber acts as an undamped vibration absorber and hence two resonant frequency peaks are obtained. Similarly, when ($\xi = \infty$), then the two masses m_1 and m_2 behave as if they are join together to become single DOF system. For damping value between 0 to ∞, the amplitudes are reduced substantially. So here, when $\xi = 0$ then amplitude of machine is infinite and when $\xi = \infty$ then also the amplitude is infinite. The reduced vibration amplitude of machine would obtain in between these two extreme ξ values. Now, the task is to obtain optimum value of ξ where the amplitude of machine vibration is obtained minimum and the frequency ratio at which this optimum ξ occurs.

4. Optimization of Damped DVA

Figure 49.2 shows the three curves for $\xi = 0$, $\xi = 0.1$ & $\xi = \infty$ intersect each other at point A and point B. At point A and B, the amplitude of machine vibration is same for $\xi = 0$ and $\xi = \infty$. To obtain the frequency ratio for point A and B, put $\xi = 0$ and $\xi = \infty$ in eq. (11) and equate the two. This yield

$$g^4 - 2g^2 \left(\frac{1 + f^2 + f^2}{2 +} \right) + \frac{2f^2}{2 +} = 0 \quad (12)$$

Eq. (12) is quadratic equation whose solution gives two values of frequency ratio (g). Lower value of g gives the frequency ratio for point A and higher value for point B. So, it has been observed that the most efficient absorber is the one where amplitudes at point A and B are equal. This condition is obtained by equating Eq. (10) by substituting conditions for point A and B. the condition obtained is given in Eq. (13)

$$f = \frac{\omega_a}{\omega_n} = \frac{1}{1 +} \quad (13)$$

So here the damped vibration absorber is efficient when both the amplitude peaks at A and B are equal. And this condition is obtained by tuning using eq. (14). Eq. (14) also gives the condition for tuned DVA, but

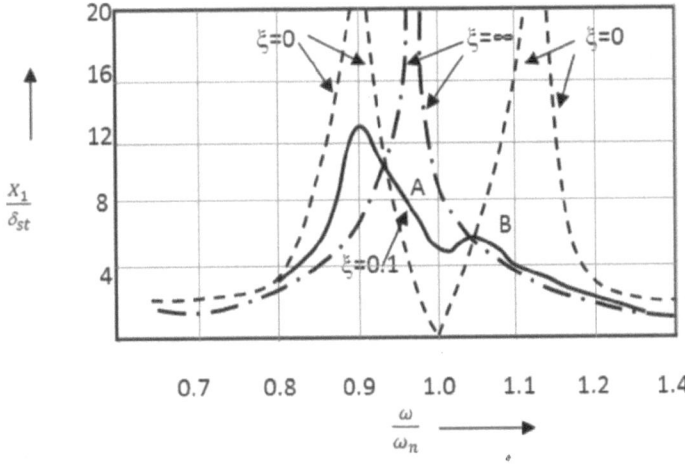

Figure 49.2: Amplitude response of primary system when damped DVA tuned to natural frequency of primary system with mass ratio of 0.05.

Source: Author.

it does not give the optimum value of the damping ratio ξ and vibration amplitude $(\frac{X_1}{\delta_{st}})$. The whole concept of damped vibration absorber is based on to eliminate the possibilities of multiple peaks in undamped DVA. The objective of whole optimization process is to obtain flat response curve of main system. In other words, optimization of amplitude can be achieved by minimizing the slope of the response curve of main system. This can be achieved by putting the value of optimum frequency ratio in eq. (11). The resulting equation gives the amplitude of main system for optimum tuned condition. Now, to obtain optimal damping ratio ξ, the modified eq. (11) is differentiated w. r. t. g to find the slope of the curve $\dfrac{d(\frac{X_1}{\delta_{st}})}{dg}$. Now equating the slope equal to zero at point A and B.

$$\xi^2 = \frac{\left\{3 - \sqrt{+2}\right\}}{8(1+)^3} \ for \ point \ A \qquad (14)$$

$$\xi^2 = \frac{\left\{3 + \sqrt{+2}\right\}}{8(1+)^3} \ for \ point \ B \qquad (15)$$

These are the two optimal damping ratios which would flat the curve at point A and B. these two damping ratios can not be implemented to one absorber system, hence an average of these two value are taken to find the final optimum damping ratio. The optimum average damping ratio is given in Eq. (16)

$$\xi^2_{optimal} = \frac{3}{8(1+)^3} \qquad (16)$$

The corresponding vibration amplitude for optimal damping ratio will be optimal amplitude which is given in Eq. (17)

$$\left(\frac{X_1}{\delta_{st}}\right)_{optimal} = \left(\frac{X_1}{\delta_{st}}\right)_{max} = \sqrt{1 + \frac{2}{\mu}} \qquad (17)$$

5. Conclusion

This paper presents design modifications in undampedDVA to make it damped

DVA. This design modification eliminates the possibility of introducing two natural frequencies in the system which further eliminates the possibility of high vibration amplitude between start to operating frequency. Damped DVA reduces the amplitude of vibration response of primary system by many folds. This paper also determines the condition of optimal frequency ratio of absorber system to primary system, damping factor and response amplitude for most efficient tuned damped DVA. The study shows that the auxiliary system absorbs most of the vibrations from the primary system and vibrates with much higher amplitudes than that of primary system. This design consideration helps designer to accommodate heavy vibrations of auxiliary system while attaching it to the primary system. Also, the spring design of auxiliary system should be capable of withstanding heavy cyclic loading and should be designed from fatigue point of view. Addition of damping in vibration absorber does not eliminate the vibration of primary system completely. So, utmost care is taken while adding the damping. Damping is to be added only in situations where operating frequency band is very narrow.

References

Jordanov, I. N., and Cheshankov, B. I. (1988). Optimal design of linear and non-linear dynamic vibration absorbers. Journal of Sound and Vibration, 123(1), 157–170.

Lee, E. C., Nian, C. Y., and Tarng, Y. S. (2001). Design of a dynamic vibration absorber against vibrations in turning operations. Journal of Materials Processing Technology, 108(3), 278–285.

Martins, L. A., Lara-Molina, F. A., Koroishi, E. H., and Cavalini Jr., A. A. (2020). Optimal design of a dynamic vibration absorber with uncertainties. Journal of Vibration Engineering & Technologies, 8, 133–140.

Noori, B., and Farshidianfar, A. (2013). Optimum design of dynamic vibration absorbers for a beam, based on H∞ and H 2 Optimization. Archive of Applied Mechanics, 83, 1773–1787.

Roseiro, L. M., Neto, M. A., Amaro, A. M., Alcobia, C. J., and Paulino, M. F. (2016). Hand-arm and whole-body vibrations induced in cross motorcycle and bicycle drivers. International Journal of Industrial Ergonomics, 56, 150–160.

Shen, Yongjun, Xiaoran Wang, Shaopu Yang, and Haijun Xing. (2016). Parameters optimization for a kind of dynamic vibration absorber with negative stiffness. Mathematical Problems in Engineering, 2016, 1–10.

Younes, M. F. (2018). Optimal design of dynamic vibration absorber for rolling systems. *The International Conference on Applied Mechanics and Mechanical Engineering*, vol. 18, no. 18th International Conference on Applied Mechanics and Mechanical Engineering, Military Technical College, 1–16.

50 Design modification of knuckle for "Formula Student" vehicle

Vimal Chand Sontake[a], Sanchit Fulzele[b], Abhishek Yuonate[c], Ashish Mathew[d], Harsh Bhandarkar[e], and Harshal Jambhulkar[f]

Department of Mechanical Engineering, St. Vincent Pallotti College of Engineering and Technology, Nagpur, India

Abstract

The upright of the formula student vehicle is the pivotal and load bearing component. It plays a significant role in supporting the transmitting forces through the suspension assembly and between the wheels and chassis. It improves the overall performance of Formula Student Vehicle (FSAE). This work presents a thorough analysis of the design and engineering considerations involved in the development of an upright system for an FSAE Vehicle. Important aspects covered in this work include material selection, weight optimization, and structural integrity by maintaining a balance between performance and safety. The thorough research with extensive finite element analysis (FEA) simulations have been done to assess the upright's structural integration. Moreover, the study discusses the significance of fine-tuning of the upright design to improve vehicle handling characteristics competitiveness. The proposed upright design demonstrates significant improvement in terms of performance and weight reduction which in turn contributes to vehicle's competitiveness.

Keywords: Upright assembly, formula vehicle student, pivotal components, finite element analysis, structural integrity

1. Introduction

An upright in a formula vehicle student, also known as a hub carrier or upright assembly, is a critical component of the vehicle's suspension system. It is typically located at each wheel and serves several important functions within the vehicle's overall architecture. Here are its primary roles:

1. Wheel Attachment: The upright serves as the mounting point for the vehicle's wheels. It contains wheel bearings and studs that secure the wheel in place.
2. Suspension Linkage: The upright connects to the suspension components, such as control arms, wishbones, and shock absorbers. These connections allow the wheel to move up and down, providing support and damping to the vehicle's suspension system.
3. Steering Linkage: In most vehicles, the upright is also involved in the steering system. It can contain steering knuckles

---

[a]vimalsontake44@gmail.com, [b]sanchitfulzele2002@gmail.com, [c]abhishekyuonate2000@gmail.com, [d]ashishm.me20@stvincentngp.edu.in, [e]harshbhandarkar6280@gmail.com, [f]harshaljambhulkar29@gmail.com

DOI: 10.1201/9781003567653-50

and provide attachment points for tie rods, enabling the wheel to turn and steer the vehicle.

4. Load Bearing: The upright bears the load and forces experienced by the wheel, including the vehicle's weight and the forces generated during braking, acceleration, and cornering. It plays a crucial role in distributing these forces effectively and maintaining stability.

5. Wheel Alignment: The upright's design and geometry influence the vehicle's wheel alignment, including camber (tilt of the wheel), caster (tilt of the steering axis), and toe (alignment of the wheels with respect to the vehicle's centerline). Proper alignment is essential for optimal handling and tire wear.

6. Impact Absorption: In the event of a collision or impact with an obstacle, the upright may also play a role in dissipating and absorbing some of the energy, contributing to the vehicle's safety.

2. Literature Review

Marek et al. (2020) in their work focused on the design of an upright using topological optimization. This optimization was achieved by conventional methods. The optimized upright played a crucial role in the development of a fully operational prototype for the student formula vehicle in the "Formula Student Competition". Through the application of finite element analysis, its performance under different load cases was accurately predicted. Employing various principles of topology optimization, the team defined a concept for simulating topology, ensuring the utmost efficiency in the design process. They did design iterations on various materials like AISi10Mg and Ti6AI4V. Later, they focused on the creation of design and setting the non- design parts for different sets of load and supports. The aim was to increase stiffness and weight. In addition, they concluded that utilizing a heavy metal such as maraging steel officered excellent strength yield characteristics.

However, its high density posed a challenge in achieving a model weighing under 1kg while maintaining a factor of safety above 1.5.

Vijay Kumar et al. (2019) designed a specialized upright for formula vehicles. This component was designed to meet the specific needs of the vehicle and was optimized through thorough analysis to enhance its performance and withstand various loads experienced during acceleration, braking, cornering, and repetitive loading. The updated design resulted in a reduction in weight compared to the previous upright design, which weighed 2 kg. Despite the weight reduction, the strength of the upright was maintained, and its adaptability to suit different vehicle designs was improved. The design also ensured a factor of safety of 2, ensuring reliability under operational conditions.

Dai et al. (2016) studied how the angle of a car's wheel (called camber) can be adjusted to help reduce skidding when the car turns. This adjustment is needed because when a car turns, the weight shifts and can cause the car to slide sideways. The camber of the wheel can be changed by adjusting the angle of the steering axis, which is represented by the caster or lean angles. By changing these angles, the camber can be adjusted accordingly. The researchers found that by finding the right way to adjust the caster angle, along with using a smart caster mechanism, they could make the car safer, more stable, and easier to maneuverer.

Ayush Garg (2017) investigated, how mechanical loads change over time for vehicle components. Understanding the lifespan of these parts is crucial to ensuring the vehicle performs well, stays reliable, and lasts long. It was observed that parts with higher stress during static analysis tend to have shorter lifespans due to fatigue. The study predicted how many laps such components could endure before needing replacement. It was concluded that analyzing fatigue in components could lead to significant improvements in reliability, performance prediction, weight reduction, and other design aspects of vehicles.

Few more studies (were reported on the design and analysis of front and back upright for formula vehicle.

3. Methodology

In the construction of the front upright, this study systematically navigated through several phases to optimize the design (Ashish & Srivastava, 2018; Babannavar & Deshpande, 2021). This ensured that final structure of the upright was capable of meeting dynamic conditions effectively. The following phases (Figure 50.1) were the integral to the processes:

1. Research Phase: In this phase, examination of the existing designs was done and relevant literature review was done to establish a comprehensive understanding of upright structures in the context of FASE Vehicles.
2. Calculation Phase: Detailed calculations on the specific area were done. In addition, the extreme dynamics were considered for the processes that could deform upright
3. Design and Analysis Phase: Based on the research and calculations, the design process was initiated by formulating the model for the upright of the FASE Vehicle. Advanced methodologies were utilized for structural and deformation analysis.
4. Iterative Optimization: A series of iterations were executed to optimize the geometry, material selection, and overall structural composition of the upright.
5. Manufacturing Phase: Translated the finalized design into a manufacturable product. Considered practical aspects, such as manufacturability within cost constraints, utilizing appropriate materials, and quality standards.

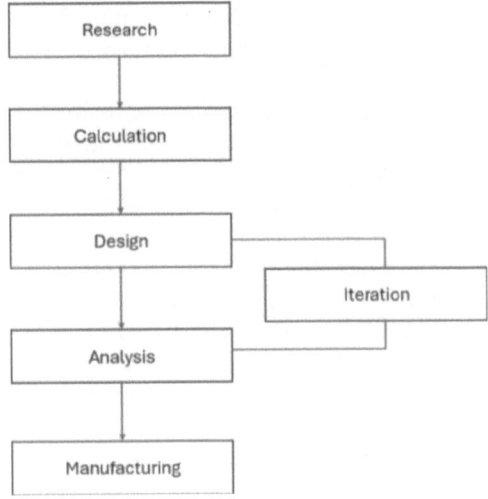

Figure 50.1: Flow diagram of various phases in development of upright.

Source: Author.

4. Material Selection

The upright of the FASE vehicle is, generally, subjected to various mechanical load and environmental conditions during its operation (Jixiong et al., 2020). In the design and operation of a FASE vehicle, several types of forces act on its upright components, such as the weight of the chassis or frame, load transfer, Centripetal Force, and Bump force. Material considerations play a significant role in determining how effectively the vehicle's structure can withstand these forces while maintaining performance, safety, and durability. Here are the selected materials with their properties.

To overcome all the forces in dynamic conditions and maintain structural integrity, various types of materials are selected. Analysis is performed to get a clear idea of which materials are compatible in dynamic conditions. Based on the graph, two materials were selected: AL 7075 T6 and Al 6061 T6. Both materials exhibited similar properties.

Therefore, Aluminum Alloy 7075-T6 was chosen for the component due to its

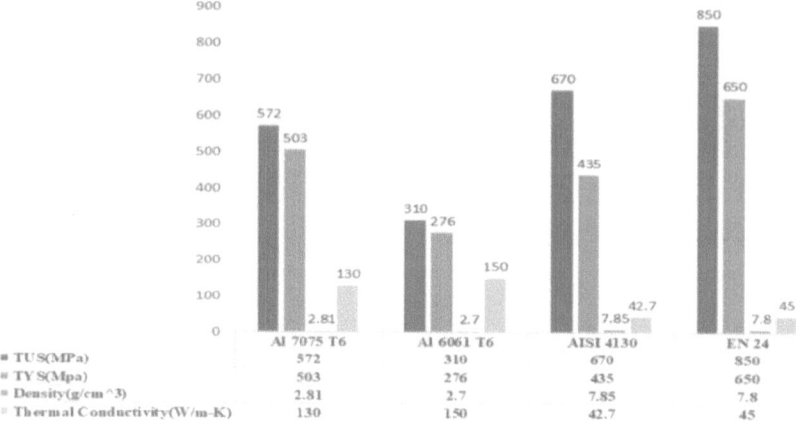

Figure 50.2: Matrix of materials.

Source: Jixiong et al., 2020.

superior strength-to-weight ratio and excellent fatigue properties compared to lighter aluminum alloys like 6061-T6. Al 7075-T6 is well suited to various manufacturing processes, including CNC machining and forging. This material resists the corrosion to ensure long term durability and reliability. By leveraging the unique properties of Al 7075-T6, the FASE vehicle can achieve enhanced performance, durability and sustainability. The characteristics utilized for the analysis are shown in Tables 50.1 and 50.2 and Figure 50.3.

5. Calculations

Following are Specifications of Formula student vehicle for which upright was designed (Kritika and Kanishka, 2020).

- Mass of Vehicle (Including Driver Mass 60 kg) = 260kg.
- Mass at front = 104kg
- Mass At rear = 156kg
- Wheelbase = 1550mm
- Trackwidth Front = 1150mm
- Trackwidth Rear = 1130mm
- Height of Centre of Gravity = 228.559mm
- Radius of Wheel = 228.6 mm
- Scrub Radius = 28 mm

Table 50.1: Material properties Al 6061-T6 and Al 7075-T6

Properties	A16061	A17075
Elastic modulus (GPa)	70–80	70–80
Density (g/cc)	2.7	2.81
Poisson's ratio	0.33	0.33
Hardness (HB500)	30	60
[Tensile strength (MPa)	115	115

Source: Jixiong et al., 2020.

Table 50.2: Material properties Al 6061-T6 and Al 7075-T6

Properties	
Ultimate Tensile Strength	572 MPa
Tensile yield strength	503 MPa
Elongation on Break	11%
Fatigue strength	159 MPa
Fracture toughness	25 MPa$^{m1/2}$
Shear Module	26.9 GPa
Shear Strength	331 MPa
Machinability	70%

Source: Jixiong et al., 2020.

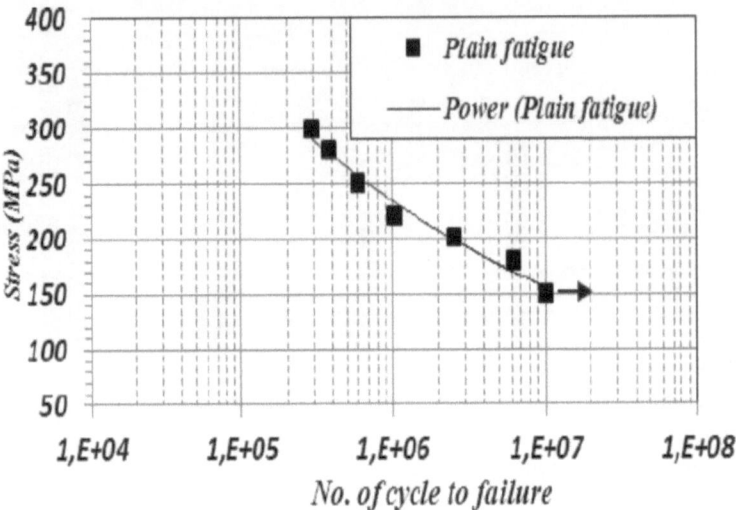

Figure 50.3: Fatigue (S/N Curve) Al 7075-T6.

Source: Azmeer et al., 2017.

- Length Of Steering arm = 56.277 mm
- Radius Of Pinion = 28.93 mm
- Braking Torque = 210 Nm
- Effective Radius = 64 mm

A. *Force on Steering Arm*

The steering arm experiences force from the push and pull of the tie rod, caused by the linear motion of the rack, in response to the steering wheel input. This force operates at a right angle to the steering arm.

Force of Friction (on one
wheel) = $\mu \times g \times$ load transfer   (1)
= 0.6 × 9.81 × 32.4
= 190N

The torque
due to = Force of friction × frictional
force scrub radius   (2)
= 190 × 28.93
= 5496.7 Nmm

This torque will be equal to lateral push from tie rods

Force on
tie rods = Torque due to lateral push /
Steering arm length   (3)
= 5496.7 × 56.27 = 97.67 N

B. *Braking Torque*

The braking torque exerted on the wheel results from the frictional force occurring at the contact patch of the tire. Braking
torque (Tb) = Frictional force ×
Radius of tire(R)   (4)
= 3843.71456 × .064
= 200.8980 N-m

C. *Bump Force*

For a quick change in road surface height and a sustained loading situation, the 3g bump condition case is used.

Bump Force = 3g × load on front
wheel   (5)
= 3 × 9.81 × 52
= 1530.36 N

D. *Vertical Load*

The load of vehicle on the single wheel

FV = Mass on wheel × 9.81   (6)
= 52 × 9.81 = 510.12N

6. Simulation and Analysis

In the design process, **Ansys Workbench** 19.2 was used to assess the front Upright, a key component of a Formula-Style Vehicle.

This analysis was crucial because it simulated real-world forces, helping validate the design (Ahmad et al., 2020). In simple terms, a static structural analysis was carried out by fixing the spindle and applying different forces to specific locations on the Upright. Subsequently, the equivalent stress, equivalent elastic strain, total deformation, factory safety, and fatigue life were determined, and the adjustment were made to the design for the next iteration, aiming to enhance its strength and reliability. During the analysis, following data used:

1. Meshing size – 1mm
2. Element order – Quadratic
3. Method – Tetrahedrons
4. Node – 3,754,78
5. Element – 2,246,62

7. Fatigue Analysis

The aim of the fatigue analysis was to check how the proposed upright designs would hold up over time (Singh and Srivastava, 2018). This involved applying specific condition and load to see how they affected the stressful life of the design. To do this, the

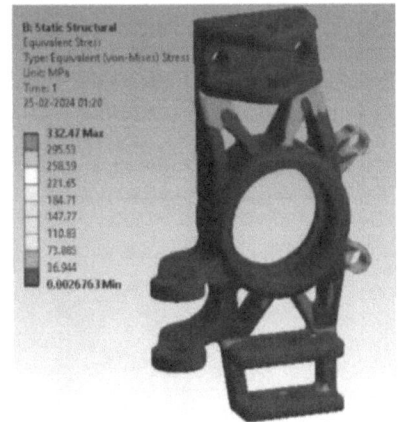

Figure 50.5: Total deformation MPa (0.0 Min – 0.2245 Max) mm.

Source: Author.

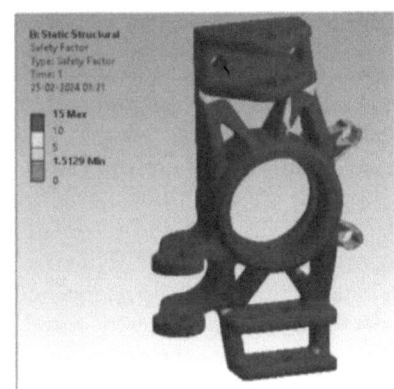

Figure 50.6: Safety factor is 1.5129.

Source: Author.

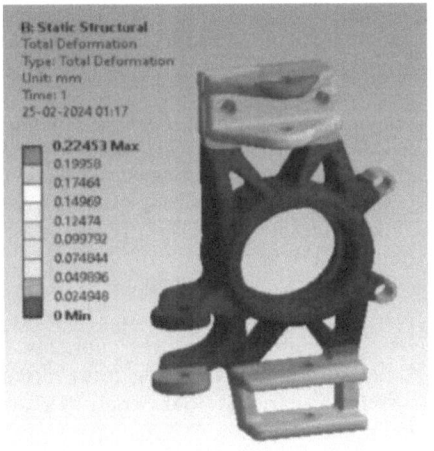

5.   Element – 2,246,62

Figure 50.4: Equivalent (von-mises) stress (0.0026 Min - 332.47Max).

Source: Author.

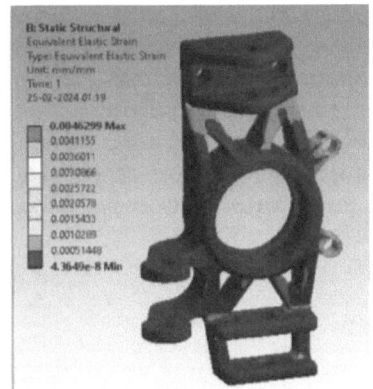

Figure 50.7: Equivalent elastic stain (4.3e -8 Min – 0.04629 Max) mm/mm.

Source: Author.

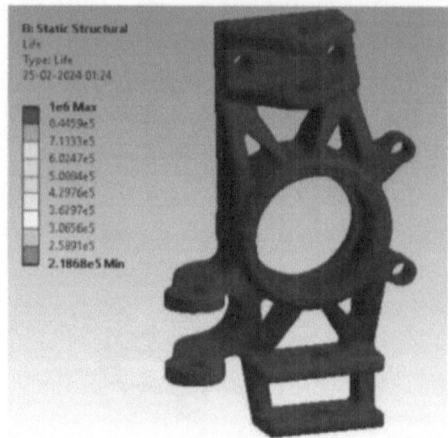

Figure 50.8: Fatigue Life (2.18e5 Min–1e6 Max) life.

Source: Author.

model was divided into six parts to apply the forces. Lateral Forces, Longitudinal Forces, Vertical Forces, Braking Torque, Steering Force and Bump Force are the forces that were applied on the front upright. The standard load was used and recorded how the load changed over time (Hunar et al., 2020; Singh and This help Gabel, 2020). to understand how the different forces and moments could impact the durability of the proposed design.

It was checked that every part of the upright to make sure the vehicle can handle forces during dynamic conditions for a long time. Design step were adjusted step by step, considering things like stress, strain, deformation, safety, and how long it can endure fatigue. Result were closely looked at how the forces change over time, making the upright stronger and more dependable. This means it can perform better and be safer in real-world situations.

8. Conclusion

The motive of this paper was to design the Front Upright by calculating the dynamic forces. Upright was engineered with suspension points and dynamic force application

to consider the factor of safety. Topological optimization was used in analysis, as an innovative approaches.

In this upright, the upper ball joint can be adjusted as per the various dynamic conditions. Because of it could be possible to increase the performance of the vehicle as per the track and change in the alignment of the wheel. Following thorough structural and fatigue analysis, it has been established that the upright can endure static and cyclic loads without encountering premature failure. Consequently, based on these findings, it is concluded that the component is deemed suitable and safe for its intended use.

The material selection for the upright was based on the dynamic forces applied to the component. Through careful consideration of various metal mechanical and thermal properties, Al 7075 T6 was chosen as the optimal material for the upright. Due to its exceptional strength-to-weight ratio and favorable stress-life curve.

Overall, the study showcased the significant improvement in various aspect in terms of structural integrity, fatigue life, enhancement in diverse conditions, vehicle reliability and other design condition that are achieved by carrying out the static and fatigue analysis of the component.

References

Ahmad, A., Hassaan, M., Nayeem, F., and Alam, M. (2020). Design & analysis of suspension system for a formula student vehicle. International Research Journal of Engineering and Technology (IRJET), 7(3), 1629–1635. doi: 10.13140/RG.2.2.22070.68164.

Azmeer, M., Basha, M. H., Hamid, M. F., Rahman, M. T. A., and Hashim, M. S. M. (2017). Design optimization of rear uprights for UniMAP automotive racing team formula SAE racing car. IOP Conf. Series: Journal of Physics: Conf. Series, 908, 012051.

Babannavar, O., and Deshpande, A. M. (2021). Design, analysis, and optimization of an FSAE upright. IOP Conf. Series: Materials Science and Engineering, 1166 (2021) 012043. 10.1088/1757- 899X/1166/1/012043.

Garg, A. (2017). Fatigue analysis and optimization of upright of a FSAE vehicle. International Journal of Science and Research (IJSR), 6(9), 1983–1988. doi: 10.21275/ART20176999

Hunar, M., Jancar, L. Krzikalla, D., Kaprinay, D., and Srnicek, D. (2020). Comprehensive view on racing car upright design and manufacturing. Symmetry, 12(6), 15–16. doi: 10.3390/sym12061020

Li, Jixiong, Jianliang Tan, and Jianbin Dong (2020). Lightweight design of front suspension upright of electric formula car based on topology optimization method. World Electric Vehicle Journal, 11(1), 15. doi: 10.3390/wevj11010015

Marek Hunar, Lukas J., David Krzikalla (2020). Design reviews on racing car upright design and manufacturing system. Symmetry, 12(6), 1–14.

Singh, Ashish Kumar, and Srivastava, T. (2018). Design of upright assembly for SUPRA Vehicle. International Journal of Applied Engineering Research, 13(6), 323–326.

Singh, K., and Gabel, K. (2020). Calculation of dynamic forces and analysis of front upright for ATV. International Research Journal of Engineering and Technology (IRJET), 7(4), 6314–6319.

Vijaya Kumar, B., Ruchitha, B., Goud, E. N., and Kumar, H. (2019). Design and analysis of upright for formula vehicle. International Journal of Trend in Scientific Research and Development (IJTSRD). 3(3), 1185–1189. https://www.ijtsrd.com/papers/ijtsrd23303.pdf.

Vo, D. Q., Marzbani, H., Fard, M., and Jazar, R. N. (2016). Caster–Camber relationship in vehicles. nonlinear approaches in engineering applications: Advanced analysis of vehicle related technologies. 63–89. doi: 10.1007/978-3-319-27055-5_2